Soil Physics: Agriculture and Environmental Applications

Soil Physics: Agriculture and Environmental Applications

Editor:

Dr. Hazem Fouda, Ph.D.

Delve Publishing

www.delvepublishing.com

Soil Physics: Agriculture and Environmental Applications

Editor: Dr. Hazem Fouda, Ph.D.

708 3rd Avenue, 6th Floor
New York, NY 10017
United States of America

www.delvepublishing.com

ISBN: 978-1-68095-778-5

Library of Congress Control Number: 2016951534

About the Editor

Dr. Hazem Fouda, Ph.D.

Hazem Shawky Fouda has a PhD. In Agriculture Sciences, obtained his PhD. From the Faculty of Agriculture, Alexandria University in 2008, He is working in Cotton Arbitration & Testing General Organization (CATGO).

Contents

List of Contributors

Mili Mandal
Division of Plant & Soil Sciences, West Virginia University, Morgantown, WV 26506-6108, USA
Department of Pharmacology and Toxicology, Ernest Mario School of Pharmacy, Rutgers University, Piscataway, NJ 08854, USA

Rakesh S. Chandran
Division of Plant & Soil Sciences, West Virginia University, Morgantown, WV 26506-6108, USA

John C. Sencindiver
Division of Plant & Soil Sciences, West Virginia University, Morgantown, WV 26506-6108, USA

Amir M. González-Delgado
Post-doctoral Researcher at New Mexico State University, Las Cruces, NM, US.

Manoj K. Shukla
Professor of environmental soil physics at New Mexico State University, Las Cruces, NM, US.

Blair Stringam
Professor in the Plant and Environmental Science Department of New Mexico State University, Las Cruces, NM, US.

Mehran Parsheh
Principal Engineer for Sapphire Energy Inc.

B. Dobrzański
Institute of Agrophysics Polish Academy of Sciences, Lublin, Poland

S. Grundas
Institute of Agrophysics Polish Academy of Sciences, Lublin, Poland

A. Stępniewski
Institute of Agrophysics Polish Academy of Sciences, Lublin, Poland

Ayodele O. Adelana
Institute of Agricultural Research Training, Obafemi Awolowo University, P.M.B. 5029, Moor Plantation, Ibadan, Nigeria.

Gabriel A. Oluwatosin
Institute of Agricultural Research Training, Obafemi Awolowo University, P.M.B. 5029, Moor Plantation, Ibadan, Nigeria.

Celinah Agunbiade
Department of Agronomy, University of Ibadan, Ibadan, Nigeria.

Kayode S. Are
Institute of Agricultural Research Training, Obafemi Awolowo University, P.M.B. 5029, Moor Plantation, Ibadan, Nigeria.

Olateju D. Adeyolanu
Institute of Agricultural Research Training, Obafemi Awolowo University, P.M.B. 5029, Moor Plantation, Ibadan, Nigeria.

Lívia Cristina Pinto Dias
The Federal University of Viçosa, Viçosa, MG 36570-900, Brazil

Márcia N. Macedo
The Woods Hole Research Center, 149 Woods Hole Rd, Falmouth, MA 02540, USA

Marcos Heil Costa
The Federal University of Viçosa, Viçosa, MG 36570-900, Brazil

Michael T. Coe
The Woods Hole Research Center, 149 Woods Hole Rd, Falmouth, MA 02540, USA

Christopher Neill
The Ecosystems Center, Marine Biological Laboratory, 7 MBL Street, Woods Hole, MA 02543, USA

Órlaith **Ní Choncubhair**
Teagasc Environmental Research Centre, Johnstown Castle, Co. Wexford, Ireland
UCD School of Biology & Environmental Science, University College Dublin, Dublin 4, Ireland

Bruce Osborne
UCD School of Biology & Environmental Science, University College Dublin, Dublin 4, Ireland
UCD Earth Institute, University College Dublin, Dublin 4, Ireland

John Finnan
Teagasc Crops Research Centre, Oak Park, Carlow, Ireland

Gary Lanigan
Teagasc Environmental Research Centre, Johnstown Castle, Co. Wexford, Ireland

Hebah I. Hasan
Research and teaching assistant (master degree), Faculty of agriculture/The University of Jordan, Amman, Jordan

Anwar M. Battikhi
Professor of Soil Physics, Faculty of agriculture/The University of Jordan, Amman, Jordan

Mostafa M. Qrunfleh
Professor of Horticulture, Faculty of agriculture/The University of Jordan, Amman, Jordan

Aweke M. Gelaw
Ethiopian Agricultural Transformation Agency, Off Meskel Flower Road across Commercial Graduates, Addis Ababa, Ethiopia

B. R. Singh
Norwegian University of Life Sciences, 1432 Ås, Norway

R. Lal
Carbon Sequestration and Management Center, The Ohio State University, Columbus, OH 43210, USA

Ate Poortinga
Soil Physics and Land Management Group, Wageningen University and Research Center, Wageningen, The Netherlands

Joep G.S. Keijsers
Soil Physics and Land Management Group, Wageningen University and Research Center, Wageningen, The Netherlands

Jerry Maroulis
Soil Physics and Land Management Group, Wageningen University and Research Center, Wageningen, The Netherlands
Faculty of Science, Health, Education and Engineering, University of the Sunshine Coast, Maroochydore DC, Queensland, Australia

Saskia M. Visser
Team Soil Physics and Land Use, Alterra, Wageningen University and Research Center, Wageningen, The Netherlands

Giovanni Bitella
School of Agriculture, Forestry, Food and Environmental Sciences, University of Basilicata, Potenza 85100, Italy

Roberta Rossi
Consiglio per la Ricerca e la sperimentazione in Agricoltura—Research Unit for Cropping Systems in Dry Environments (CRA-SCA), Bari 70125, Italy

Rocco Bochicchio
School of Agriculture, Forestry, Food and Environmental Sciences, University of Basilicata, Potenza 85100, Italy

Michele Perniola
School of Agriculture, Forestry, Food and Environmental Sciences, University of Basilicata, Potenza 85100, Italy

Mariana Amato
School of Agriculture, Forestry, Food and Environmental Sciences, University of Basilicata, Potenza 85100, Italy

Isaurinda Baptista
Instituto Nacional de Investigação e Desenvolvimento Agrário (INIDA), CP 84, 7600 Praia, Cape Verde
Soil Physics and Land Management group, Wageningen University (WUR), 6700 AA Wageningen, The Netherlands

Luuk Fleskens
Soil Physics and Land Management group, Wageningen University (WUR), 6700 AA Wageningen, The Netherlands
Sustainability Research Institute, School of Earth and Environment, University of Leeds, Leeds LS2 9JT, UK

Coen Ritsema
Soil Physics and Land Management group, Wageningen University (WUR), 6700 AA Wageningen, The Netherlands

António Querido
Environment, Energy & Natural Disaster Prevention Unit, Office of United Nations, PO Box 62, 7600 Praia, Cape Verde

Jacques Tavares
Instituto Nacional de Investigação e Desenvolvimento Agrário (INIDA), CP 84, 7600 Praia, Cape Verde

António D. Ferreira
CERNAS, Escola Superior Agrária de Coimbra, Bencanta, P-3040-316 Coimbra, Portugal

Eduardo A. Reis
Instituto Nacional de Investigação e Desenvolvimento Agrário (INIDA), CP 84, 7600 Praia, Cape Verde

Samuel Gomes
Instituto Nacional de Investigação e Desenvolvimento Agrário (INIDA), CP 84, 7600 Praia, Cape Verde

Anabela Varela
Direcção geral de Agricultura e Desenvolvimento Rural (DGADR/DSER), CP 115, 7600 Praia, Cape Verde

Ranjan Bhattacharyya
Centre for Environment Science & Climate Resilient Agriculture, NRL Building, Indian Agricultural Research Institute, New Delhi 110 012, India

Birendra Nath Ghosh
Central Soil & water Conservation Research & Training Institute, Dehradun 248 195, India

Prasanta Kumar Mishra
Central Soil & water Conservation Research & Training Institute, Dehradun 248 195, India

Biswapati Mandal
Bidhan Chandra Krishi Viswa-Vidayala, Kalyani, West Bengal 741 235, India

Cherukumalli Srinivasa Rao
Central Research Institute on Dryland Agriculture, Hyderabad, Telangana 500 059, India

Dibyendu Sarkar
ICAR Research Complex for North Eastern Hill Region, Imphal, Manipur 795 004, India

Krishnendu Das
National Bureau of Soil Survey & Land Use Planning, Kolkata Regional Center, Kolkata 700 091, India

Kokkuvayil Sankaranarayanan Anil
National Bureau of Soil Survey & Land Use Planning, Bangalore Regional Center, Bangalore 560 024, India

Manickam Lalitha
National Bureau of Soil Survey & Land Use Planning, Bangalore Regional Center, Bangalore 560 024, India

Kuntal Mouli Hati
Division of Soil Physics, Indian Institute of Soil Science, Bhopal 462 038, India

Alan Joseph Franzluebbers
USDA-ARS, Plant Science Research Unit, Raleigh, NC 27695, USA

Ronald A. Fleming
University of Kentucky

Richard M. Adams
Oregon State University

David E. Ervin
Portland State University

Ilaria Zambon
Department of Agricultural and Forestry Sciences (DAFNE), Tuscia University, Via San Camillo de Lellis snc, Viterbo 01100, Italy

Fabrizio Colosimo
Department of Civil and Environmental Engineering, University of Strathclyde, 75 Montrose Street, Glasgow G1 1XJ, UK

Danilo Monarca
Department of Agricultural and Forestry Sciences (DAFNE), Tuscia University, Via San Camillo de Lellis snc, Viterbo 01100, Italy

Massimo Cecchini
Department of Agricultural and Forestry Sciences (DAFNE), Tuscia University, Via San Camillo de Lellis snc, Viterbo 01100, Italy

Francesco Gallucci
Council for Agricultural Research and Agricultural Economy Analysis (CREA) Research Unit for Agricultural Engineering, Via della Pascolare 16, Monterotondo, Rome 00015, Italy

Andrea Rosario Proto
Department of Agriculture, Mediterranean University of Reggio Calabria, Feo di Vito, Reggio Calabria 89122, Italy

Richard Lord
Department of Civil and Environmental Engineering, University of Strathclyde, 75 Montrose Street, Glasgow G1 1XJ, UK

Andrea Colantoni
Department of Agricultural and Forestry Sciences (DAFNE), Tuscia University, Via San Camillo de Lellis snc, Viterbo 01100, Italy

Shane R. Brockhoff
Iowa State University

Nick E. Christians
Iowa State University

Afranio Almir Righes
Environmental and Sanitary Engineering of Franciscan University, Santa Maria, Brazil

Galileo Adeli Buriol
Environmental and Sanitary Engineering of Franciscan University, Santa Maria, Brazil

Valduino Estefanel
Environmental and Sanitary Engineering of Franciscan University, Santa Maria, Brazil

Mareike Ließ
Department of Soil Physics, Helmholtz Centre for Environmental Research–UFZ, Halle (Saale), Germany
Department of Geosciences/ Soil Physics Division, University of Bayreuth, Bayreuth, Germany

Johannes Schmidt
Department of Physics and Geosciences/ Institute of Geography, University of Leipzig, Leipzig, Germany

Bruno Glaser
Department of Soil Biochemistry, Martin-Luther-Universität Halle-Wittenberg, Halle (Saale), Germany

Preface

Soil physics is the study of soil physical properties and processes. It is applied to management and prediction under natural and managed ecosystems. Soil physics deals with the dynamics of physical soil components and their phases as solid, liquids, and gases. It draws on the principles of physics, physical chemistry, engineering, and meteorology. It is especially important in this day and age because most farmers require an understanding of agro-ecosystems. Soil physics applies these principles to address practical problems of agriculture, ecology, and engineering. Soil Physics Agriculture and Environmental Applications consists of seventeen chapters. First chapter aims to evaluate the changes in physical and chemical properties of subsoil as a result of using CPL as a soil amendment. The main objective of second chapter was to determine the effect of different levels of compaction and polyacrylamide polymer application treatments (soil sealant) on the Ks of soils. Third chapter aims at ecological use of agricultural ecosystem to assure best quality of agriculture products and the preservation of agricultural landscapes. Fourth chapter aims to investigate Cd and Pb distribution in different types of wetlands, their variation with depth, and characterize the geochemical fractions of Cd and Pb in these wetlands. Fifth chapter presents a combination of field measurements and model simulation results for the Upper Xingu focused around three main goals: (1) analysis of water yield data collected in tropical forest and soybean catchments to understand the effects of land cover change on regional water dynamics; (2) determination of the best set of soil hydrological properties for simulating the partitioning of annual precipitation into evapotranspiration and discharge in forest and soybean catchments; and (3) investigation of the differences in annual evapotranspiration and streamflow in tropical forest, cerrado, pasture and soybean catchments in the Upper Xingu. Sixth chapter presents the comparison of Miscanthus and RCG to investigate comparative differences in crop development, leaf longevity and ecosystem-scale C fluxes from initial establishment to near maturity. Seventh chapter aims to evaluate the short-term influence of applying TWW on some soil chemical and physical properties and to study the yield and quality of Gladiolus communis, irrigated with TWW. Eighth chapter presents an approach on soil quality for evaluating small agricultural land use and the aim of ninth chapter was to characterize aeolian mass flux from wind tunnel and field data by comparing passive trap and high-frequency saltiphone data. The objective of tenth chapter the design of a low cost "open hardware" platform for the measurement, recording and wireless transmission of data on water content at different depths, air, soil and canopy temperatures. Eleventh chapter aims to corroborate the importance of SLM measures for Cape Verde, the strategies undertaken to address desertification, and the impact of those measures on the environment and rural livelihood. Twelfth chapter highlights on the challenges and solutions of soil degradation. The purpose of thirteenth chapter is to assess the potential for soil tests to improve ambient groundwater quality and producer profit. Fourteenth chapter aims to determine the opportunity to create an innovative supply chain from agricultural waste biomass, especially regarding olive (Olea europaea) and hazelnut (Corylus avellana) in order to evaluate the production of biochar from their pruning residues. The objectives of fifteenth chapter were to quantify soil water retention capabilities, determine water infiltration rates, and measure rooting depths of creeping bentgrass in sand and biochar soils. Sixteenth chapter reviews on water infiltration and time to recharge the profile of soils, and spatial prediction of soil organic carbon stocks is assessed in last chapter.

Editor
Dr. Hazem Fouda, Ph.D.

INTRODUCTION

Soil Physics describes the physical properties of soils and how these properties affect agriculture and the environment. Soil physics is the study of soil physical properties and processes, including measurement and prediction under natural and managed ecosystems. The science of soil physics deals with the forms, interrelations, and changes in soil components and multiple phases. The typical components are: mineral matter, organic matter, liquid, and air. Three phases are solid, solution and gas, and more than one liquid phase may exist in the case of no aqueous contamination. Physical edaphology is a science dealing with application of soil physics to agricultural land use. The study of the physical phenomena of soil in relation to atmospheric conditions, plant growth, soil properties and anthropogenic activities is called physical edaphology. Study of soil in relation to plant growth is called edaphology, whereas study of soil's physical properties and processes in relation to plant growth is called physical edaphology. Thus, physical edaphology is a branch of soil physics dealing with plant growth.

Soil physics is a young and emerging branch of penology, with significant developments occurring during the middle of twentieth century. It draws heavily on the basic principles of physics, physical chemistry, hydrology, engineering and micrometeorology. Soil physics applies these principles to address practical problems of agriculture, ecology, and engineering. Its interaction with emerging disciplines of geography (geographic information system or GIS), data collection (remote sensing), and analytical techniques (fuzzy logic, fractal analysis, neural network, etc.) has proven beneficial in addressing practical problems in agriculture, ecology, and environments. Indeed, soil physics plays a pivotal role in the human endeavor to sustain agricultural productivity while maintaining environment quality.

Agricultural production depends very much on the physical properties of the soil, and mainly on those related to the soil's water holding and transmission capacities. These properties affect the availability of water to crops and may, therefore, be responsible for crop yields. The knowledge of the physical properties of soil is essential in defining and/or improving soil water management practices to achieve optimal productivity for each soil/climatic condition. In many parts of the world, crop production is also severely limited by the high salt content of soils and water. Such soils, classified either as saline or sodic/saline depending on their alkalinity, are capable of supporting very little vegetative growth.

Soil Physics and Agricultural Sustainability

Agricultural sustainability implies non-negative trends in productivity while preserving the resource base and maintaining environmental quality. The role of physical edaphology in sustaining agricultural production while preserving the environment cannot be overemphasized. While the economic and environmental risks of soil degradation and desertification are widely recognized, the underlying processes and mechanisms are hardly understood. It is in this connection that the application of soil physics or physical edaphology has an important role to play in:

- preserving the resource base,
- improving resource use efficiency,
- minimizing risks of erosion and soil degradation, and restoring and reclaiming degraded soils and ecosystems, and
- enhancing production by alleviation of soil/weather constraints through development and identification of judicious management options

Notable applications of soil physics include control of soil erosion; alleviation of soil compaction; management of soil salinity; moderation of soil, air, and water through drainage and irrigation; and alteration of soil temperature through tillage and residue management. It is a misconception and a myth that agricultural productivity can be sustained by addition of fertilizer and/or water per se. Expensive inputs can be easily wasted if soil physical properties are suboptimal or below the critical level. High soil physical quality plays an important role in enhancing soil chemical and biological qualities. Applications of soil physics can play a crucial role in sustainable management of natural resources. Fertilizer, amendments, and pesticides can be leached out, washed away, volatilized, miss the target, and pollute the environment under adverse soil physical conditions. Efficient use of water and nutrient resources depends on an optimum level of soil physical properties and processes. Soil fertility, in its broad sense, depends on a favorable interaction between soil components and phases that optimize soil physical quality. Soil physical properties important to agricultural sustainability are texture, structure, water retention and transmission, heat capacity and thermal conductivity, soil strength, etc.

These properties affect plant growth and vigor directly and indirectly. Important soil physical properties and processes for specific agronomic, engineering, and environmental functions are outlined. Soil structure, water retention and transmission properties, and aeration play crucial roles in soil quality.

Soil Physics and Environment Quality

In the context of environment quality, soil is a geomembrane that buffers and filters pollutants out of the environment. It is also a vast reactor that transforms, deactivates, denatures, or detoxifies chemicals. Soil physical properties and processes play an important role in these processes. The environmental purification functions of soil are especially important to managing and moderating the quality of air and water resources. Soil physical properties and processes influence the greenhouse effect through their control on emission of radioactively-active gases (e.g., CO_2, CH_4, N_2O, and NO_x). A considerable part of the 80 ppmv increase in atmospheric CO_2 concentration since the industrial revolution has come from C contained in world soils. Soil physical properties and processes determine the rate and magnitude of these gaseous emissions. Formation and stabilization of soil structure (i.e., development of secondary particles through formation of organ mineral complexes), is a prominent consequence of C sequestration in soil. Air quality is also influenced by soil particles and chemicals (salt) airborne by wind currents. Management of soil structure, control of soil erosion, and restoration of depleted soils are important strategies of mitigating the global climate change caused by atmospheric enrichment of CO_2.

Fresh water, although renewable, is also a finite quantity and a scarce resource especially in arid and semiarid regions. Soil, a major reservoir of fresh water, influences the quality of surface and ground waters. The pedosphere processes (e.g., leaching, erosion, transport of dissolved and suspended loads in water) interact with the biosphere and the atmosphere to influence properties of the hydrosphere. Soil physical properties important to the hydrosphere, in terms of the quality and quantity of fresh water resources, are water retention and transmission properties of the soil, surface area and charge properties, and composition of inorganic and organic constituents.

The approach that integrates knowledge is very important in Agriculture, including farmers, extensionists, researchers and professors. The specialists, including the soil physicists, must have a global view of the crop production system. Therefore, their expertise can be useful for the society. The Essence of scientific knowledge is its practical application. The soil physics is a subarea of Agronomy. There are many examples of this specific subject related to Agriculture. This paper will focus, in general, the following cases: (i) erosion, environmental pollution and human health, (ii) plant population and distribution, soil fertility, evapotranspiration and soil water flux density, and (iii) productivity, effective root depth, water deficit and yield.

Plant Population and Distribution, Evapotranspiration, Soil Fertility, and Soil Water Flux Density

The understanding of relationship of plant population and distribution, soil fertility, evapotranspiration and soil water flux density is fundamental to optimize the soil resources. This implies that soil physicists, with agricultural system global vision, could develop techniques to make the crop production system more adequate for each specific environment. Increasing of plant population demands more water and better plant distribution. Increasing evapotranspiration requires more soil water flux density. The soil fertility depends on the soil volume per plant (plant population and distribution), evapotranspiration (leaf area, species, wind, air temperature and relative humidity, mainly) and soil water flux density. The maximum water requirement occurs at flowering. The correct plant population is defined as function of the probability of soil water flux density to be equal or superior to maximum evapotranspiration any day in the whole crop cycle. For high population, the plant distribution becomes more important. The soil physicist must minimize intra specific competition for water and nutrients. The better plant distribution maximizes the soil volume per plant, and the critical content values for all nutrients (soil fertility) are lower. Consequently, the fertilizer requirement decreases. The corn grain production per plant is constant when there is no intra specific competition for water and nutrients, and the grain production per area has linear increment with the increasing of plant population (phase A – Figure 1). The corn grain production per plant decreases when there is intra specific competition for water and nutrients, and the grain production per area has potential (less than linear) increment with the increasing of plant population (phase B – Figure 1). The grain production per plant decrement rate is lower than the plant population increment rate, then the grain production per area increases. The corn grain production per plant decreases when there is intra specific competition for water, nutrients and light, and the grain production per area decreases with the increasing of plant population (phase C – Figure 1). The grain production per plant decrement rate is higher than the plant population increment rate, then the grain production per area decreases. The point AB (Figure 1) shows when the intra specific competition for water and nutrients starts. The correspondent plant population can be larger in better plant distribution, and the maximum grain production per area also can be larger in higher plant population (point BC – Figure 1), when the intra specific competition for light starts. The soil resources (physical and chemical attributes) optimization (plant distribution) is a subject of soil physicists (see Plant population below). Soil physicists should start with a dynamic focus and replace the traditional static emphasis. An example is the critical value for potassium (soil fertility) (see Plant distribution still further below).

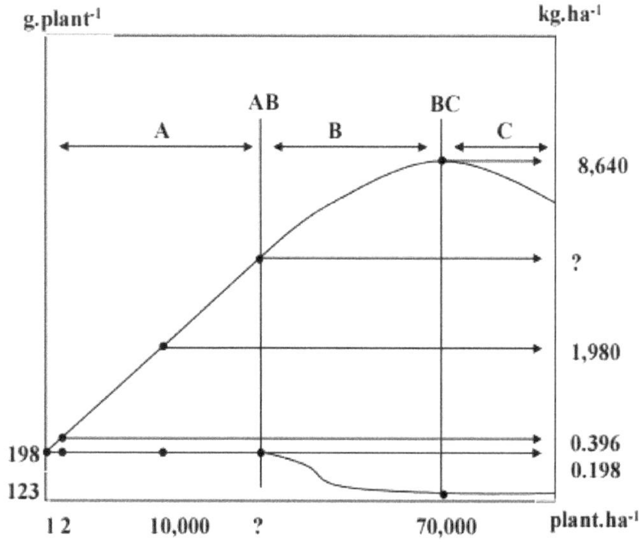

Figure 1. Corn grain production per plant (g.plant-1) and per area (kg.ha-1) as function of plant population.

The beginning of soil physics

The importance of soil physical properties has been recognized since the early days of agriculture, as evidenced by the use of tillage tools for land preparation, for example. It is, however, generally accepted that the scientific study of the physical properties of soils started in Europe in the early nineteenth century. Most of the early soil physics research focused on the effects of soil physical properties on crop growth, although the discovery of Darcy's law was a direct result of the application of soil physics to water pollution problems in France. German scientists Schübler, Schumacher, and Wollny were among the first to study soil physical properties for crop production in the nineteenth century. Schübler determined specific gravity, bulk density, water-holding capacity, swelling/shrinkage, hygroscopic water content, heat capacity, latent heat of wetting, and electrical conductivity of soils. Schumacher's interest was more in the movement of air and water in the soil. Schumacher introduced the concept of capillary-saturation capacity (or 'field capacity') of soils, and suggested various management practices to improve the structure of soils. Wollny studied the effects of plant growth and soil management on soil physical properties. He founded one of the first soil physics journals, and is therefore considered by many to be the father of soil physics in Europe. He did experiments to measure soil temperature, water and gaseous composition, how these properties were affected by soil management and earthworms, and how this affected crop productivity. Also in Europe, French scientist Schloesing and Dutch scientist Van Bemmelen wrote about the colloidal properties of clay in the late nineteenth century, including its ability to hold water when air-dry. Mitscherlich researched the heat of wetting of soil and refined the method to determine hygroscopic water content. Darcy did his experiments in the nineteenth century in France that lead to the formulation of Darcy's law of saturated flow, published in 1856. Darcy worked for the city of Dijon that hired him to develop a treatment system for its surface water that was highly polluted by its famous mustard industry. Darcy's law was later independently derived by Hagen and Poiseuille from Newton's law of viscosity and the water flux in capillary tubes. In America, scientists Hillgard, Johnson, King, and Slichter investigated various aspects of soil physics in the nineteenth century. Hillgard studied the particle size of soils and laid the foundations for modern soil survey. Johnson researched the effects of tillage on soil physical properties and capillary movement of water. King, considered to be the father of soil physics in the United States, researched the water holding capacity of soils, the water requirements of different crops, air porosity, and the effects of tillage on soil. King published a book 'Physics of Agriculture' that remained the authoritative text on soil physics for several decades. Slichter did fundamental research into steady-state movement of water in soils and introduced the term 'potential' into soil physics literature. Briggs and Shantz developed the concept of maximum available moisture, defined as the difference between wilting coefficient ('wilting point') and moisture equivalent (roughly 'field capacity'). Their study 'The wilting coefficient for different plants and its indirect determination' was published in 1912. Buckingham, in 1907, published 'Studies on the Movement of Soil Moisture', in which he recognizes that water flow in unsaturated soils is proportional to matric potential. Unfortunately, the importance of this work was only recognized many years later.

Contemporary soil physics

Theory of soil physics

Solid phase

A large fraction of the solid phase of the soil consists of a size distribution of inorganic particles. Various scientific disciplines have classified inorganic particles into groups, called soil separates, based on their size. One common classification, used by the

USDA in the United States, is shown in Table 1. Soil texture is defined as the relative proportion of sand, silt, and clay in the < 2.0-mm inorganic fraction of the soil.

Table 1. The USDA classification of soil particles and selected properties.

Soil Separate	Size Range (mm)	Shape	Surface Area (m^2g^{-1})
Rock Fragments	> 2.0	Irregular, rounded, to thin and flat	up to 0.001
Sand	0.05 – 2.0	rounded, irregular	0.001 to 0.04
Clay	0.002 – 0.05	rounded, irregular	0.04 to 1.0
Silt	< 0.002	platelike	5 to 870

The organic fraction of the solid phase of the soil typically represents 10 to 50 g kg^{-1} of solids in the A horizon of an inorganic soil. Lower horizons in the soil profile typically contain < 10 g of organic matter per kg of solids. Soils are classified as organic soils if the proportion of organic materials exceeds 200 g kg^{-1} of solids. Organic matter in a soil ranges from relatively inert particles of plant and animal material to their decomposition residues, called humus. Most of the humus, the active fraction of the organic matter, is bonded to the inorganic fraction of the soil. Humus decomposes very slowly and some of it has been in the soil over 1000 years. The < 0.002-mm fraction of the soil, the clay and humus portion, is the physically and chemically active part of the soil. The mineralogy and the proportion of the < 0.002-mm fraction of the soil determine to a large extent the physical and chemical properties of a soil. Most of the < 0.002-mm fraction of the soil is composed of sheets of layer silicates. The properties of these minerals are shown in Table 2. The solid phase is not a completely disorganized jumble of materials. We know that in the top meter or so, most of the material that is smaller than 2.0 mm in size is usually organized into units, called peds or soil structural units, as shown in Figure 1. These units are formed from numerous adjacent particles that are weakly glued together by the highly active clay and humus material. Normally only one of these shapes is dominant in any one horizon of the soil. The genesis of peds is not completely understood, but it involves the types of layer silicate clays, iron and aluminum oxides, shrinking and swelling of the solid phase, the wetting and drying of the particles, the activity of animals and plants, and the availability of organic exudates and decay products.

Table 2. Properties of some common layer silicate clays.

Layer Silicate	Surface Area (m^2g^{-1}	Cation Exchange Capacity ($cmol_c kg^{-1}$)
Kaolinite	10 – 20	3 - 15
Illite	65 – 100	0 - 40
Vermiculite	40 – 870	100 - 150
Smectite	50 – 840	80 - 150
Chlorite	80	10 – 40

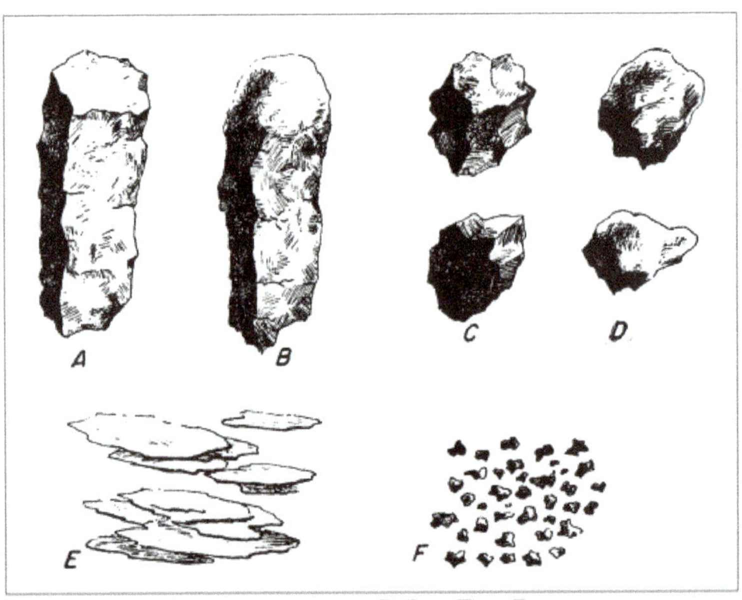

Figure 2. Drawings illustrating some of the types of soil structure: A. prismatic, B. columnar, C. angular blocky, D. subangular blocky, E. platy, and F. granular.

The solid phase of the soil is characterized as well by its density. Individual particles have a density similar to their primary minerals ranging from 1.3 Mg m-3 for organic particles to 3.5 Mg m-3 or greater for soils with high iron content. The average density of all particles in the soil is called the particle density. Most soils have a particle density close to that of quartz and a value of 2.65 Mg m-3 is frequently used. Since soils contain pore space as well as solid-filled space, a second density, the bulk density, is used to represent the mass of solids divided by the total volume occupied by the solids and pores. Soils vary widely in this property, but values of bulk density typically are between 1.0 and 2.0 Mg m-3 with values in the range of 1.2 to 1.5 Mg m-3 being optimum for most plant growth.

CHAPTER 1

AMENDING SUBSOIL WITH COMPOSTED POULTRY LITTER-I: EFFECTS ON SOIL PHYSICAL AND CHEMICAL PROPERTIES

Mili Mandal, Rakesh S. Chandran and John C. Sencindiver

Keywords: Amendment; improving turf medium; turfgrass; urban lawn establishment

INTRODUCTION

Topsoil on sloping landscapes in the Appalachian region is often thin, rocky and infertile [1]. The topsoil is usually lost or is mixed with subsoil during the construction process. Lawns established under such conditions are not healthy and may be vulnerable to pest problems. Amending such soils with suitable composts may improve the medium for turfgrass growth. Composted poultry litter (CPL) contains several plant macronutrients and micronutrients and has been considered as a valuable soil amendment to enhance soil fertility, organic matter, and tilth [2,3,4,5,6]. Improper use and disposal of poultry litter may have detrimental effects on environment and human health. Instances of diseases such as methemoglobinemia, cancer, and respiratory illness have occasionally been attributed to NO_3^- leaching from poultry litter [5,7,8]. Other undesirable attributes such as eutrophication, production of phytotoxic substances, air pollution, and emission of ozone depleting gases have also been associated with improper disposal of this waste product [7].

Citation: Mili Mandal, Rakesh S. Chandran, and John C. Sencindiver, Amending Subsoil with Composted Poultry Litter-I: Effects on Soil Physical and Chemical Properties, doi:10.3390/agronomy3040657.

Composting may mitigate the environmental problems associated with the application of raw poultry litter and may provide a beneficial method for handling this waste product of the poultry industry. Along with the reduction in odor, pathogens, and weeds, composting also immobilizes highly mobile NO_3^- in the litter to its slow-release form [9,10,11]. It is also easier to handle, store, transport, and apply litter that has been composted [12]. Composting also reduces total phosphorus (P), due to dilution of manure with woodchips and straw [13].

Considerable data on the effect of various organic materials on soil physical and chemical properties are available. Sewage sludge reduced bulk density and increased soil water content, available water, CEC, and soil pH [14,15]. Municipal solid waste increased nutrient and water retention capacity, total pore space, and aggregate stability, and decreased cone penetration resistance and bulk density [16,17]. Cattle feedlot litter increased soil organic carbon and water stable aggregates, and decreased bulk density and hydraulic conductivity of disturbed soils [18]. Application of poultry litter on pasture soils over a 20-yr period altered soil properties (pH, electric conductivity, and bulk density) macrocation levels (Ca, K, Mg and Na levels), and increased hay yields [19]. In this study, the authors also noted that the labile form of phosphorus increased following long-term sequential application of poultry litter. Similar conclusions were drawn by Gascho and Hubbard [20] when broiler litter was used to amend soils over a 5-yr period.

The effects of different types of compost materials as an amendment for lawns and other types of turf have been well documented [21,22,23,24,25]. While sequential applications of poultry litter over multiple years may result in undesirable effects to the environment, single incorporation of the same as compost to the subsoil may be beneficial. Limited information is available on the effect of composted poultry litter (CPL) on infertile subsoil as it relates to turf production. The objective of this experiment was to evaluate the changes in physical and chemical properties of subsoil as a result of using CPL as a soil amendment. In a related field experiment, the effects of amending a subsoil using CPL on the growth attributes of Kentucky bluegrasss (*Poa pratensis* L.) and weed populations were investigated [26]. Turfgrass was chosen as the plant material due to its ubiquitous nature in the landscape typically established in housing and other developments with poor soil quality and a minimal A-horizon.

RESULTS AND DISCUSSION

Soil Physical Properties

Total Water Content and Soil Organic Matter

The subsoil was determined to be a silt loam. The effects of compost rates on total water content, θ_T, are presented inFigure 1. For both seeded and sodded treatments, percent soil water content was higher in compost-treated plots than in fertilized and control plots and increased linearly with percent compost (sodded, $R^2 = 0.95$; seeded, $R^2 = 0.73$). There was an increase in organic matter content due to initial addition of CPL into the subsoil (sodded $R^2 = 0.94$; seeded $R^2 = 0.98$) but differences in the same could not be determined with subsequent incorporation rates of CPL (Figure 2).

The increases in θ_T may be associated with the increase in organic matter and its effect on soil physical properties. Organic matter is capable of absorbing up to 20 times its weight of water [27]. This ability is related to the large number of hydrophilic aromatic and aliphatic compounds in the organic matter structure [28]. Organic matter also acts as a cementing agent between soil particles, thereby improving aggregation and increasing porosity [27]. This process of aggregation improves water infiltration, soil tilth, and subsequently plant growth especially in loamy soils.

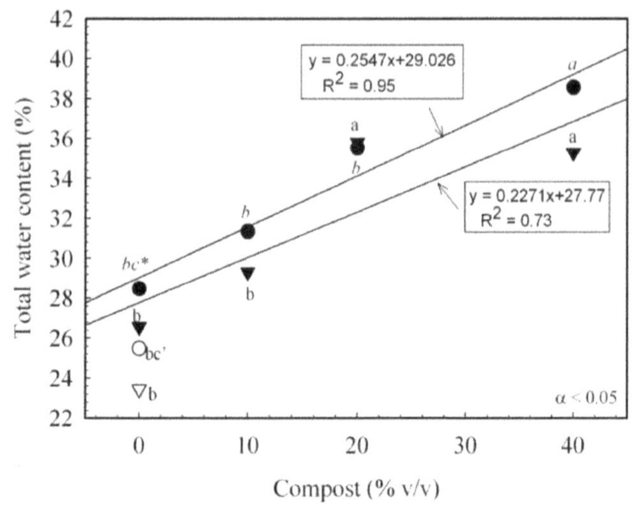

Figure 1. Effect of incorporating composted poultry litter into top 12.7 cm of subsoil on percent total water content (θ_T) by volume for • sodded compost; ○ sodded fertilizer; ▼ seeded compost; and ▽ seeded fertilizer plots; * Means with same letters are statistically insignificant based on Fisher's Protected LSD test ($P \leq 0.05$) (italicized statistics correspond to sodded treatments); † values for standard error were small ranging between 0.085 (min.) to 0.553 (max.) and hence are not shown in figure.

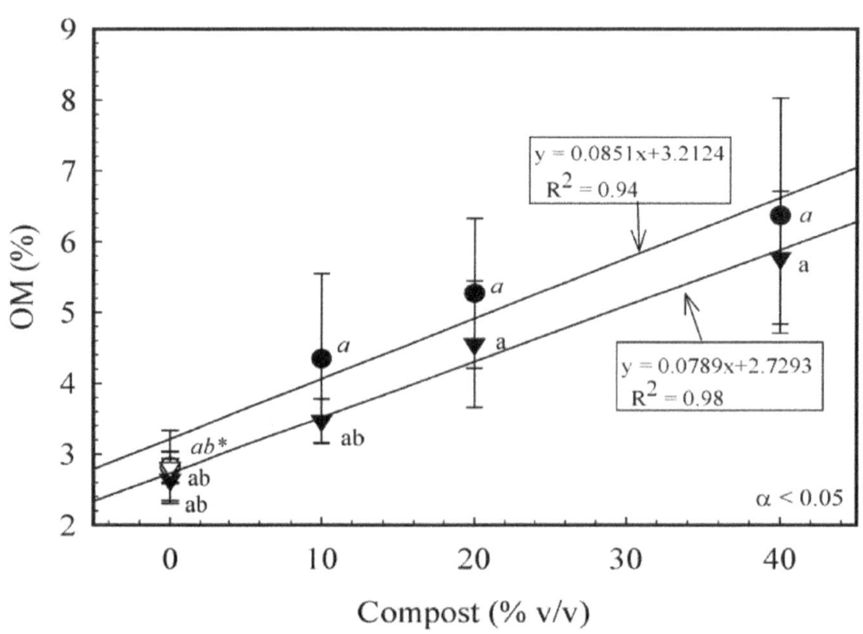

Figure 2. Effect of incorporating composted poultry litter into top 12.7 cm of subsoil on organic matter for • sodded compost; ○ sodded fertilizer; ▼ seeded compost; and ▽ seeded fertilizer plots; * Means with same letters are statistically insignificant based on Fisher's Protected LSD test ($P \leq 0.05$) (italicized statistics correspond to sodded treatments).

Available Water Capacity and Soil Bulk Density

A linear negative relationship (sodded, $R^2 = 0.89$; seeded, $R^2 = 0.97$) was observed between percent compost and available water, A_v (Figure 3). As percent compost increased, A_v decreased (Figure 3) whereas the total amount of water in soil increased (Figure 1). This suggests that the higher the compost rate, the larger is the total volume of water held by soils and lower is the volume available for plant uptake which indicates that more water is retained by soil particles.

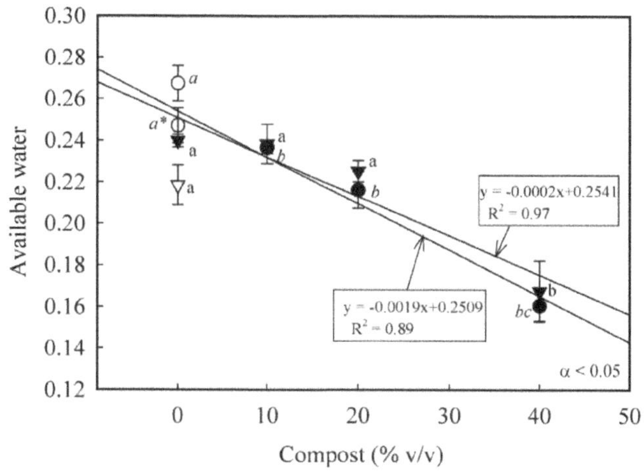

Figure 3. Effect of incorporating composted poultry litter into top 12.7 cm of subsoil on available water, A_v for • sodded compost; ○ sodded fertilizer; ▼ seeded compost; and ▽ seeded fertilizer plots; * Means with same letters are statistically insignificant based on Fisher's Protected LSD test ($P \leq 0.05$) (italicized statistics correspond to sodded treatments).

In this study, we determined both θ_T, and A_v, since each measured a different pool of water. Soil water content measures the total amount of water held by soil ($\theta_T = A_v + Unavailable$) whereas A_v determines the amount of water that can be extracted by plants. This water is held between field capacity and wilting point by capillary forces [29].

Organic matter addition has been reported to increase A_v only in coarse textured soils with <15% clay. With soils having >15% clay such as the soil used in this study (24% clay), factors other than organic matter determine A_v [30]. Reports on the effects of compost on A_v ranged from increase, decrease, and no change of A_v with compost additions, and, were related to variation in the soil physical properties such as texture, pore size distribution [5], percent clay and structure [30], and bulk density [31,32].

Angle *et al.*, [33] reported an increase in θ_T of soil with increased rate of sludge compost but observed no effect on A_v. Jamison and Kroth [34] attributed the changes in A_v to the increase in either percent silt or clay and their effect on pore size distribution. They concluded that the clay particles in silt loam soils with more than 20% clay reduced the water storage by diluting the soil mass and shifting the pore size distribution towards smaller pores, where more water is held in the unavailable range. Metzger and Yaron [35] observed an increase in A_v of sandy soils, whereas that of a loamy soil remained unchanged upon addition of sludge. MacRae and Mehuys [36] reported a decrease in A_v in silt loam following the addition of peat at different rates.

Stevenson [37] explained that the net result of addition of organic material on A_v depends on the relative changes in percent water by weight of soil (W_{15}–$W_{1/3}$) and decreases of bulk densities dictated by soil type. For silt loam, he associated the decrease in A_v (at 0.3×10^5 Pa) to a consistent decrease in soil bulk density with peat moss addition. These observations agree with our results (Figure 4). We recorded a decrease in A_v of a silt loam from addition of CPL along with a decrease in bulk density (Figure 3 and Figure 4).

Soil bulk density decreased linearly, up to 42% with 40% compost, with increasing rates of compost (sodded, $R^2 = 0.98$; seeded, $R^2 = 0.97$) (Figure 4). Decrease in bulk density is generally associated with the low particle density of organic matter, which when mixed with the mineral fractions of soils greatly improves aggregation and porosity [36]. The reduction in the soil bulk density with organic material addition is also related to factors such as compost application rate, soil type, and extent of soil compaction [37]. Wei *et al.* [38] also observed a 6.3% decrease in bulk density of silty clay loam by the application of sludge compost at 1.34×10^{-8} kg m^{-2}.

Soil Chemical Properties

The subsoil was determined to have a pH of 6.0 and CEC of 11 cmol$_c$ kg^{-1}. The addition of CPL to subsoil increased its pH linearly (sodded, $R^2 = 0.97$, seeded, $R^2 = 0.84$) from 6.0 (control) to 7.4 (0.4 cm-compost/cm-soil) (Figure 5). An increase in the availability of soil basic cations also was observed as a result of compost addition (Table 1). These effects may be explained by the initial high pH of 8.3 of CPL and its high basic cation content (Table 1) and a possible increase in availability of inherent cations associated with the pH change. Increases in soil pH through compost application (poultry litter) were also reported by Rasnake and Murdock [39]. They attributed this increase to the high Ca content of poultry litter and its effect on exchangeable soil Ca. However, in a more recent study, application of broiler litter had no effect on total soil pH nor was there any interaction between broiler litter rate and depth [20].

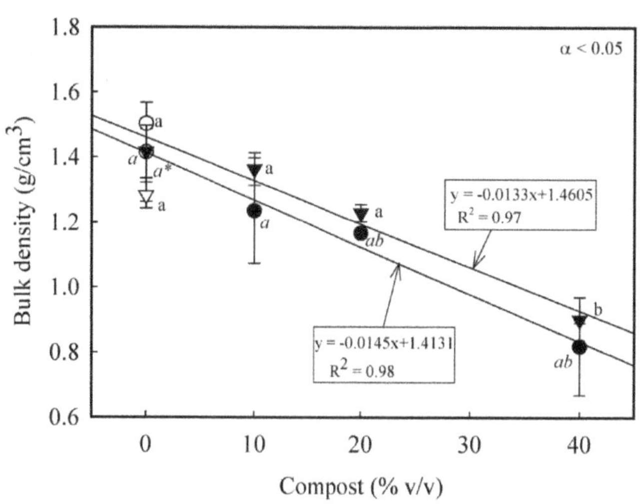

Figure 4. Effect of incorporating composted poultry litter into top 12.7 cm of subsoil on soil bulk density for • sodded compost; ○ sodded fertilizer; ▼ seeded compost; and ▽ seeded fertilizer plots; * Means with same letters are statistically insignificant based on Fisher's Protected LSD test ($P \leq 0.05$) (italicized statistics correspond to sodded treatments).

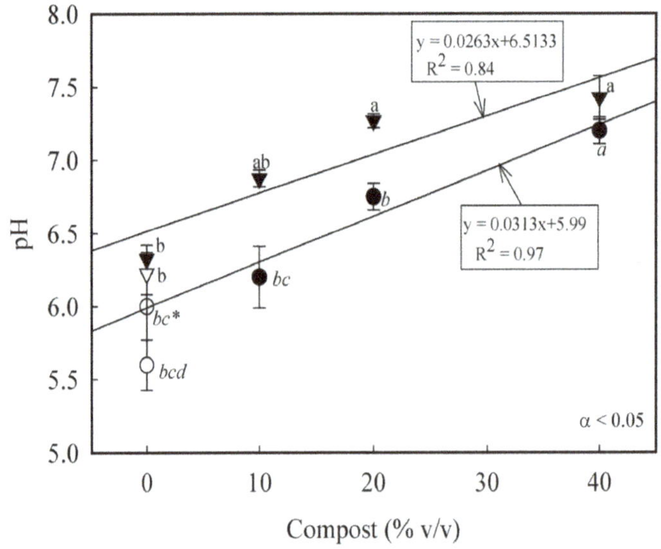

Figure 5. Effect of incorporating composted poultry litter into top 12.7 cm of subsoil on soil pH for ● sodded compost; ○ sodded fertilizer; ▼ seeded compost; and ▽ seeded fertilizer plots; * Means with same letters are statistically insignificant based on Fisher's Protected LSD test ($P \leq 0.05$) (italicized statistics correspond to sodded treatments).

Table 1. Effect of incorporating composted poultry litter into top 12.7 cm of subsoil on its mineral nutrient content; * Means with same letters are statistically insignificant based on Fisher's Protected LSD test ($P \leq 0.05$).

Treatment (cm-compost/cm-soil)	P	K	Ca	Mg
		Seeded		
		--------- ppm -----------		
Compost (0.1)	0.227 a*	1.306 ab	4.879 bc	0.620 b
Compost (0.2)	0.232 a	1.915 a	7.606 b	0.919 b
Compost (0.4)	0.241 a	2.337 a	9.087 a	1.671 a
Fertilizer (0.0)	0.110 b	1.28 b	2.661 bcd	0.255 bc
Control (0.0)	0.101 b	0.245 b	2.708 bcd	0.288 bc
		Sodded		
		--------- ppm -----------		
Compost (0.1)	0.227 a	0.738 b	4.442 b	0.581 bc
Compost (0.2)	0.237 a	1.236 b	7.044 a	0.946 b
Compost (0.4)	0.240 a	2.291 a	7.929 a	1.530 a
Fertilizer (0.0)	0.091 b	0.158 bc	2.156 bc	0.264 bcd
Control (0.0)	0.055 bc	.259 bc	2.266 bc	0.250 bcd

Control plots had higher pH (6.0) compared to fertilized plots (5.9) (Figure 5). The decrease in pH of fertilized plots can be attributed to the application of fertilizer with a composition of NH_4-N (4.7%), urea (8.3%), other water soluble N (6.4%), and water insoluble N (0.6%). NH_4-N fertilizers are known to decrease pH through leaching of basic cations and production of H^+ as a result of nitrification [40]. In this study, leaching of basic cations and the resultant exchange with NH_4^+ ions was most likely responsible for the lowering of pH in fertilized plots. Our results are consistent with those of Rasnake and Murdock [39] and Eghball [2] who reported a decrease in pH from 6.7–6.3 and 6.4–5.6 respectively, following NH_4-N fertilizer applications.

Soil CEC increased from 11 $cmol_c\ kg^{-1}$ (control) to an average of 28 $cmol_c\ kg^{-1}$ with increasing rates of compost application (Figure 6). This can be explained by both the increase in the amount of organic matter due to CPL addition and the concomitant increase in pH and available cations (Figure 5). The CEC of organic matter ranges from 100–300 $cmol_c\ kg^{-1}$, [29]. The effect of pH on CEC is most prominent in soils with a significant amount of pH-dependent clays. In our study, the mineralogical composition of soil was not determined, but based on the low CEC of control plots (11 $cmol_c\ kg^{-1}$) we believe that it is mostly formed of pH-dependent clays. Gao and Chang [3] also observed increases in CEC following cattle feedlot manure applications and attributed this increase mainly to the increase in organic matter.

In general, the addition of CPL to subsoil improved its nutrient content (Table 1). Higher concentrations of P, K, Mg and Ca were observed at the high (40%) compost rate for both seeded and sodded plots. This increase in nutrient content also reflects the rich nutrient composition of composted poultry litter. Levels of basic cations (Ca, K, and Mg) increased linearly and consequently increased base saturation. Gagnon [41] reported an increase of 60% in available soil P treated with poultry litter as compared with other on-farm and industrial composts. Significantly larger increases in soil P (from 52%–240%) were reported by Balkcom *et al.* [42]. These increases are due to relatively high content of total P in poultry litter. In studies by Gascho and Hubbard [20], soil P and Ca levels also increased with broiler litter application, while soil K levels decreased. Decrease in soil K was attributed to crop uptake. Our studies demonstrate that composted poultry litter can be used successfully to improve soils where turfgrasses are established in poor subsoil, typically encountered in suburbia.

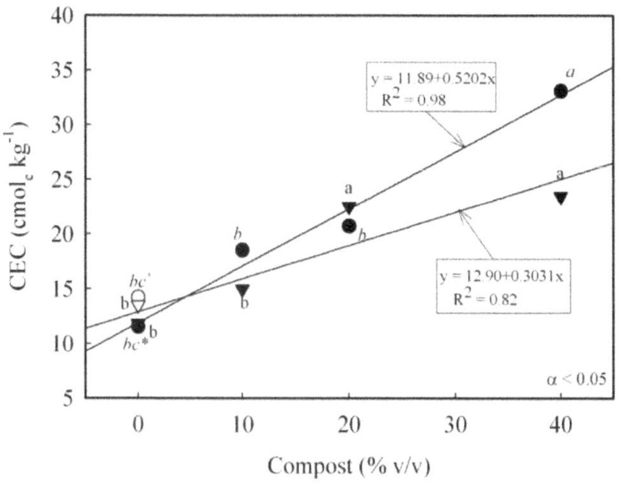

Figure 6. Effect of incorporating composted poultry litter into top 12.7 cm of subsoil on cation exchange capacity (CEC) for • sodded compost; ○ sodded fertilizer; ▼ seeded compost; and ▽ seeded fertilizer plots; * Means with same letters are statistically insignificant based on Fisher's Protected LSD test ($P \leq 0.05$) (italicized statistics correspond to sodded treatments); † values for standard error were small ranging between 0.065 (min.) to 0.443 (max.) and hence are not shown in figure.

Positive linear increases in θ_T, organic matter, pH, CEC, and mineral nutrient content were observed with a linear decrease in soil bulk density and A_v by composted poultry litter incorporation. Improvement in physical and chemical properties of subsoil could be attributed to its dual role of supplying organic matter and essential nutrients, and raising the soil pH thereby releasing more cations. Both the amount and composition of organic matter added to the soil affect the extent to which its physical and chemical properties are modified. These results demonstrate that composted poultry litter can be used to enhance the soil physical and chemical properties of eroded or disturbed soils. Amending soils with organic wastes can have two advantages; apart from managing the disposal problem it can also improve soil nutrient and organic matter content.

EXPERIMENTAL SECTION

Experimental Site: Field experiments were established in fall 2003 at the West Virginia University, Agronomy Farm, Morgantown (39.66N, 79.90W). The soil was a Dormont silt loam (fine-loamy, mixed, superactive, mesic Oxyaquic Hapludalfs). To simulate construction disturbance existing topsoil containing the A-horizon was removed (to a depth of 20 cm) and the subsoil was exposed. The exposed subsoil was tilled uniformly to a depth of 12.7 cm using a rototiller and plots 4.6 m × 3.1 m separated by 0.6 m alleyways were delineated. The compost utilized for the experiments was derived from poultry litter with wood chips and cardboard as a primary source of carbon with a ratio of 2:2:2 (N, P_2O_5, K_2O) (Borderline LLC, Baker, WV). Treatments consisted of CPL incorporated to a depth of 12.7 cm at rates of 0.1 cm-compost/cm-soil, 0.2 cm-compost/cm-soil, and 0.4 cm-compost/cm-soil (equivalent to 4.3 kg m^{-2}, 8.75 kg m^{-2}, and 17.5 kg m^{-2}, respectively), control (untreated), and fertilized plots (N, P_2O_5, K_2O fertilizer 20-27-5; applied at 50 × 10^{-4} kg m^{-2}). The three compost treatments provided N at 0.086, 0.172, and 0.34 kg m^{-2}, whereas the fertilizer treatment provided N at 0.001 kg m^{-2}. The fertilizer was applied using a drop spreader and incorporated lightly (1 cm) into the soil using a rake. Based on the different depths to which the compost and the fertilizer treatments were incorporated, it was estimated that the compost provided N at 6.77× 10^{-7} kg cm^{-3} whereas the fertilizer provided N at 10^{-7} kg cm^{-3}

The experimental design was a randomized complete block with turf type (seeded or sodded) and amendment treatments arranged factorially. All the treatments were replicated four times. Twenty plots were seeded on 7 October 2003 with Kentucky bluegrass seeds (Scotts "Classic"; 97.75% Kentucky bluegrass, 0.05% other crop seeds, 2.11% inert matter, 0.09% weed seeds) at the rate of 50 × 10^{-4} kg m^{-2} using a drop spreader. Seeded plots were covered with straw mulch in order to retain soil moisture and to offer protection from predators. The 20 remaining plots were sodded with Kentucky bluegrass manually on 10 October 2003 with sod (from Rich Farm, Smithfield, PA, USA). Plots were irrigated uniformly as needed during establishment. After establishment, the turf was maintained at a height of 7.6 cm using a reel-mower and clippings removed. Rainfall data were collected during the months of May through September 2004 (Table 2).

Table 2. Precipitation recorded from May to September in 2004.

Month	Precipitation amount (cm)
May	13.36
June	12.45
July	9.02
August	16.23
September	17.07
Total	68.12

Soil Analysis: Soil samples were collected twice at a depth of 20.3 cm from each plot during the study using a soil probe. At each sampling time, 7–8 subsamples were collected from random locations within each plot and were mixed thoroughly to make one composite sample. Samples collected during February 2004 were air dried, machine ground and sieved through a 2-mm screen, and were analyzed for CEC and levels of extractable Ca, P, Mg, and K. Soil samples were collected again as described above during January 2005 and were manually cleaned for any pebbles and debris and then air dried, hand ground, sieved through a 2-mm screen was used to analyze available water and organic matter content. The pH in water was measured with a glass electrode using a 1:1 soil/water ratio [43]. Soil organic matter (OM) was determined by the loss on ignition method (LOI) [44]. The LOI was corrected for dehydroxylation of inorganic constituents through regression analysis prior to determination of the actual OM content. The following equation derived from the Walkley Black method was used to correct for soil types in Northeast and North Central states [44].

$$OM = [(0.8 \times LOI\%) - 0.23]$$

The CEC was determined by ammonium acetate–Kjeldahl method (1.0 N NH_4OAc, pH 7.0) [45]. Calcium, Mg, K and, P were extracted using the Mehlich 1 procedure [46]. Calcium, P, and Mg were determined using inductively coupled plasma (ICP). Potassium was measured with atomic absorption spectrophotometry (Perkin Elmer Model 100, Perkin Elmer Corp., Norwalk, CT, USA).

Soil texture was determined using the pipette method [47]. Non-disturbed soil cores of known volume (3.47×10^{-4} m^3) were collected to determine bulk density using a hammer-driven core sampler. Bulk density was determined by measuring the mass (dried at 105 °C) and volume of soil in each core. To determine available water (A_v), soil samples were saturated with distilled water and desorbed at 3.3×10^4 and 1.5×10^6 Pa pressure using the pressure plate method [48]. To determine total water content, fresh soil samples of known volume (3.47×10^{-4} m^3) were taken from the field and their moist weights were recorded immediately. Samples were kept in an oven at 105 °C for 48 h and the dry weight was recorded.

Data Analysis: Analysis of variance was used to analyze the data using PROC GLM™ procedure of the Statistical Analysis System (SAS Institute, 2002). Fisher's Protected Least Significant Difference (LSD) test, with a probability value of 0.05, was used to separate treatment means.

CONCLUSIONS

In conclusion, our research determined that physical and chemical characteristics of subsoil, that govern turfgrass growth, could be improved significantly by composted poultry litter incorporation. Various field studies have demonstrated the benefits and risks of poultry litter application to soils used to grow food and fiber crops. In our experiment, composted poultry litter incorporation into a silt-loam subsoil, used to grow turfgrasses, improved its physical and chemical properties. Specifically, increases in total water content, organic matter, pH, and cation exchange capacity were noted following incorporation of the compost material. Overall, compost treatments were able to significantly improve soil physical and chemical properties compared to conventionally fertilized and control plots.

ACKNOWLEDGMENTS

The experiment was funded through USDA Hatch Project no. 438. The authors wish to thank Borderline LLC., for their generous donation of compost materials used in this experiment.

CONFLICTS OF INTEREST

The authors declare no conflict of interest.

REFERENCES

1. Daniels, W.L.; Amos, D.F. Generating productive topsoil substitutes from hard rock overburden in the southern Appalachians. *Environ. Geochem. Health* 1984, *7*, 8–15.
2. Eghball, B. Soil properties as influenced by phosphorus and nitrogen based manure and compost applications. *Agron. J.* 2002, *94*, 128–135.
3. Gao, G.; Chang, C. Changes in CEC and particle size distribution of soils associated with long term annual applications of cattle feedlot manure. *Soil Sci.* 1996, *161*, 115–120.
4. Nyakatawa, E.Z.; Reddy, K.C.; Sistani, K.R. Tillage, cover cropping, and poultry litter effects on selected soil chemical properties. *Soil Tillage Res.* 2001, *58*, 69–79.
5. Tyson, S.C.; Cabrera, M.L. Nitrogen mineralization in soils amended with composted and uncomposted poultry litter.*Commun. Soil Sci. Plant Anal.* 1993, *24*, 2361–2374.
6. Sharpley, A.N.; Smith, S.J.; Bain, W.R. Nitrogen and phosphorus fate from long-term poultry litter application to Oklahoma soils. *Soil Sci. Soc Am. J.* 1993, *57*, 1131–1137.
7. Kelleher, B.P.; Leahy, J.J.; Henihan, A.M.; O'Dwyer, T.F.; Sutton, D.; Leahy, M.J. Advances in poultry litter disposal technology—A review. *Bioresour. Tech.* 2002, *83*, 27–36.
8. Haapapuro, E.R.; Barnard, N.D.; Simon, M. Animal waste used as livestock feed: Dangers to human health. *Prev. Med.*1997, *26*, 599–602.
9. Chang, C.; Janzen, H.H. Long-term fate of nitrogen from annual feedlot manure applications. *J. Environ. Qual.* 1996,*25*, 785–790.
10. Goyne, K.W.; Jun, H.J.; Anderson, S.H.; Motavalli, P.P. Phosphorus and nitrogen sorption to soils in the presence of poultry litter-derived dissolved organic matter. *J. Environ. Qual.* 2008, *37*, 154–163.
11. Paul, E.A. Soil microbiology, ecology, and biochemistry in perspective. In *Soil Microbiology and Biochemistry*, 3rd ed.; Paul, E.A., Ed.; Academic Press: San Diego, CA, USA, 1996; pp. 13–17.
12. Millner, P.D.; Sikora, L.J.; Kaufman, D.D.; Simpson, M.E. Agricultural uses of biosolids and other recyclable municipal residues. *Agricultural Uses of Municipal, Animal, and Industrial Byproducts*; Wright, R.J., Kemper, W.D., Millner, P.D., Power, J.F., Korcak, R.F., Eds.; USDA Agriculture Research Service: Washington, DC, USA, 1998; pp. 9–44.
13. Vadas, P.A.; Meisinger, J.J.; Sikora, L.J.; McMurtry, J.P.; Sefton, A.E. Effect of poultry diet on phosphorus in runoff from soils amended with poultry manure and compost. *J. Environ. Qual.* 2004, *33*, 1845–1854.
14. Epstein, E.; Taylor, J.M.; Chaney, R.L. Effects of sewage sludge and sludge compost applied to soil on some soil physical and chemical properties. *J. Environ. Qual.* 1976, *5*, 422–426.
15. Tester, C.F. Organic amendment effects on physical and chemical properties of a sandy soil. *Soil Sci. Soc. Am. J.* 1990,*54*, 827–831.
16. Avimelech, A.; Cohen, A.; Shkedi, D. The effect of municipal solid waste compost on the fertility of clay soils. *Soil Tech.* 1990, *3*, 275–284.

17. Shiralipour, A.; McConnell, D.B.; Smith, W.H. Physical and chemical properties of soils as affected by municipal solid waste compost application. *Biomass Bioenerg.* 1992, *3*, 261–266.

18. Tiarks, A.E.; Mazurak, A.P.; Chesnin, L. Physical and chemical properties of soil associated with heavy applications of manure from cattle feedlots. *Soil Sci. Soc. Amer. Proc.* 1974, *38*, 826–830.

19. He, Z.; Tazisong, I.A.; Senwo, Z.N.; Zhang, D. Soil properties and macro cations status impacted by long-term applied poultry litter. *Commun. Soil Sci. Plant Anal.* 2008, *39*, 858–872.

20. Gascho, G.J.; Hubbard, R.K. Long-term impact of broiler litter on chemical properties of a coastal plain soil. *J. Soil Water Conserv.* 2006, *61*, 65–74.

21. Barker, A.V. Evaluation of composts for growth of grass sod. *Comm. Soil Sci. Plant Anal.* 2001, *32*, 1841–1860.

22. Harrell, M.S.; Miller, G.L. Composted yard waste affects soil displacement and roadside vegetation. *HortScience* 2005,*40*, 2157–2163.

23. Linde, D.T.; Hepner, L.D. Turfgrass seed and sod establishment on soil amended with biosolid compost. *HortTechnology* 2005, *15*, 577–583.

24. O'Brien, T.A.; Barker, A.V. Evaluation of field-applied fresh composts for production of sod crops. *Compost Sci. Utlzn.*1995, *3*, 53–65.

25. Wright, A.L.; Provin, T.L.; Hons, F.M.; Zuberer, D.A.; White, R.H. Compost impacts on sodicity and salinity in a sandy loam turf grass soil. *Compost Sci. Utlzn.* 2008, *16*, 30–35.

26. Mandal, M.; Chandran, R.S.; Balasko, J. Amending subsoil with composted poultry litter-II: Effects on Kentucky bluegrass (*Poa pratensis* L.) establishment, root growth, and weed populations. *Agronomy* 2013. in press.

27. Stevenson, F.J. *Humus Chemistry: Genesis, Composition, Reactions*; John Wiley & Sons: Hoboken, NJ, USA, 1994; pp. 1–24.

28. Evangelou, V.P. *Environmental Soil and Water Chemistry-Principles and Applications*; John Wiley & Sons, Inc.: New York, NY, USA, 1998; pp. 24–38.

29. Hillel, D. *Introduction to Environmental Soil Physics*; Elsevier-Academic Press: San Diego, CA, USA, 2004; pp. 363–407.

30. Jamison, V.C. Changes in air-water relationships due to structural improvement of soils. *Soil Sci.* 1953, *76*, 143–151.

31. Stevenson, E. Influence of peat moss on soil water retention for plants. *Can. J. Soil Sci.* 1974, *54*, 109–110.

32. Reeve, M.J.; Smith, P.D.; Thomasson, A.J. The effect of density on water retention properties of field soils. *J. Soil Sci.*1973, *24*, 356–367.

33. Angle, J.S.; Wolf, D.C.; Hall III, J.R. Turfgrass growth aided by sludge compost. *BioCycle* 1981, *22*, 40–43.

34. Jamison, V.C.; Kroth, E.M. Available moisture storage capacity in relation to textural composition and organic matter content of several Missouri soils. *Soil. Sc. Soc. Am. Proc.* 1958, *22*, 189–192.

35. Metzger, L.; Yaron, B. Influence of sludge organic matter on soil physical properties. *Adv. Soil Sci.* 1987, *7*, 141–163.

36. MacRae, R.J.; Mehuys, G.R. The effect of green manuring on the physical properties of temperate area soils. *Adv. Soil Sci.* 1985, *3*, 71–94.

37. McConnell, D.B.; Shiralipour, A.; Smith, W.H. Compost application improves soil properties. *BioCycle* 1993, *34*, 61–63.

38. Wei, Q.F.; Lowery, B.; Peterson, A.E. Effect of sludge application on physical properties of silty clay loam soil. *J. Environ. Qual.* 1985, *14*, 178–180.

39. Rasnake, M.; Sikora, F.; Murdock, L. Nutrient accumulation and movement in soils following the use of poultry litter. In Animal, Agricultural and Food Processing Wastes–Proceedings of the Eighth International Symposium, Des Moines, USA, 9-11 October 2000; pp. 562–567.

40. Havlin, J.L.; Beaton, J.D.; Tisdale, S.L.; Nelson, W.L. *Soil Fertility and Fertilizers, An Introduction to Nutrient Management*; Prentice Hall: Upper Saddle River, NJ, USA, 1999; p. 499.

41. Gagnon, B. Contribution of on-farm and industrial composts to soil pH and enrichment in available nutrients and metals. *Can. J. Soil Sci.* 2004, *4*, 439–445.

42. Balkcom, K.S.; Adams, J.F.; Hartzog, D.L. Peanut yield response to poultry litter and municipal sludge application.*Commun. Soil Sci. Plant Anal.* 2003, *34*, 801–814.

43. Thomas, G.W. Soil pH and soil acidity. In *Methods of Soil Analysis*; Chemical Methods-SSA Book Series No. 5; Soil Science Society of America and American Society of Agronomy: Madison, WI, USA, 1996; pp. 475–490.

44. Nelson, D.W.; Sommers, L.E. Total carbon, organic carbon, and organic matter. In *Methods of Soil Analysis*; Chemical Methods-SSA Book Series No. 5; Soil Science Society of America and American Society of Agronomy: Madison, WI, USA, 1996; pp. 961–1010.

45. Sumner, M.E.; Miller, W.P. *Methods of Soil Analysis*; Chemical Methods-SSA Book Series No. 5; Soil Science Society of America and American Society of Agronomy: Madison, WI, USA, 1996; pp. 1201–1229.

46. Mehlich, A. *Determination of P, Ca, Mg, K, Na and* NH_4; Short test methods used in soil testing division, Department of Agriculture: Raleigh, NC, USA, 1953; pp. 1–53.

47. Gee, G.W.; Bauder, J.W. Particle-size analysis. In *Methods of Soil Analysis*; Chemical Methods-SSA Book Series No. 5; Soil Science Society of America and American Society of Agronomy: Madison, WI, USA, 1986; pp. 383–409.

48. Klute, A. Water retention: laboratory methods. In *Methods of Soil Analysis*; Chemical Methods-SSA Book Series No. 5; Soil Science Society of America and American Society of Agronomy: Madison, WI, USA, 1986; pp. 635–660.

CHAPTER 2

SOIL COMPACTION AND SEALANT APPLICATION FOR COMPACTED EARTHEN LINERS

Amir M. González-Delgado, Manoj K. Shukla, Blair Stringam, Mehran Parsheh

Keywords: Saturated hydraulic conductivity; bulk density; soil texture; microalgae; evaporation ponds

INTRODUCTION

Microalgae are photosynthetic organisms with commercial and industrial applications. Different chemical compounds derived from microalgae cultures are used for cosmetics, dietary supplements and food sources. Microalgae extracts are present in face and skin care products [1]. Alternative protein sources and supplements such as pigments, antioxidants, β-carotene and fatty acids are produced from algae culture systems [2]. Microalgae are known as a natural food source of aquatic organisms, therefore algal biomass is mostly used to produce animal feed supplements for aquaculture. The microalgae biomass is commonly used for molluscs, shrimp and fish production and improvement of quality products [3]. Also, microalgae have been recognized as an alternative renewable energy source by the biofuel industry because of their high lipid content and rapid biomass production [4].

There are two microalgae culture systems used for large scale biomass production. These systems are known as closed photobioreactor systems and open pond production systems [5]. However, the majority of microalgae production occurs in open pond systems due to the low cost of construction and operation compared with photobioreactors [1, 3, 6, 7]. Microalgae, are also commonly produced in artificial ponds known as raceway ponds [3]. Raceway ponds are shallow closed loop oval channels that are circulated with a paddlewheel. The paddlewheel circulation enhances the microalgae culture and the shallow depth maximizes solar energy absorption for photosynthesis. Most of these ponds are lined, with the liner making up the bulk of the cost of building the pond [8]. As this industry continues to grow, it would be critical to evaluate lower cost alternatives to these liners. However, large scale adoption of low cost liners for microalgae culture should be preceded with careful evaluation of environmental and human health risks.

Coarse textured soils with high hydraulic conductivity are not suitable for building unlined ponds [9]. Lining materials are required to prevent the loss of water and pollutants from the bottom of the pond. Otherwise human health and water resources may be negatively impacted [10]. Currently, artificial ponds are lined with concrete or compacted clay or they may be lined with plastic liners (geo-membranes and geosynthetic clay liners) [11, 12]. Concrete and plastic liners are impervious materials that are effective against seepage. However they

Citation: Amir M. González-Delgado, Manoj K. Shukla, Blair Stringam, Mehran Parsheh, Soil Compaction and Sealant Application for Compacted Earthen Liners, DOI 10.7603/s40872-015-0004-5.

may not be cost-effective for some projects due to their high installation and maintenance costs [13]. Geo-membranes are 0.75 mm to 2 mm thick flexible plastic membranes that are water impermeable. Geosynthetic clay liners are made of a dry layer of bentonite-clay between two geomembranes with a total thickness of 5 to 10 mm. Compacted clay liners are made of a layer of clay compacted to a K_s of 1×10^{-9} m/s or lower and usually have a thickness of 0.9 m to meet the U.S. Environmental Protection Agency (USEPA) regulations [14, 15]. Compacted clay liners are less expensive than other lining technologies and are used in the production of microalgae and aquaculture products. The application of polyacrylamide polymers as a lining technology to control water seepage is also being evaluated [16, 17]. Previous studies reported that polyacrylamide polymers have the capacity of reducing the hydraulic conductivity by sealing the soil pores [18, 19]. Studies are needed to evaluate the sealing capacity of polyacrylamide polymers in order to develop cost-effective lining technologies.

The main objective of this study was to determine the effect of different levels of compaction and polyacrylamide polymer application treatments (soil sealant) on the K_s of soils. The other objective was to evaluate the decrease in the soil K_s by puddling treatments. The experimental site for the potential development of open ponds is Sapphire Energy Integrated Algal Biorefinery (IABR) located in Columbus, New Mexico. This study is expected to provide useful information on treatments that likely will assist in the selection of a treatment or treatments for earthen liners in southwestern New Mexico and in areas with similar soils.

METHODS AND MATERIALS

Collection of Soil Samples

Undisturbed soil core and loose soil samples were collected from Sapphire Energy IABR site located at Columbus, NM. Six undisturbed soil core samples were collected from the bottom of a previously drained pond (the pond) at a depth of 0-5 cm. The exposed bottom of the pond allowed for the collection of samples to evaluate the existing physical soil properties at the pond. The undisturbed soil cores were covered with clear plastic wrap to prevent desiccation and soil loss. The soil core method was used to collect the undisturbed soil samples and the soil bulk densities were determined using the method described by Blake and Hartge [20]. The bulk density of soil throughout the report refers to the dry soil bulk density. The soil surrounding the impounded cores at the time of extraction was collected in plastic Ziploc bags to determine the moisture content and texture.

Two soil types were identified by reviewing the soil survey of Luna County, New Mexico and used to coordinate the collection of soil samples. Loose soil samples were collected using a soil auger from 0-25 cm and 25-55 cm depths, respectively at a second soil sampling site located approximately 300 m north of the pond. The collection of deeper loose soil samples was not possible with a soil auger because of the existence of a gravelly material that inhibited the digging with a soil auger. The third soil sampling site was located approximately 2,000 m northeast of the pond. Using a shovel, three loose soil samples were collected from 0- 17 cm depth and were stored in plastic bags. The fourth soil sampling site was located approximately 1,500 m southeast of the pond and loose soil samples were collected from 0-17 cm depth. After identifying the dominant soil texture, additional loose and undisturbed soil core samples were collected from the third sampling location (northeast site) during a second sampling event. Twelve 3-gallon buckets were filled with loose soil collected with a shovel from 0-17 cm and 17-40 cm depths. Also, two undisturbed soil core samples were collected at each of the 0-5 cm and 40-45 cm depths. The buckets were covered with plastic bags and secured to prevent contamination and loss of soil during their transportation.

Soil moisture content and Soil texture

A representative sample was obtained from each of the loose soil samples collected from the field to determine the soil moisture content immediately after returning from the sampling sites. The moisture content of the samples were determined with the soil moisture content method defined by Gardner [21]. Pre-weighed metal cans were used to dry the soil samples in an oven for 24 hours at 105° C. The difference in weight before and after drying was obtained to calculate the water loss during the drying. The ratio of water mass lost during drying and dry soil mass gave the gravimetric moisture content of the soil.

The remaining soil samples were spread on a table, large soil clods were hand crushed and the soil was air dried for three days. The air-dried soil was run through a 2 mm sieve (Dual Manufacturing Co., Chicago, Ill). Organic material, debris, and stones were removed from the soil samples. Soil clods (> 2 mm) were broken down with a rubber mallet, and the material was sieved again. The soil material <2 mm was stored in plastic bags to conduct soil texture analyses. The hydrometer method was used to determine the soil texture and identify the soil types in the experimental site [22]. A hydrometer (Fisher Scientific Inc., Pittsburgh, PA) was used to measure the density of the soil suspension that is influenced by the soil particle size [23].

Saturated Hydraulic Conductivity

In this study, treatments were applied at three different scales using aluminum cores, proctor molds and stock pots because it is reported that soil hydraulic conductivity is a function of the sample support [24]. Aluminum cores were 5 cm long and 5 cm in diameter, proctor molds were 10 cm long and 10 cm in diameter, and stock pot was 55 cm long and 60 cm diameter. Metal cores were used to repack the soil to conduct K_s experiments. The bottom of each metal core was covered with cheesecloth to prevent the loss of soil during saturation and during K_s experiments. The weight of the metal cores were recorded prior to and after repacking them with each soil type identified by hydrometer method. The same procedure was used to repack the soil in proctor molds and stock pot. After repacking, soil cores and molds were water-saturated from the bottom by placing them in a plastic water pan and slowly raising the head of water in the pan. The repacked soil cores and molds were saturated from the bottom because it is easier and more efficient to remove most of the air from the soil pores and obtain more complete saturation. The hydraulic head was kept small and water inside the core was allowed to move via capillary rise. The time allowed for saturation was at least 12 hours.

The K_s for the soil cores and molds was determined by the constant head method [25] as follows

$$K_S = \frac{Q}{\left(\frac{h+L}{L}\right) x\ A} \tag{1}$$

where Q is the volumetric rate of flow (m^3 /s), A is the cross-sectional area (m^2), h is the constant head (m) applied during the experiment and L is the height (m) of the soil column. The area (A) for the metal cores, molds and the 160-quart stock pot were 19.63, 7.85 and 2,826 cm^2 , respectively.

An empty metal core of the same diameter was attached at the top of each soil core to provide space for maintaining a constant head of water on top of the soil. Circular filter papers were placed on top of the soil samples to prevent erosion/disturbance of the soil surface. Each soil core and mold was placed on top of an iron wire mesh which was glued to the top of a funnel. A 100 ml graduated cylinder was placed under the funnel to collect all the water that ran out of the bottom of the funnel. A constant water head of 5 cm was maintained on top of the soil. The plastic Marriott bottle continuously supplied an equal amount of water into the core that came out of the bottom of the core to maintain a constant head of 5 cm of water at the top of the soil in each core. The effluents were collected in graduated cylinders at specific time intervals until three to six consistent readings were obtained with only small differences in effluent volume (< 2ml) for a given time interval. Four replications were made for each soil types identified in the study area. A similar procedure was followed for proctor molds; however, a constant head of 10 cm was maintained on top of the soil in each proctor mold and 16 cm on top of 30 cm of soil in the stock pot.

Saturated Hydraulic Conductivity Experiments with Soil Sealant

Four soil cores for each soil type were used to evaluate the reductions in the soil K_s by soil sealant (Seepage Control Inc., Chandler, AZ) application. The K_s of the soil were determined before and after adding one ml of soil sealant. A volume of one ml of soil sealant was added to the soil samples based on the recommended application rate of 3.78 liters (1 gal) of soil sealant per 7,560 liters (2,000 gal) of water (0.0005 L/L). The K_s experiments were started 24 hours after applying the soil sealant to the cores under a constant head of five cm of water. Additional experiments were conducted with the same soil cores with five mL of soil sealant application. The recommended amount of soil sealant is much smaller than one mL but one mL was the minimum measurable with the graduated dropper to approximate the recommended application rate of 50 mL for 100 mL of water present in the soil cores. The experiments conducted with 5 mL of soil sealant represent 100 times more sealant application than the recommended application rate. The circular filter papers were removed from the soil cores to maximize the interaction of soil sealant with the soil. The volume of soil sealant added to the samples was measured with a graduated dropper. After adding the soil sealant, water developed a whitish color that slowly vanished over time. This is probably due to the settlement of the chemical and subsequent filling of the pores. At the cessation of the experiments, a greasy soil surface was observed.

Compaction of Soil Cores and Proctor Molds

The effect of compaction on K_s was evaluated on four soil cores for each soil. The K_s of the soil was determined under three successively increasing compactions. The compaction was created by using a standard two inch (five cm) proctor hammer. Before manual compaction, soil cores were drained to near field capacity soil moisture content. At the end of each compaction, the volume of soil in each core was determined to calculate the new dry bulk density. Compacted cores were subsequently saturated from the bottom prior to the determination of the K_s. As the last step of the K_s determination, one mL and five mL of soil sealant were added to each of the compacted soil cores. Similar experiments were conducted using repacked proctor molds. The K_s was determined at the end of each of the four successive compaction treatments.

Puddling Treatments

A 160-quart stock pot was modified to work as a permeameter. The soils were repacked separately in the stock pot to simulate puddling experiments similar to that in a paddy field. First, holes were drilled through the bottom of the 160-quart stock pot and a plastic tray with a hole in the center was attached below the bottom of the pot. A funnel with a valve to turn the flow from the bottom of the pot on or off was attached. This arrangement ensured that water flowed free out of the bottom of the pot and allowed for easy determination of the rate of flow through the soil. A layer of gravel approximately four cm thick was placed at the bottom of the pot and a circular metallic mesh was placed on the top of the gravel. Cheesecloth was placed on the metallic mesh to prevent soil loss during the puddling experiments. The K_s experiments were conducted after saturating the soil from the top. After saturation was accomplished, 16 cm of water head was maintained on the soil surface, and the effluent was collected from the bottom of the pot using a 1000 ml graduated cylinder. After completing the first hydraulic conductivity experiment, the soil in the stock pot was puddled with a standard proctor hand compaction hammer (Humboldt MFG. Co., Schiller Park, Ill.). The puddling depth was about 10 cm. The wet soil was worked similarly to the way manual puddling is done on a paddy field. At the end of a puddling treatment, the soil suspension was left to settle for about 24 hours and water contained much less suspended particles (more clear) than immediately after puddling. Puddling operations were repeated three times, and the K_s was determined 24 hours after the cessation of each test. The water coming out of the bottom of the stock pot during K_s tests were mostly clear and contained little sediment.

After conducting four K_s experiments with each soil type, 58 mL of soil sealant was added to the stock pot and after 24 hours, the K_s was determined again. The stock pot was subsequently drained to collect soil samples with a push-probe. The collected soil samples were used to determine the vertical variation of bulk density of the soil in the stock pot that showed bulk density was higher in the bottom half of the stock pot. It was difficult to collect an intact sample from the upper puddled layer (no data are presented).

RESULTS AND DISCUSSION

Particle Size Distribution and Soil texture

Determining the texture of different soils found at the experimental site is important for evaluating the potential of earthen liners and the viability of compaction treatments. Areas with soils containing a fine textured clay are suitable for the development of earthen liners [14]. The soil particle size analyses identified three types of soil textures present throughout the study site. The soil types were identified based on their percentages of sand, clay, and silt contents and using USDA soil textual classification (Table 1).

Table I. Location and depth of undisturbed and loose soil samples with their respective percentage of soil particle size and soil texture.

Location	Depth (cm)	n[a]	Sand (%)[b]	Clay (%)[b]	Silt (%)[b]	Soil Texture
Pond[c]	0-5	6	43.29 ± 0.88	36.59 ± 0.94	20.12 ± 0.11	Clay loam
North	0-25	3	55.12 ± 0.34	16.40 ± 0.64	28.48 ± 0.78	Sandy loam
North	25-55	3	53.84 ± 0.00	27.25 ± 0.48	18.91 ± 0.48	Sandy clay loam
Northeast	0-12	6	57.85 ± 0.41	13.36 ± 0.92	28.78 ± 1.08	Sandy loam
Northeast	0-17	3	58.21 ± 1.78	13.36 ± 2.78	28.42 ± 1.00	Sandy loam
Northeast[d]	0-17	3	59.06 ± 1.09	12.21 ± 0.54	28.72 ± 0.94	Sandy loam
Northeast[d]	17-40	3	42.40 ± 0.00	28.21 ± 0.54	29.38 ± 0.54	Loam
Southeast	0-20	3	69.83 ± 0.94	9.92 ± 0.19	20.24 ± 0.77	Sandy loam

[a] n; number of soil particle size analysis.
[b] Mean ± Standard Error of sand, clay and silt.
[c] Samples collected from pond bottom.
[d] Samples collected during the second sampling event.

The soil type identified at the surface (0-25 cm depth) was sandy loam at each sampling site. Sandy clay loam was present at the 25-55 cm depth in the northern site and loam was the soil type identified at the17-40 cm depth in the northeastern site. The particle size analysis of soil samples collected from the bottom of the pond classified the soil as clay loam. Because there is no information on the addition of clay material, it seems the compaction treatment of the soil on the pond bottom during construction caused a breakdown of sand and coarse silt-sized particles and changed the soil texture. Sandy loam and sandy clay loam soils were chosen for compaction treatments to determine Ks of soil in repacked cores, proctor molds, and stock pot. Sandy clay loam was chosen because sandy clay loam and loam are expected to provide similar values of conductivity under compaction that exceeds natural soil compaction. The texture analysis (Table 1) also showed no statistically significant difference in clay content between sandy clay loam and loam. Because loam has slightly higher silt plus clay contents than sandy clay loam, loam was selected in place of sandy clay loam for puddling treatments where dispersion and subsequent settlement of finer particles are important

Hydraulic Properties of Undisturbed Soil in the Pond

Under a constant head of 5 cm, only one of the six undisturbed cores, collected from the pond bottom, conducted water through it during the K_s experiments. During pond construction, clay layers were found in some areas near the pond. Although the clay layer is not contiguous, compaction of the soil along with likely texture modifications by heavy machinery could be the reason that no water came out of the cores during hydraulic conductivity tests. The only core that was permeable had a bulk density of 1.35 g/cm^3 and a K_s of 2.83 x 10^{-7} m/s. The average bulk density values for all the undisturbed soil cores collected from the pond bottom was 1.44 ± 0.10 g/cm^3.

Hydraulic Properties of Undisturbed Soil Cores

The bulk density of the undisturbed sandy loam sample collected at the northeastern site was 1.21 g/cm^3 and K_s of 3.25 x 10^{-5} m/s. One of the two undisturbed loam samples collected from 20-25 cm depth transmitted water when hydraulic conductivity experiments were performed. The bulk density and K_s values for this undisturbed core sample were 1.51 g/cm^3 and 2.22 x 10^{-6} m/s, respectively. The average bulk density value for the two undisturbed loam soil cores were 1.59 ± 0.07 g/cm^3 . The high soil dry bulk density was due to the weight of the stones and gravels included in the dry soil weight. Because gravels and stones can have large pores around them, hydraulic conductivity of this system can be high in spite of the high bulk density.

Hydraulic Properties of Soil Cores Treated with Sealant

The effect of compaction was evaluated in soil cores repacked with sandy loam and sandy clay loam soils, separately. At the end of the third K_s test on each soil core, two different concentrations of soil sealant were applied to the cores, and K_s was determined 24 hours after the application of the soil sealant. There was no outstanding change in K_s values between the third hydraulic conductivity test and the tests conducted after applying 1 mL and 5 mL of soil sealant, for both soils. The final average K_s values without soil sealant for sandy loam and sandy clay loam were 2.17 x 10^{-7} ± 1.47 x 10^{-7} m/s and 4.88 x 10^{-8} ± 2.45 x 10^{-8} m/s, respectively. The K_s values for sandy loam and sandy clay loam after the application of 5 mL of soil sealant were 1.99 x 10^{-7} ± 1.11 x 10^{-7} m/s and 9.86 x 10-^{8} ± 5.14 x 10^{-8} m/s, respectively. The K_s value for both soils before and after the application of soil sealant at final compactions remained within their standard deviations, and no significant differences were observed. These experiments do not show the usefulness of the soil sealant for decreasing hydraulic conductivity even at rates much higher than those recommended by the manufacturer. Therefore, experiments should be conducted in the field to assess the true potential of the soil sealant.

The experiments in Fig. 1 with soil sealant were conducted at the highest level of compaction (highest bulk density) of the soil cores. To be sure that the effect of soil sealant on reductions in K_s were not related to the high bulk density of soil in the cores (Fig. 1), more experiments were performed by packing the cores with both soils at or close to the natural bulk density of soils in the field. There was only a slight reduction in the K_s for both soils after applying 1 mL of soil sealant (Fig. 2). The application of 5 mL of soil sealant did not change the K_s of the soils. The final average K_s values for sandy loam and sandy clay loam were $5.71 \times 10^{-6} \pm 7.07 \times 10^{-7}$ m/s and $7.88 \times 10^{-6} \pm 6.18 \times 10^{-7}$ m/s, respectively.

The current recommended application rate for the study site is 0.20 L of soil sealant/m² (0.005 gal/ft²). The soil sealant's manufacturer recommends an application rate of 3.78 liters (1 gal) of soil sealant per 7,560 liters (2,000 gal) of water. Soil in the cores (Fig. 1 and Fig. 2) had a surface area of 19.63 cm² (0.021 ft²), and the volume of ponded water was 98.13 cm³ (0.0259 gal). Thus, based on any of the above recommendations, the cores should have been treated with 4.35×10^{-3} L/m² (1.06×10^{-4} gal/ft²) and 0.05 mL (1.30×10^{-5} gal/gal). However, soil cores were treated with 1 mL and 5 mL of soil sealant because of the ease of measuring and to determine if there were appreciable reductions in the hydraulic conductivity of the soil. The amount of 1 mL and 5 mL of soil sealant were 2.50 and 12.52 times greater than the application rate used in the study site. The data in Fig. 2 indicate that K_s decreased with increasing amounts of soil sealant, but it is highly unlikely that addition of amounts greater than 5 mL of soil sealant will be enough to reduce the K_s to USEPA levels; moreover costs associated with soil sealant may make its application impractical. Although laboratory experiments conducted by Young et al. [18] showed K_s reduction in the soil with increasing concentration of polyacrylamide polymer that promoted the clogging of soil pores.

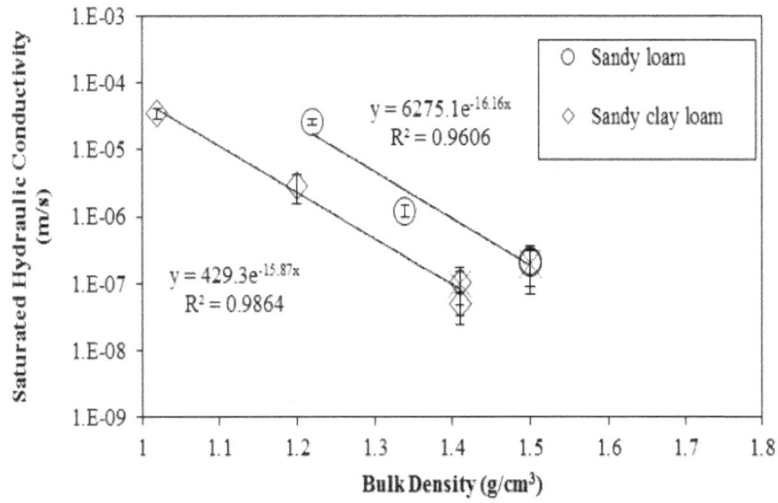

Figure 1. Saturated hydraulic conductivity of soil cores compacted at different levels and subsequently treated with 1 mL and 5 mL of Soil Sealant after the third compaction. Experiments conducted with the soil sealant are marked with the dashed-X symbols. The vertical axis has a logarithmic scale, and error bars are included for each data point. Error bars represent standard error

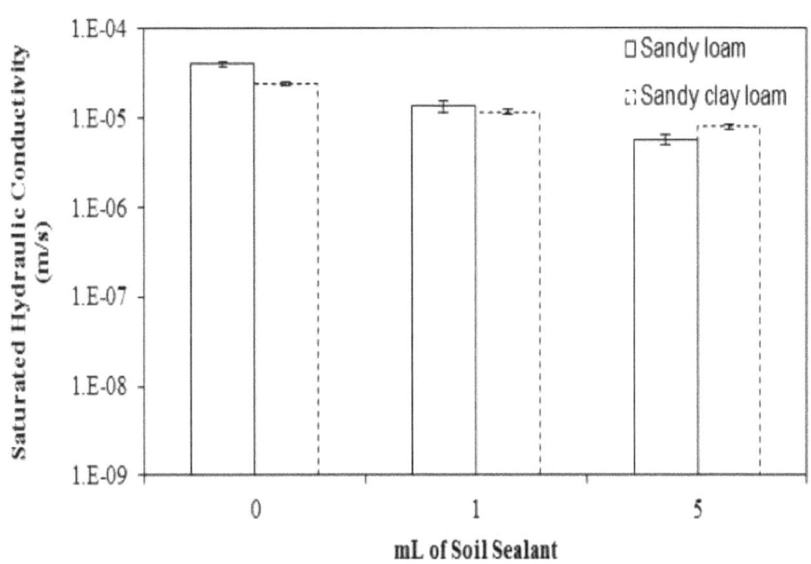

Figure 2. Saturated hydraulic conductivity of soil cores determined 24 hours after the application of 0 mL, 1 mL and 5 mL of soil sealant. Error bars represent standard error.

Relationships between Compaction and Soil Volume

The soil was repacked in cores and each core was slightly compacted by using a standard proctor hammer (2 inch diameter) prior to the saturation to determine hydraulic conductivity. Soil in each core was subsequently drained below field capacity prior to the next compaction obtained by hammering the soil four times with the proctor hammer. The change in volume of the soil in each core was recorded at the end of each compaction treatment (Fig. 3A and Fig 3B). Each core was saturated again and the hydraulic conductivity tests were repeated. The procedure was repeated for each core and soil four times. Soil volume decreased with increasing compaction, and the average decrease in soil volume was 19% for sandy loam soil and 28% for sandy clay loam. The relative compaction expressed as the ratio of initial bulk density and final bulk density (not the maximum bulk density) was 72% for sandy loam soil and 73% for sandy clay loam soil at a soil water content below field capacity (below 0.33 bars or 4.8 psi). A similar decrease in soil volume with increasing compaction also was observed in proctor molds. Average decreases in the soil volume were 29% for sandy loam soil and 25% for sandy clay loam soil (Fig. 3C and Fig. 3D). The relative compaction expressed as the ratio of the initial bulk density and the final bulk density were 71% for sandy loam and 75% for sandy clay loam soils.

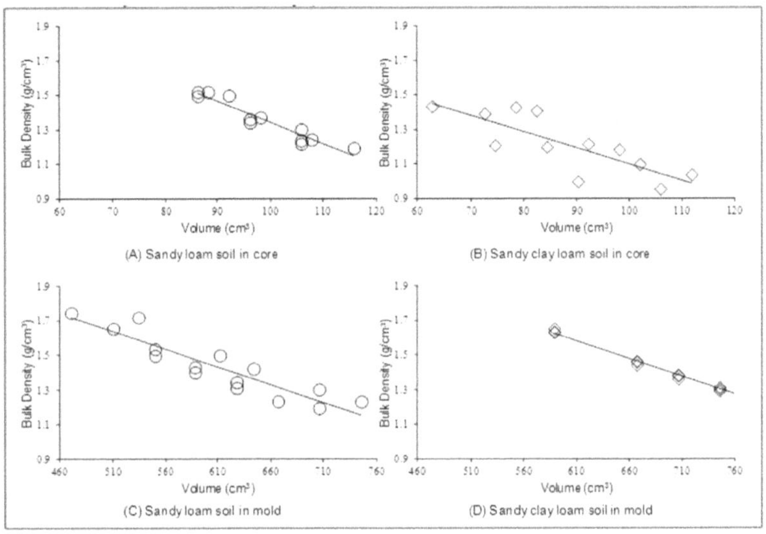

Figure 3. The relationship between soil bulk density and soil volume after compaction using a standard 2 inch proctor hammer in soil cores (A and B) and proctor molds (C and D) for sandy loam and sandy clay loam soils.

Relationships between Hydraulic Conductivity and Bulk Density

It is evident from Fig. 1 that for both soil types, K_s decreased with increasing soil bulk density in soil cores. The K_s for sandy clay loam soil was consistently lower than for those obtained for sandy loam soil at each level of compaction. A negative exponential relationship between K_s and compaction was obtained for both soils. Previous studies have reported a similar relationship between K_s and compaction [26, 27, 10]. The exponential relationship suggests that the sandy loam and sandy clay loam should be compacted to 1.82 and 1.69 g/cm^3 , respectively to meet the USEPA limit of 1 x 10^{-9} m/s for compacted clay liners.

Six proctor molds, three of them repacked with sandy loam and three with sandy clay loam soils, were used to determine the effect of compaction on the K_s (Fig. 4). In general, with increasing bulk density, K_s decreased in proctor molds. A decrease in the K_s of proctor molds packed with sandy clay loam soil was clearly evident, and at a bulk density of about 1.4 g/cm^3 , the average $_K$s value was 3.04 x 10^{-7} ± 5.04 x 10^{-8} m/s. However, the K_s for proctor molds packed with sandy loam soil was 1.62 x 10^{-6} ± 1.01 x 10^{-6} m/s at a bulk density of 1.72 g/cm^3 .

The compaction process decreases the movement of water through the porous media as a result of reduction of soil pore size and pore connectivity [28, 29]. Fig. 4 presents a negative exponential relationship between K_s and compaction for both soils in proctor molds. The exponential relationship suggests that the sandy loam soil should be compacted to 2.4 g/cm^3 and sandy clay loam to 1.59 g/cm^3 to meet the USEPA limit of 1 x 10^{-9} m/s for compacted clay liners. It is highly unlikely that sandy loam soil can be compacted to 2.4 g/cm^3 to yield USEPA specified K_s for compacted clay liners. There is a potential of utilizing the sandy clay loam soil present in the area and compacting it further to yield a USEPA specified K_s for compacted clay liners. This is further supported by laboratory work where repacked sandy clay loam soil molds did not transmit water when compacted to a bulk density ≥ 1.45 g/cm^3 .

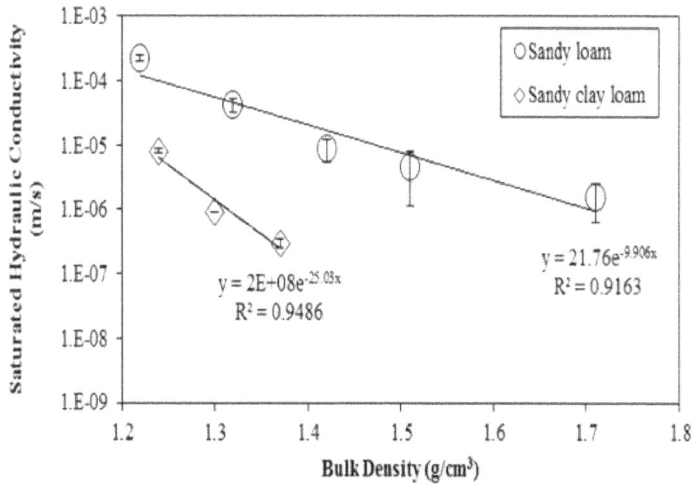

Figure 4. Saturated hydraulic conductivity of soil molds for different bulk densities. The vertical axis has a logarithmic scale and error bars are included for each data point. Error bars represent standard error.

Puddling Treatments

The puddling treatments that simulate a paddy field situation were conducted with sandy loam and loam soils because these two represent the most contrasting soil textures. The bulk densities of the sandy loam and loam soils packed to conduct puddling experiments were 1.25 g/cm^3 and 1.07 g/cm^3 , respectively. During puddling, the soil at and below the depth of puddling is usually compacted, and the fine soil particles within the puddling depth get dispersed and become suspended. At the termination of puddling operations, the suspended particles began to settle slowly, further clogging the pores and decreasing the hydraulic conductivity of the puddled layer (Fig. 5). Rezaei et al. [30] have also reported soil aggregate breakdown due to puddling in a paddy field with attendant reduction of macropore but increase of micropore volumes. The elimination of soil aggregates and modification of soil pores through the puddling treatment contributed to the formation of a blocked soil surface that decreased the K_s. Also they reported that the bulk density increased with depth for each puddling intensity level.

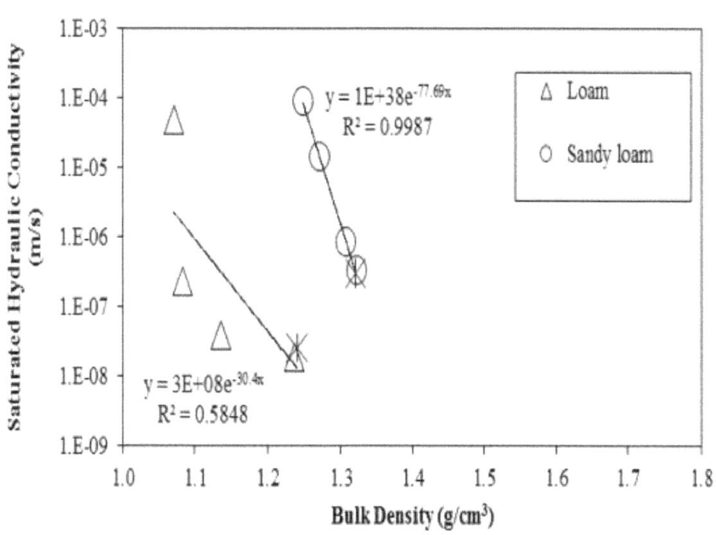

Figure 5. Saturated hydraulic conductivity of soil after puddling treatments to simulate paddy field conditions under different bulk densities. The saturated hydraulic conductivity of the porous media also was determined after 24 hours of adding 58 mL of soil sealant at the end of the last paddy field simulation. Experiments conducted with the soil sealant are marked with the dashed-X symbols. The vertical axis has a logarithmic scale.

The puddling treatments with sandy loam and loam soils showed a decreasing trend for K_s with increasing compaction. Lower K_s values were obtained from loam than from sandy loam for each bulk density. The lowest K_s obtained at the end of the last puddling was 1.90 x 10^{-8} m/s for loam soil that was compacted to a bulk density of 1.24 g/cm^3 . This clearly indicates that larger amounts of finer particles clog the soil pores and further decrease the hydraulic conductivity of the soil.

Similar to other experiments conducted in soil cores, soil sealant was also applied to the stock pot at the end of the fourth K_s experiment. The stock pot had a crosssectional area of 2,826 cm^2 (3.14 ft^2). Based on the recommended application rate used on the existing pond site in Columbus, NM, 58 mL of soil sealant was added to the stock pot. The K_s experiments conducted about 24 hours later showed that the K_s did not change substantially (Fig. 5). The final K_s value for the soil sealant application was only 13% smaller in sandy loam soil; however, the value was almost 28% higher in loam soil. These experiments further demonstrate that application of soil sealant alone will not decrease the K_s to the USEPA mandated limit for compacted clay liners. The puddling experiments of Fig. 5 were conducted when water was ponded on the soil surface. Because the average bulk density of the puddled loam soil was only 1.24 g/cm^3 and K_s was 1.90 x 10^{-8} m/s, there is a possibility that further increases in compaction by conducting a puddling treatment when soil water content is below the field capacity may decrease the K_s to the EPA mandated limit. Previous studies have shown that puddling treatment can increase or decrease the bulk density depending on the soil moisture content [31, 32]. The cohesion between soil aggregates is greater when the soil is at a moisture content below saturation. Therefore, the puddling of soils at moisture content below saturation produces higher soil bulk density values than those at saturated moisture content. Ahmad [33] reported that the bulk density increased with decreasing water-soil ratio after conducting a study to determine the influence of water-soil ratio on puddling efficiency.

CONCLUSION

The soil texture analysis showed three different types of soils present in the study area. The soil near the surface at all of the sampling locations was classified as sandy loam while soil at deeper depths (below 17 cm) was sandy clay loam or loam. Experiments in soil cores, proctor molds, and stock pot (three different scales) did not support the use of soil sealant for decreasing K_s to USEPA mandated criteria for compacted clay liners. Although soil sealant did not seem to decrease the K_s of the soil, field experiments should be conducted before ruling out the sealing potential of the soil sealant. As compaction increased, hydraulic conductivity of the soil decreased, and a negative exponential relationship were obtained between K_s and compaction for both soils. The exponential relationship suggested the sandy loam and sandy clay loam should be compacted to 1.82 and 1.69 g/cm^3 , respectively in soil cores and 2.40 and 1.59 g/cm^3 , respectively in soil molds, to meet the USEPA limit. The puddling experiments with sandy loam and loam soils showed a decreasing trend for K_s with increasing compaction. Puddling in loam produced consistently lower K_s values than sandy loam. The lowest K_s obtained at the end of the puddling was 1.90 x 10^{-8} m/s for loam soil with a bulk density (compaction) of 1.24 g/cm^3 . These results indicate that with a further increase in the bulk density, puddling treatment has the potential to reduce the hydraulic conductivity of the soil to the USEPA-specified limit. The results were derived from experiments conducted with soil samples without gravel therefore experiments under field conditions will be required to evaluate the effect of gravel on the K_s of compacted porous media. The three types of soils present in the experimental site contain enough clay content for the development of compacted earthen liners for open ponds, therefore, it could be useful to conduct K_s experiments under different compaction treatments after mixing these soil types in the field.

ACKNOWLEDGEMENT

Authors thank New Mexico State University (NMSU) Agricultural Experiment Station, Las Cruces, NM, USA for support. Sincere thanks are for Sapphire Energy Inc., Las Cruces, NM, USA for providing the financial support and valuable assistance during sampling.

REFERENCES

1. P. Spolaore, C. Joannis-Cassan, E. Duran, and A. Isambert, "Commercial applications of microalgae," Journal of Bioscience and Bioengineering, 2006, vol. 101: pp. 87-96.
2. T.M. Mata, A.A. Martins, and N.S. Caetano, "Microalgae for biodiesel production and other applications: A review," Renewable & Sustainable Energy Reviews, 2010, vol. 14: pp. 217-232.
3. L. Brennan, and P. Owende, "Biofuels from microalgae-a review of technologies for production, processing, and extractions of biofuels and co-products," Renewable & Sustainable Energy Reviews, 2010, vol. 14: pp. 557-577.
4. I. Priyadarshani, and B. Rath, "Commercial and industrial applications of micro algae – a review," Journal of Algal Biomass Utilization, 2012, vol. 3: pp. 89-100.
5. D. Chaumont, "Biotechnology of algal biomass production - a review of systems for outdoor mass-culture," Journal of Applied Phycology, 1993, vol. 5: pp. 593-604.
6. R. Slade, and A. Bauen, "Micro-algae cultivation for biofuels: Cost, energy balance, environmental impacts and future prospects," Biomass & Bioenergy, 2013, vol. 53: pp. 29-38.
7. P. Letcher, S. Lopez, R. Schmieder, P.A. Lee, C. Behnke, M. Powell, and R. McBride, "Characterization of amoeboaphelidium protococcarum, an algal parasite new to the cryptomycota isolated from an outdoor algal pond used for the production of biofuel", PLOS ONE, 2013, 82 e56232. DOI: 10.1371/journal.pone.0056232.
8. R.C. McBride, and D.S. Merrick, "Catalyzing innovation. Innovation in open pond algae agriculture for biofuel production," 2014, vol. 10(3) DOI:10.1089/Ind.2013.1614.
9. E.R. Venteris, R.C. McBride, A.M. Coleman, R.L. Skaggs, and M.S.Wigmosta, "Siting algae cultivation facilities for biofuel production in the United States: trade-offs between growth rate, site constructability, water availability, and infrastructure," Environmental Science and Technology, 2014, vol. 48 (6): pp.3559-3566.
10. A.M. Gonzalez-Delgado, M.K. Shukla, A.L. Ulery, A.S. Bawazir, and P.V. Brady, "Saturated hydraulic conductivity of self-sealing lining materials for desalination evaporation ponds," Desalination and Water Treatment, 2011, vol 29: 187- 195.
11. T.J. Lundquist, I.C. Woertz, N.W.T. Quinn, and J.R. Benemann, "A realistic technology and engineering assessment of algae biofuel production," Energy Biosciences Institute. University of California Berkeley. 2010.
12. Y. Chisti, "Biodiesel from microalgae," Biotechnology Advances, 2007, vol. 25: pp. 294–306.

13. O. Sivrikaya, "Models of compacted fine-grained soils used as mineral liner for solid waste," Environmental Geology, 2008, vol. 53: pp. 1585-1595.
14. C.H. Benson, and J.M. Trast, "Hydraulic conductivity of thirteen compacted clays," Clays and Clay Minerals, 1995, vol. 43: pp. 669-681.
15. J.P. Giroud, K. BaduTweneboah, and K.L. Soderman, "Comparison of leachate flow through compacted clay liners and geosynthetic clay liners in landfill liner systems," Geosynthetics International, 1997, vol. 4: pp. 391-431.
16. R.E. Sojka, D.L. Bjorneberg, J.A. Entry, R.D. Lentz, and W.J. Orts, "Polyacrylamide in agriculture and environmental land management," Advances in Agronomy, 2007, vol. 92: pp. 75- 162.
17. B.T. Story, M.A. Urynowicz, and G.F. Vance, "Field demonstration of polyacrylamide in an unlined irrigation canal," Journal of Irrigation and Drainage Engineering-ASCE, 2013, vol. 139: pp. 399-404.
18. M.H. Young, E.A. Moran, Z. Yu, J. Zhu, and D.M. Smith, "Reducing saturated hydraulic conductivity of sandy soils with polyacrylamide," Soil Science Society of America Journal, 2009, vol. 73: pp. 13-20.
19. J. Zhu , and M.H. Young, "Sensitivity of unlined canal seepage to hydraulic properties of polyacrylamide-treated soil," Soil Science Society of America Journal, 2009, vol. 73: pp. 695-703.
20. G.R. Blake, and K.H. Hartge, "Bulk density, In A. Klute (Ed), Methods of Soil Analysis," Part I, ASA Monograph No. 9, Madison, WI, 1986, pp. 363-375.
21. W.H. Gardner, "Water content. In A. Klute (Ed), Methods of Soil Analysis," Part I, ASA Monograph No. 9, Madison, WI, 1986, pp. 493-544.
22. G.W. Gee, and J.W. Bauder, "Particle size analysis, In A. Klute (Ed), Methods of Soil Analysis," Part I, ASA Monograph No. 9, Madison, WI, 1986, pp. 337-382.
23. J.M. Keller, and G.W. Gee, "Comparison of American society of testing materials and soil science society of America: Hydrometer methods for particle-size analysis'" Soil Science Society of America Journal, 2006, vol. 70: pp. 1094-1100.
24. M.K. Shukla, "Introduction to soil hydrology. In M.K. Shukla (Ed), Soil hydrology, land use and agriculture: Measurement and modeling," CAB International, Wallingford, U.K., 2011, pp. 1-23.
25. A. Klute, and C. Dirksen, "Hydraulic conductivity and diffusivity: Laboratory methods, In A. Klute (Ed), Methods of Soil Analysis," Part I, ASA Monograph No. 9, Madison, WI, 1986, pp. 687-734.
26. S. Assouline, "Modeling the relationship between soil bulk density and the hydraulic conductivity function," Vadose Zone Journal, 2006, vol. 5: pp. 697-705.
27. B.K. Behera, "Effect of puddling on puddled soil characteristics and performance of self-propelled transplanter in rice crop, In B. P. Varshney (Ed)," Agricultural Engineering International: the CIGR Ejournal, 2009, vol. 10: pp. 1-18.
28. D. Dec, J. Dorner, O. Becker-Fazekas, and R. Horn, R "Effect of bulk density on hydraulic properties of homogenized and structured soils," Revista De La Ciencia Del Suelo Y Nutricion Vegetal, 2008, vol. 8: pp. 1-13.
29. J. Lipiec, and R. Hatano, "Quantification of compaction effects on soil physical properties and crop growth," Geoderma, 2003, vol. 116: pp. 107-136.
30. M. Rezaei, R. Tabatabaekoloor, S.R. Mousavi Seyedi, and N. Aghili Nategh, "Effects of puddling intensity on the in-situ engineering properties of paddy field soil," American Journal of Agricultural Economics, 2012, vol. 3: pp. 22-26.
31. G.B. Bodman, and J. Rubin, "Soil puddling," Soil Science Society of America Journal, 1949, vol. 13: pp. 27-36.
32. P.K. Sharma, and S.K. De Datta, "Effects of puddling on soil physical properties and processes," Advances in Soil Science, 1986, vol. 5: pp. 139-178.
33. D. Ahmad, "Puddling efficiency for rice root growth in a cohesive soil," Pertanika Journal of Science & Technology, 1993, vol. 1: pp. 93-102.

CHAPTER 3 INTRODUCTION TO SCIENTIFIC DISCIPLINE AGROPHYSICS

B. Dobrzański, S. Grundas and A. Stępniewski

INTRODUCTION

Agrophysics's Definition and Scope

From the very beginning of its existence the definition of Agrophysics as a science was described a number of times. In general, Agrophysics is a branch of natural and agricultural sciences which applies physics into agriculture. Therefore sometimes it is also called agricultural physics. It explores agricultural materials and processes to describe their physical properties in order to assure best quality of agricultural products or raw material for industry, taking into account the role of environment and other factors. As a field of science, Agrophysics is of interdisciplinary scope and it is closely related to Biophysics. It is however limited strictly to the agricultural environment, i.e., soil, plants and animals and also takes into account the knowledge of Agronomy and Agriculture Engineering.

Agrophysics deals with physical processes in the soil-plant-atmosphere system, taking into account various external factors (climate, impact of the machinery, pollution) and issues related to the growth, harvest, transport, storage and processing of agricultural materials.

Some examples of the wide scope of agrophysical investigation are: developing systems for monitoring and controlling the condition of soil (moisture, salinity etc.) and plant growth (maturity), evaluation of the soil's susceptibility to water and wind erosion, monitoring and diagnosis of soil biological activity, determination of pollution in agricultural products (fruits, vegetables etc.), the assessment of the technological value of grain, evaluation of quality of fruits and vegetables during their storage and changes of their nutrition value during storage.

Citation: B. Dobrzański, S. Grundas and A. Stępniewski (2013). Introduction to Scientific Discipline Agrophysics, Advances in Agrophysical Research, Prof. Stanisław Grundas (Ed.), InTech, DOI: 10.5772/56982.

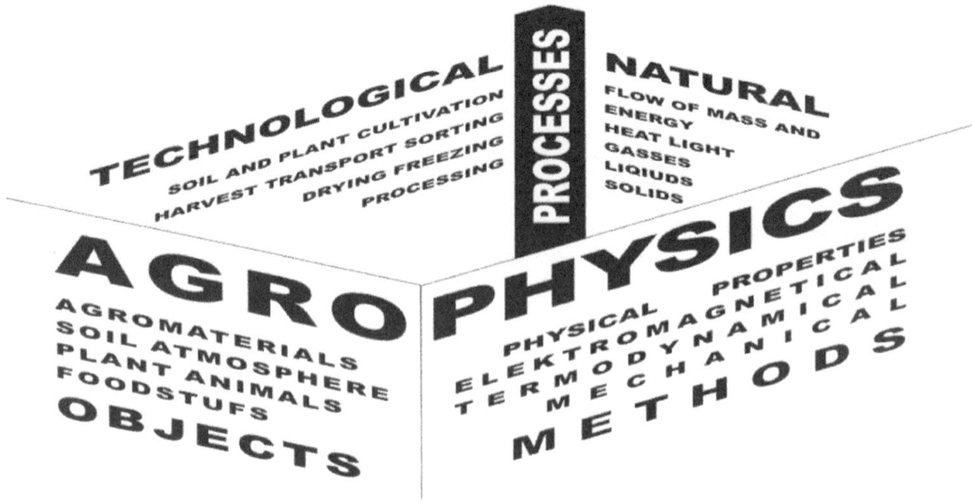

Figure 1. Graphical description of the definition of Agrophysics.

The following definition of Agrophysics was recently included in the "Encyclopedia of Agrophysics":

"Agrophysics is a science that studies physical processes and properties affecting plant production. The fundaments (nutrients) and energy (light, heat) transport in the soil– plant–atmosphere and soil–plant–machine–agricultural products–foods continuums and way of their regulation to reach biomass of high quantity and quality with the sustainability to the environment. The knowledge of physical phenomena in agricultural environment allows increasing efficiency of use of water and chemicals in agriculture and decreasing biomass losses during harvest, transport, storage, and processing."

Agrophysics therefore aims at ecological use of agricultural ecosystem to assure best quality of agriculture products and the preservation of agricultural landscapes.

HISTORY

In the end of XIX century scientists recognized the need of application of physics to agriculture. The pioneer in this field of research was probably Franklin Hiram King. He was born in 1848 in Whitewater Wisconsin USA. He graduated from Cornell University and served as professor of natural sciences at River Falls State Normal School and after that as professor of agricultural physics at the University of Wisconsin. In 1888 the University of Wisconsin called him to the Chair of Agricultural Physics, the first of its kind in America. F.H. King was interested in a wide range of scientific problems but he made his major contributions to the applications of physics to agriculture. Most attention was given to soil physics, e.g., he studied water-holding capacities, moisture requirements of plants, aeration, movement of water in soils, movement of groundwater, and the drafts of plows. During his last years in Madison he also began studies of soil fertility. He has been called the "father" of soil physics in the USA.

King published some books, the titles of which prove his scientific interest:

- Elementary lessons in physics of agriculture. F.H. King, 1894. Madison, WI.
- The soil: Its nature, relations and fundamental principles of management. Macmillan, 1895. New York.
- Irrigation and drainage. Macmillan, 1898. New York.
- Physics of agriculture. F.H. King, 1901. Madison, WI.
- Ventilation for dwellings, rural schools and stables. F.H. King, 1908. Madison, WI
- Farmers of forty centuries. F.H. King, 1911. Madison, WI.
- Soil management. Orange-Judd, 1914. New York.

F.H. King's observations and investigations could be nowadays regarded as a contribution to the so called sustainable agriculture. He died in 1911 in Madison, Wisconsin and his last two books were completed and published after his death.

Beginnings of application of physics in agriculture in Europe were in Soviet Russia. Agrophysical research was commenced in Russia in the middle of XX century. Russian physicist Abram Fieodorovich Ioffe (1880–1960) is regarded as the initiator of this branch of science. He studied measuring methods and used them in agriculture, biology and agrochemistry. He studied electromagnetism, radiology, crystals, high-impact physics, thermoelectricity and photoelectricity. He established research laboratories for radioactivity, superconductivity, and nuclear physics. In 1932 Ioffe organized the Institute of Physics and Agriculture in Leningrad (Sankt Petersburg at present) and became its first director. A.F. Ioffe published two fundamental works entitled: Physics and Agriculture (1955), and Physics for Agriculture (1959). His successors included: A.F. Chudnovsky (1910–1985), F.E. Koliasev (1898–1958);P.V. Vershinin (1909–1978); I.B. Revut (1909–1978), S.V. Nerpin (1915–1993), E.I. Ermakov (1929–2006), N.F. Bondarenko's (1928–2003); I.S. Lisker; A.M. Globus, V.P. Yakushev and others. The research topics undertaken by the above mentioned scientists were: mathematical modeling of agricultural production, simulation of agrophysical systems and processes; interaction of biological objects with different physical fields (light, gravity, magnetic, electromagnetic, acoustic, electrostatic), information technologies of production management in arable farming and plant growing, agrophysical instrumentation, elaboration of vegetative systems in controlled climate, a new type of coordinate precision agriculture as the first step to the creation of the "electronic farmer".

The second scientific Institute whose main field of research is Agrophysics was established in Lublin, Poland. Created in 1968 by prof. Bohdan Dobrzański, the Institute of Agrophysics of the Polish Academy of Sciences was soon recognized as a leading research centre not only in Eastern and Central Europe, but also in the global research. Numerous professors and younger researchers from the Institute frequently visited leading universities from Western Europe, Japan and the USA, bringing home the knowledge of recent trends and novel methods in the research field. In turn, the high quality of IA staff attracted a number of doctoral students and young researchers from Poland and abroad.This attractiveness, especially to young researchers made the Institute become also a leading educational centre at an advanced level in numerous research areas connected with environment, agriculture and food sector. The results of the research activities undertaken in the Institute were introduced into industry and gave basis for new products and technologies. The outstanding scientists of that period were: I. Dechnik (1929 - 2003), R. Walczak (1943 - 2003), M. Malicki (1939 - 2009), J. Stawiński (1942 - 2005), B. Szot (1933 – 2012), J. Gliński, W. Stępniewski, J. Lipiec, K. Konstankiewicz,

Figure 2. Coworkers of prof. B. Dobrzański from the Instite of Agrophysics at the Conference in Prague 1985 (from left; in the front row: R. Walczak, B. Szot, G. Skubisz, B. Dobrzański, J. Gliński; in the secondrow: M.Molenda, A. Pukos, W. Stępniewski, W. Woźniak, A. Kuczyński, K. Konstankiewicz, J. Lipiec).

The main fields of scientific activity of the Institute of Agrophysics PAS were as follows: investigation of physical and physical–chemical processes of mass and energy exchange in the soil – plant – atmosphere system, physical properties of agricultural materials and processes affecting plant production as well as processes related to gathering, transport and storage of agricultural materials.

The main feature of these studies was the elaboration of new theoretical and experimental research methods, developing of physical – mathematical models and their experimental verification, processing of data, taking into account their variabilityin time and space. Data bases and thematic maps created in this process can be used in practice for agricultural and environmental protection.

At present, there are more than 100 employees of interdisciplinary character. The scientific staff of the institute constitutes an interdisciplinary team of physicists, chemists, agronomists, horticulturists, biologists, engineers, geographers and mathematicians.

Agrophysics initially focused mainly on the study of soil and plant materials. Over time, this research area started to gradually expand, on the one hand including ever more elements of the soil-plant-atmosphere system, on the other hand, more and more focused on the process of food production and its quality - from the stage of agricultural production through the period of storage of agricultural products, their processing until the final product.

Agrophysics began also to be useful in the wider environment, being a part in the study of not only the degradation of soils, but also researching marshy land formation and greenhouse gas emissions. Currently Agrophysics concentrates on a number of agricultural specialties, it is used to interpret interactions, design, control and optimization of processes. It is also widely used in environmental protection, pedology, tillage and plant engineering, agriculture, agri-food technology, and others.

AGROPHYSICS IN OTHER COUNTRIES

Agrophysics was developed also in other countries, initially in Eastern European countries. In Czechoslovakia (Prague) R. Řezniček together with his coworkers (J. Blahovec, J. Pecen, P. Hnilica) conducted some important investigations at the Chair of Physics of Prague Agricultural University. They concentrated on physical properties of cereal grains and developed some interesting testing methods. In Czech M. Kutilek from Czech Technical University in Prague conducted advanced studies within soil science too. Also in Hungary (Gödöllö and Budapest) researchers carried out studies on physical properties of agricultural materials: soil, seed, grains, vegetables and fruits. In this context the following names should be mentioned: G. Sitkei, I. Husar, G. Várallyay, I. Farkas and others. Also German researchers joined international agrophysical society. The universities in Hohenheim (H-D. Kutzbach, E. Schlihting, K. Stahr), Bonn (K.-H Kromer), Kiel (R. Horn), Berlin (G. Wessolek) and in Potsdam (J. Helebrand) were engaged in application of physics in agriculture. The events called "Agrofizik Tagung" were organized on a regular base in Germany. These symposiums gathered scientists interested in this topic from Germany. Agrophysics was researched also in Spain. Spanish scientists were interested both in soil (J. Moreno, M. Aranda and D. de la Rosa) and fruit properties (M. Ruiz-Altisant, J. Caniavate). Belgium has also undertaken serious research activities to become a strong agrophysical centre. Catholic University in Louvain (J. DeBaerdemaker, B. Nicolai) and University of Ghent (M. De Boodt, D. Gabriëls) were main centers of agrophysical research there. In France agrophysics was developed in INRA Montfavet (S. Auber, P. Varoquaux), while in Italy the University of Torino (A. Ferero) was the centre of agrophysical studies. Same investigation were also conducted in Belarus and a leader there was I.I. Lishtvan. In Austria prof. W.E.H. Blum interested in some agrophysical aspects of soil sceinces. From Sovakia it is necessarry to mentione Jech.

Agrophysics was developed not only in Europe. A fundamental publication on this topic "Physical Properties of Agricultural Products" has been written by N.N. Mohsenin – scientist of Iranian origin who worked in the USA. Prof. A. Tabatabaeefar is a continuator of Mohsenin agrophysical research in Iran. There were also other scientists interested in agricultural physics i.e. S. Gunasekaran, O.R. Kunze, J.I. Ross, G. Brusevitz, F. McClure, Y.A. Pachepsky, S.A. Thompson, P.P. Chen and others. Some topics of applied physics in agriculture were undertaken in Japan (R. Hatano) and China (T. Ren,).

In Canada at the Guelph University physics in agriculture was practiced by number of researchers (W.K. Bilanski, R.L. Kushwaha). Also at University of Saskatchewan F. Sosulski. Agrophysics was developed even in New Zealand by C.J. Studman.

In Israel some aspects of agrophysics were studied by I. Shmulevich, E. Bresler, K. Peleg.

The above mentioned names and centres do not exhaust the list of places and persons who conduct their investigations in the field of applied physics in agriculture. These are only examples of people who cooperated with scientists from the Institute of Agrophysics PAS proving that agrophysics is a world-wide recognized science discipline.

AGROPHYSICAL CONFERENCES

Conferences on Agrophysics have always been an occasion for long discussionson all aspects of physical, physicochemical and biological processes of mass and energy exchange in soil-plant-atmosphere system and of plant production, as well as characteristics of agricultural products and materials, agrophysical measuring methods, soil degradation and remediation problems.

Since Agrophysics became a widely practiced discipline of science there were already a number of agrophysical conferences organized. They gathered scientists which main research interests focus on physics in agriculture. The conferences were organized as follows:

- Lublin, Poland 1976, B. Szot;
- Gödöllö, Hungary 1980, I. Husár;
- Prague, Czech Republic 1985, R. Řezniček;
- Rostock, Germany 1989, H.-J. Hellebrand;
- Bonn, Germany 1993, K.-H. Kromer;
- Lublin, Poland 1997, J. Gliński;
- Prague, Czech Republic 2001, J. Blahovec;
- Louven, Belgium 2004, J. De Baerdemaeker;
- Lublin, Poland 2005, R. Walczak;
- Lublin, Poland 2011, J. Horabik;

A number of scientific papers that were submitted and presented during the above mentioned conferences were published both in conference materials as well as in special issues of scientific journals.

AGROPHYSICAL JOURNALS

Institute of Agrophysic of the Polish Academy of Sciences is also the publisher of an outstanding scientific journal entitled "**International Agrophysics**". The editorial board states that: "the journal focuses on physical properties and processes affecting biomass production and processing. The main topics are: mass (water, air, plant nutrients) and energy (light, heat) transport in the soil-plant-atmosphere continuum, ways of their regulation in order to reach biomass of high quantity and quality. The description of new methods and devices for measurements of the physical properties of agro- and biomaterials are published. The journal is also open to wider aspects of environmental and agricultural physics". The present Editor-in-Chief is prof. J. Gliński.

International Agrophysics is indexed by Journal Citation Reports with 1,574 impact factor.

A second journal published by the Institute of Agrophysics is "Acta Agrophysica". Acta Agrophysica has been published since 1993. At the beginning it contained mainly monographies and dissertations which were published irregularly. From 2012 Acta Agrophysica is a quarterly. It publishes papers presenting the results of fundamental and applied studies from the field of application of physics for the solution of problems relating to the management and protection of the natural environment, sustainable agriculture, and food processing. Papers can be published both in Polish and English. The present Editor-in-Chief is prof. J. Horabik.

The third journal published by the Institute of Agrophysics is "Acta Agrophysica Monographiae" which publishes reviewed papers based on original research results as well as monographs pertaining to the field of agrophysics. The monographs are published in Polish or English. The present Editor-in-Chief is prof. J. Horabik.

The above mentioned journals are available in electronic versions on the internet.

DICTIONARIES AND BOOKS, MAPS

In order to facilitate international collaboration and allow unification of terminology, a number of bilingual and multilingual dictionary of agrophysical terms (nomenclature) was prepared under editorial supervision of prof. Ryszard Dębicki and prof. Jan Gliński. The following languages were taken into consideration: English, Russian, French, Spanish, German.

"Atlas of the Redox Properties of Arable Soils in Poland" prepared by Ostrowski, J.; Stępniewska, Z.; Stępniewski, W.; Gliński, J. was published 1996. It contains a wide range of data on soils in Poland presented in the form of cartographic maps. The Atlas gives a comprehensive information on soils in Poland and can be regarded as a research tool for studying spatial characteristics of agriculture. Practical application of the database lays in the power to generate, for the first time in the world, the thematic maps on spatial differentiation of the redox soil properties throughout the country. A set of these maps has been presented in the published atlas.

The last but most fundamental publication on Agrophysics is the "Encyclopedia of Agrophysics" edited by J. Gliński, J. Horabik and J. Lipiec by Springer in 2011. This book provides an up-to-date information on the physical properties and processes affecting the quality of the environment and plant production. Encyclopedia of Agrophysics is a publication complementary to the Encyclopedia of Soil Science, (November 2007) which has been published in the field of Earth sciences series of Springer. The Encyclopedia presents a set of about 250 informative articles and ca 400 glossary terms covering all aspects of Agrophysics. It contains 450 illustrations on more than 1000 pages.

DATA BANKS (SOIL PROBE BANK)

One of the comprehensive accomplishment of the Institute of Agrophysics PAS is the bank of soil samples collected all over Poland. The collection consists of over a thousand profiles (3 levels) and their full characteristics. This is a unique set of samples which can enable scientists to monitor changes of agriculture environment in Poland.

AGROPHYSICS — EUROPEAN CENTRE OF EXCELLENCE

The Centre Of Excellence For Applied Physics In Sustainable Agriculture "AGROPHYSICS" was created on February 28-th. 2003, within the scheme of the 5-th. Framework Programme of the European Union. The project was realized for three years and ended on February 28th2006. The main goal of the project was to support the research potential of the region and strengthen its integration with the European Research Area.

The main points of the action plan of the CoE were:

- to develop programmes aimed at significant technological applications, while preserving high standard of fundamental research,
- to concentrate on modern research fields with potential applications in industry,
- to expand the Institute's activities towards market-oriented research,
- to participate in European projects within European research programmes,
- to intensify activities in attracting funds from the State Committee for Scientific Research (KBN; now – the Ministry of Science and Higher Education) under research grants,
- to start co-operation with Polish industry and SMEs,
- to develop a scientific network with universities,
- to make the IA more attractive for students by combining post-graduate and doctoral studies with a contribution to research projects,
- to establish a Foundation for Development of Agrophysical Research in order to be able to obtain more funds for research.

The prolongation of the existence of the CoE and the consequence of this prolongation was a possibility for dynamic development of the structural base of the Institute. The full renovation of buildings as well as thorough exchange of research equipment was accomplished thanks to European funds.

POLISH SOCIETY OF AGROPHYSICS

Polish Society of Agrophysics (PSA) was founded in 1996 in Lublin. The initiator and the first president of PSA was prof. Bogusław Szot (1933 - 2012). He has been managing the Society until his death in October 2012. During his presidency ten branches in major scientific centers in Poland were established, where agrophysical investigations are performed (Fig. 3.). The Society is very active and every year it organizes meetings, seminars and symposia. Today PSA comprises over 360 members, among whom over 150 hold the title of full professor.

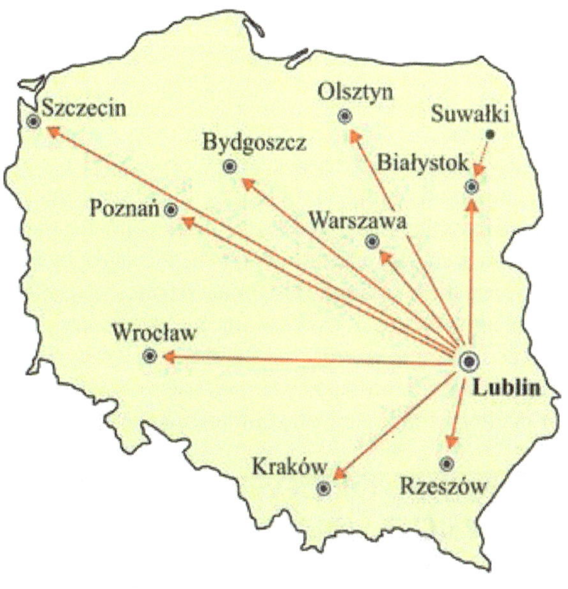

Figure 3. Location of PSA branches in Poland.

The main goal of the Society is to agglomerate researchers who conduct their research in the field of agrophysics. The Society has organized nine national and international conferences with over 1400 participants (115 of them from 29 countries outside Poland). Every conference has its own book of abstract, where the most recent results are presented. The most important papers are published in International Agrophysics and Acta Agrophysica – both journals edited at the Institute of Agrophysics PAS. Current information on the activities of PSA is distributed through its Information Bulletin.

SCIENTIFIC COMMITTEE OF AGROPHYSICS OF POLISH ACADEMY OF SCIENCES

The Committee of Agrophysics of Polish Academy of Sciences was established in 1981.

The Scientific Committees of the Polish Academy of Sciences are self-governing, nationwide representation of various disciplines or groups as well as interdisciplinary scientific problems integrating scholars throughout Poland. The scientific committee includes the Members of the Polish Academy of Sciences of the relevant specialty (outstanding scientists), eminent researchers representing universities, institutes of the Polish Academy of Sciences and the scientific institutes and research departments, as well as representatives of other institutions, including economic and social organizations. The scientific committees of the Polish Academy of Sciences are the most representative group of experts in the discipline.

The main task of the Scientific Committee of Agrophysics is to promote research of physical and physicochemical properties of the natural environment with particular emphasis on the system: soil-plant-atmosphere-machine-crops, agricultural-machine-food products. At present the chairman of the Committee is prof. Bohdan Dobrzański Jr.

Actually there are three sections operating within the structure of the Committee:

- Section of Physics Application in Engineering of Agricultural Production and in Food Technology,
- Section for Physical Measurement Techniques for the Agricultural Environment Protection,
- Section of Physical Methods of Evaluation of the Quality of Agricultural Products.

Since the beginning of its existence the Committee's special attention was directed to the integration of the scientific community and its involvement in the process of creatively solving important problems in agricultural research. The Committee in its activities focused also on identifying the main areas of research and science policy in the agrophysics. In previous years the Committee also financed research projects in the field of agrophysics. It has developed extensive cooperation with other committees working in the field of agricultural sciences. It also promoted international cooperation as a partner institution in organization of international conferences.

FOUNDATION FOR DEVELOPMENT OF AGROPHYSICAL SCIENCES (FRNA)

Foundation for Development of Agrophysical Sciences was established for the purpose of supporting and promoting the development of agrophysical research and activity of the Institute of Agrophysics PAS. The implementation of the main goals of the Foundation is achieved by publishing (Scientific Publishing House of FRNA) in the field of natural sciences and application of physics in agriculture. According to the up-to-date techniques, the Foundation proposed e-files i.e. electronic publications on CD and DVD instead traditionally printed books. The Foundation also sponsors various initiatives and initiates actions toward popularization of agrophysics as science.

PROTOTYPE APPARATUS AND INVESTIGATION METHODS

For many years a workshop for prototype apparatus production was being organized at the Institute of Agophysics PAS in Lublin. A number of pioneer measuring techniques were elaborated and used in innovative apparatus. Both construction and production of this equipment was realized at the Institute. The founder and the first manager of the workshop was M. Grochowicz, PhD.

Under this activity a measuring systems for determination of moisture content and salinity of soil and other porous materials has been elaborated. The measuring techniques are based on the Time-Domain Reflectometry (TDR) which is worldwide patented. In the last years an innovative silo for drying and safe storage of rapeseed was constructed. The silo allows automated post-harvest drying, cooling, and storage of rapeseeds in neutral gas atmosphere to assure best quality of raw material for oil production. It prevents the development of heat processes as well as moulds and fungi growth in stored seeds. Also some prototype measuring apparatus were elaborated in order to evaluate quality parameters of fruit tissue especially apple tissue.

All the above mentioned activities, institutions and persons prove the necessity and usability of Agrophysics as a branch of applied science, which can help to solve current problems in agriculture and which can benefit in progress of this part of economy.

REFERENCES

1. B Dobrzanski, 1986Możliwości rozwoju agrofizyki. Wykład z okazji nadania tytułu Doktora Honoris Causa ART w Olsztynie, Kaseta video, ART, Olsztyn.
2. B Dobrzanski, 1981Badania w zakresie fizyki i fizykochemii gleb. Kosmos, 2135139
3. B Dobrzanski, I Dechnik, J Glinski, 1979Rozwój badań agrofizycznych w Polsce. Post. Nauk Roln., 5.
4. B Dobrzanski, J Glinski, B Szot, 1988Agrophysical investigations in Poland at present and in future. Physical properties of agricultural materials and products. Hemisphere Publ. Corp., New York, 873976
5. J Glinski, J Horabik, J Lipiec, 2011Encyclopedia of Agrophysics. Springer.
6. J Haman, J Horabik, A Pukos, 1985Mechanical investigations of agricultural materials in the Institute of Agrophysics. Zesz. Probl. Post. Nauk Roln., z. 304916
7. B Szot, Dobrzański jr B., 1995Agrofizyka dla techniki rolniczej. Zesz. Probl. Post. Nauk Roln., z. 4247986
8. C. B Tanner, R. W Simonson, 1993Franklin Hiram King- Pioneer Scientist. Soil Science Society of America Journal 571
9. http://www.agrophys.ru/
10. http://www.ipan.lublin.pl/
11. http://www.international-agrophysics.org/
12. http://www.acta-agrophysica.org/
13. http://www.acta-agrophysica-monographiae.org/
14. http://www.komagrof.pan.pl/

CHAPTER 4 DISTRIBUTIONS OF CADMIUM AND LEAD IN PERI-URBAN WETLANDS AS INFLUENCED BY SOIL ORGANIC MATTER, CLAY FRACTION, AND MOISTURE CONTENT

Ayodele O. Adelana, Gabriel A. Oluwatosin, Celinah Agunbiade, Kayode S. Are and Olateju D. Adeyolanu

Keywords: Heavy metals, distribution pattern, chemical speciation, hydraulic conditions,

INTRODUCTION

Within an agro-ecosystem, soil plays a major role in food production, and it stands as the interface between human activities and other parts of the environment. Soil, not only serves as source of essential nutrients for crops but also functions as sink for heavy metal pollutants. This ecological function of soil is most pronounced in wetlands. Wetlands are usually low areas where heavy metals are more easily accumulated due to changes in natural environment and influences of human activities (May & Edwards, 2001), and this may also be due to a variety of physicochemical processes such as adsorption, ligand exchange, and sedimentation. In wetlands, heavy metals could exist mainly in water, mud at the bottom, and in plants or other organisms. The occurrence of wetland in Nigeria according to Okusami and Rust (1992) has been associated with three landform types, which are inland depression, alluvial plains, and coastal plains. They are used by peasant farmers in sub-Sahara Africa (SSA) for agricultural production and water sources for domestic consumption as well as dump sites for urban domestic wastes (Binns, Maconachie, & Tanko, 2003; Mbabazi et al., 2010). Because wetlands are important sources of food and water for human beings in SSA, they also constitute health hazard due to heavy metal deposits. It is therefore pertinent to protect soil resources while detail research studies on the distribution, transfer, and forms of heavy metals in wetlands are of great importance.

It has been documented that the total heavy metal contents in soils cannot determine their availability and toxicity (Clozel, Ruban, Durand, & Conil, 2006; Pagnanelli, Moscardini, Giuliano, & Toro, 2004). It is the physicochemical forms of any metal that determine its potential availability and toxicity. Thus, the speciation and distribution study of heavy metals in soils

Citation: Adelana, A. O., Oluwatosin, G. A., Agunbiade, C., Are, K. S., & Adeyolanu, O. D. (2016). Distributions of cadmium and lead in peri-urban wetlands as influenced by soil organic matter, clay fraction, and moisture content. Cogent Food & Agriculture, 2(1), 1159406. http://doi.org/10.1080/23311932.2016.1159406.

has become one of the most important areas in environmental research. The chemical speciation of metal ions in soil involves the fractionation of its total content into exchangeable, acid-extractable, reducible, oxidizable, and residual forms (Christine, Sylvaine, & Michel, 2002). The exchangeable and acid-extractable fractions are mobile fractions that are considered to be easily available. The oxidizable and reducible forms will be leached out only under extreme conditions, while the residual fraction is almost inert. The sequential extraction method proposed by Krishnamurti and Naidu (2000) provides information about the distribution of heavy metal associated with different specific solid phases in soils. This contributes to the understanding of the distribution and availability of Cd and Pb in some selected wetlands. Research has shown that the stabilization of heavy metals is being controlled by different physicochemical properties (Cui, Yang, Yang, & Zhang, 2009; Ip, Li, Zhang, Wai, & Li, 2007). Research by Ip et al. (2007) reported that the presence of organic matter, divalent iron, and clay enhances the presence of heavy metals in soils; while Cui et al. (2009) identified carbonate, phosphate, and Fe/Mn oxides as factors aiding their immobilization and deposition. The processes of decalcification, desalinization, and oxidation of reduced Fe, Mn, and sulfur compounds can induce changes in redox conditions, pH value, mobility of phosphorus, and behavior of Fe and other metals in the soil (Giani, Ahrens, Duntze, & Irmer, 2003). Within a given wetland type, these physicochemical properties vary and therefore aid the release and occurrence of heavy metal forms in varying proportion. Previous studies (Ngole, 2011; Oluwatosin et al., 2010) presented that the accumulation of heavy metals, such as Cd and Pb, is significantly affected by anthropogenic activities. However, the distribution and forms of Cd and Pb in different soil depths in the arrays of wetlands in southwest Nigeria have not been investigated. Understanding Cd and Pb distribution, forms, and accumulation in inland depression and alluvial plain wetlands plays a crucial role in ecological risk assessment. Thus, this study was set out to investigate Cd and Pb distribution in different types of wetlands, their variation with depth, and characterize the geochemical fractions of Cd and Pb in these wetlands.

MATERIALS AND METHODS

Site Description

Different types of wetlands, specifically, inland depression and alluvial plain wetlands, were identified within Ibadan metropolis, southwest Nigeria. Two alluvial plain wetlands and one inland depression wetland were selected for this study (Figure 1). The alluvial plain wetlands are located at Omi Adio (AP_1) (7° 23.8′N and 3° 45.2′E) covering an area of 5.1 ha and Ojoo Barrack (AP_2) (7° 26.8′N and 3° 56.6′E) with an area of 3.2 ha, which are drained by Omi and Ajibode rivers, respectively. The inland depression wetland is located within the University of Ibadan (ID) (7° 26.9′N and 3° 53.8′E) with an area of 2.2 ha. The three sites are peri-urban wetlands with a tropical humid climate characterized by high humidity, temperature, and precipitation. The annual precipitation in the area is 1,300 mm, with humidity ranging between 60 and 90%, and minimum and maximum temperatures of 17–21°C and 29–36°C, respectively.

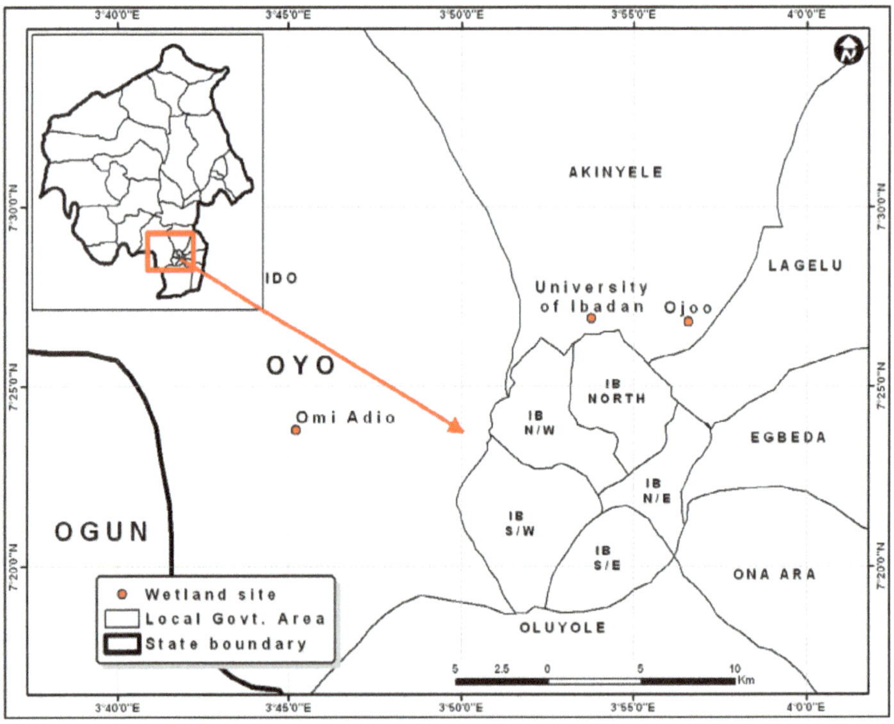

Figure 1. Map of Oyo state southwest Nigeria showing the wetland sites.

The study sites are characterized by overlaying Precambrian Basement complex rocks consisting of granites, gneisses, and quartzite. The soils of Omi Adio and Ojoo Barrack are classified as Eutric Fluvisol (FAO/UNESCO) and Aeric Fluvaquent (USDA), while the wetland soils of University of Ibadan are classified as Eutric Gleysol (FAO/UNESCO) and Typic Tropaquent (USDA). All the wetland sites receive runoff from adjacent build-up areas where household waste are generated in varying degrees. The wetlands are usually cultivated with vegetable and maize during the dry season

(November–April). During this period, the water table might have been lowered, while little or no farming activity takes place during the raining season.

Soil Sample Collection and Analyses

Soil samples were collected during the dry season in December 2013. Each of the wetland sites was divided into four quadrants representing four sampling plots. From each sampling plot, five random points were selected for soil sampling. Two techniques were used to obtain samples from the plots: (1) a 5 cm by 5 cm core sampler was used to collect undisturbed core samples for bulk density and soil moisture determination, and (2) composite samples were collected from 0–20, 20–40, 40–60, 60–80, and 80–100 cm depths, using a stainless steel soil auger. The composite soil samples were air dried at room temperature and allowed to pass through a 2-mm nylon sieve to remove gravels. The samples were stored in polythene bags prior to soil analysis.

Particle size distribution was determined using a modified Boyoucos hydrometer method as described by (Gee & Or, 2002). The undisturbed cores were used to determine soil bulk density (BD)

(Grossman & Reinsch, 2002). Moisture content was determined using the procedure described by Lowery, Arshad, Lal, and Hickey (1996). Soil pH was determined in distilled water with a pH meter in a 1:1 soil to water suspension ratio after 1 h of equilibration. Exchangeable cations (calcium-Ca, magnesium-Mg, potassium-K, and sodium-Na) were determined by extracting with 1 M NH_4 OAc (ammonium acetate) solution as described by Okalebo, Gathua, and Woomer (1993). The amounts of exchangeable Na and K in the extract were determined by flame photometry (Jenway PFP 7), while Ca and Mg were measured by atomic absorption spectrophotometry (AAS) (BUCK 211). The effective cation exchange capacity (ECEC) of the soil was determined by addition of the values of exchangeable cations and hydrogen (H^+) (Okalebo et al., 1993). Soil organic carbon (SOC) was determined by loss on ignition and the SOC content was converted to soil organic matter content (SOM) by multiplying SOC by a Van Bemmelen factor of 1.724. Active carbon (AC) was determined using permanganate oxidation following the procedure of Weil, Islam, Stine, Gruver, and Samson-Liebig (2003). Crystalline iron (Fed) and aluminum (Ald) were determined by dithionite–citrate–bicarbonate extraction (Coffin, 1963) and amorphous iron (Fe_o) and aluminum (Al_o) by ammonium oxalate extraction (Schwertmann, 1964). The total concentrations of Cd and Pb in the soils were determined by AAS after being digested with aqua regia and $HClO_4$.

Sequential extraction of soil Cd and Pb

The chemical fractionation technique of Krishnamurti and Naidu (2000) was used in the sequential extraction. The procedure separates heavy metals into seven different fractions as: exchangeable (F1), carbonate bound (F2), metal–organic complex or easily reducible metal bound (F3), organic bound (F4), amorphous Fe–Mn oxide bound (F5), crystalline Fe–Mn oxide bound (F6), and detrital (residual) (F7). The stepwise extraction is presented in Table 1. Following each successive extraction, the mixture was centrifuged at 14,000 rpm for 30 min. The supernatants were removed with a pipette, filtered with 0.2 µ nucleopore polycarbonate membrane filters, and analyzed for Cd and Pb concentration using AAS. The extraction was done in triplicates and two independent technologists performed the entire process of digestion and analysis while a high level of reproducibility was attained with R2 of 0.95 for Cd and 0.98 for Pb between the two analyses.

Statistical Analyses

The data obtained were analyzed using the analysis of variance (ANOVA) with statistical application software (SAS Institute, 2002). Significant means were separated using least significance difference (LSD) at 5% significant level of probability. The relationships among soil parameters were evaluated using Pearson correlation analysis to determine whether there are significant correlations among the pairs.

Table 1. Procedure for sequential extraction of Cd and Pb in the soils

Fraction	Species	Reagent	Shaking time/temp
F1	Exchangeable	10 ml of 1 m NH_4NO_3 (pH 7)	4 h at 25°C
F2	Carbonate bound	25 ml of 1 m CH_3COONa	6 h at 25°C
F3	Metal-organic complex bound	20 ml of 0.1 m $NH_2OH{\cdot}HCl$ in 0.1 m HNO_3	30 min at 25°C
F4	Organic bound	(1) 5 ml of 30% H_2O_2 (pH 2) 3 ml of 0.02 m HNO_3	2 h at 25°C
		(2) 3 ml of 30% H_2O_2 (pH 2) 1 ml of 0.02 m HNO_3	2 h at 25°C
		(3) Cool, add 10 ml of 2 m NH_4NO_3 in 20% HNO_3	30 min at 25°C
F5	Amorphous Fe-Mn oxide bound	20 ml of 0.2 m $(NH_4)_2C_2O_4$/2 m $H_2C_2O_4$ (pH 3)	25°C (dark)
F6	Crystalline Fe-Mn oxide bound	25 ml of 0.2 m $(NH_4)_2C_2O_4$ (pH 3) in 0.1 m ascorbic acid	30 min at 95°C
F7	Detrital (residual)	Digestion with HNO_3:$HClO_4$ (2:1)	

RESULTS AND DISCUSSION

General Properties of Soils

Mean and median values of variables measured across the plots revealed that soil properties in the wetlands varied widely. The chemical properties showed more variability when compared with the physical properties except for soil pH (Table 2). For example, the coefficient of variation (CV) obtained from exchangeable Ca, Mg, and ECEC exceeded 30%, compared with 10.9, 13.0, and 15.3% for BD, clay, and sand content, respectively (Table 2). Also, concentrations of Pb and Cd were highly variable with CVs of 55.6 and 94.5%, respectively. When compared with the study of Aiyelari and Oshunsanya (2008) on wetland soils in Ibadan metropolis, BD measurements and soil texture values measured generally fell within the ranges reported in these studies. In contrast, Ogban and Babalola (2009) reported lower BD values ranging from 0.78 to 1.36 Mg m^{-3} in some valley bottom soils of southwestern Nigeria. Low BD reported in this study may be a consequent of organic matter content of the soils. Moreover, frequent cultivation of the land made the soil loose and ultimately contributed to the low BD.

Although, soil chemical properties in the wetlands generally had high CVs, soil pH measured had a relatively low CV of 5.3% with a range of 5.3–6.6 (Table 2). It is evident that the concentration of Pb (ranging from 112.0 to 490.7 mg kg^{-1}) was high in the wetlands, followed by Cd (0.11–4.90 mg kg^{-1}) with a median value of 1.06 mg kg^{-1}. The CV obtained for both Pb and Cd were higher than 50% indicating high variability amongst the wetlands. Oluwatosin, Adeyolanu, Dauda, and Akinbola (2008) reported similar high variability in heavy metal concentration of some valley bottom soils of urban cities in south west Nigeria. These variations could be attributed to the amount of organic matter and clay particles in the wetlands. Also, hydraulic conditions could have played a part in the distribution of Pb and Cd. However, the total extractable heavy metal content in the wetlands cannot predict the availability of the metals for plant uptake. Thus, further investigation was carried out on the samples to understand the chemical speciation of Pb and Cd in the wetlands.

Table 2. Descriptive statistics of soil properties sampled to a depth of 20 cm (n = 60)

Soil property	Mean	Median	Minimum	Maximum	SE	CV (%)
pH	5.9	6.0	5.3	6.6	0.04	5.3
Organic matter (g kg^{-1})	65.3	76.5	18.4	119.3	3.42	36.0
Active carbon (g kg^{-1})	6.7	6.5	2.6	11.6	0.27	31.4
Bulk density (Mg m^{-3})	1.28	1.29	1.19	1.58	0.02	10.9
Sand (g kg^{-1})	577	590	377	771	1.14	15.3
Clay (g kg^{-1})	243	228	144	498	0.97	13.0
Silt (g kg^{-1})	180	185	145	305	0.74	11.9
Exchangeable Ca (cmol kg^{-1})	1.65	0.93	0.64	3.38	0.12	37.3
Exchangeable Mg (cmol kg^{-1})	0.94	0.82	0.58	2.08	0.06	47.1
Exchangeable Na (cmol kg^{-1})	0.49	0.42	0.24	0.66	0.02	26.8
Exchangeable K (cmol kg^{-1})	0.21	0.18	0.10	0.28	0.006	24.3
ECEC (cmol kg^{-1})	3.43	2.75	2.01	5.34	0.12	30.9
Acid extractable Cd (mg kg^{-1})	1.76	1.06	0.11	4.90	0.22	94.5
Acid extractable Pb, mg kg^{-1}	312.6	280.1	112.0	490.7	5.61	55.6
Crystalline Fe (g kg^{-1})	8.21	8.89	2.86	14.9	0.34	31.8
Amorphous Fe (g kg^{-1})	0.35	0.34	0.23	0.45	0.006	13.3
Crystalline Al (g kg^{-1})	4.43	3.70	2.82	9.04	0.34	28.5
Amorphous Al (g kg^{-1})	0.35	0.38	0.13	0.49	0.001	21.7

Notes: *n* is number of samples; SE is standard error; CV is coefficient of variation; ECEC is effective cation exchange capacity; Crystalline Fe and Al extracted by dithionite-citrate-bicarbonate; Amorphous Fe and Al extracted by oxalic acid + ammonium oxalate solution.

Surface Properties of Soils between Wetlands

The mean values of some physical and chemical properties of the wetland soils are presented in Table 3. All wetland soils were dominated by moderate sand content of 570 and 700 g kg^{-1} for ID and AP_2, respectively. The clay and silt contents in the different wetlands varied from 159 to 289 g kg^{-1} for clay and from 141 to 144 g kg^{-1} for silt. The soils were slightly acidic with pH values varying from 5.8 to 6.1, with the highest soil pH observed under ID wetland (Table 3). Figure 2 shows moisture content of the wetlands at various depths. There were significant differences ($p \leq 0.05$) in moisture content amongst the wetlands at 0–20 and 20–40 cm depths. The observed moisture content is a reflection of the drainage conditions in the wetlands. The well-drained AP wetlands had lesser moisture content than ID wetland. The improved drainage observed under alluvial plain (AP) wetland soils was enhanced by the higher sand fraction obtained from these soils.

The AC, a measure of the fraction of SOM that is readily available as carbon and energy source for the soil microbial community, was significantly higher ($p \leq 0.01$) under the alluvial plain wetlands (AP_1 and AP_2) than ID, and it was in the order of $AP_2 > AP_1 >$ ID. This perhaps increased the soil aeration under AP_1 and AP_2, and thus increased microbial activities, when compared with ID with poorer drainage. Wan, Song, Guo, Wang, and Huang (2008) reported that wetlands with water stagnation or continuous flooding did not favor the growth and multiplication of microbes leading to reduction in enzyme activities. Although, ECEC of the soils were generally low, they were significantly different ($p \leq 0.05$) from each other (Table 3). The significantly lower values of ECEC under AP_1 and AP_2 can be attributed to the leaching of exchangeable cations that occurred under the better drained alluvial plain wetland soils.

Surface Distribution of Metals in Wetlands

The surface distribution of crystalline- Fe and Al (Fe_d and Al_d) and amorphous- Fe and Al (Fe_o and Al_o) in the wetlands is summarized in Table 3. Generally, Fe_d and Al_d were higher than their Fe_o and Al_o .

Table 3. Mean surface soil properties (0 – 20 cm) for different wetlands

Wetland	Location	pH	AC	OM	Exchangeable cations				ECEC	Total		Fe_d[a]	Fe_o[b]	Al_d[a]	Al_o[b]	Particle size		
					Ca	Mg	Na	K		Cd	Pb					Clay	Silt	Sand
		1:1 H_2O	(g kg⁻¹)		(cmol kg⁻¹)					(mg kg⁻¹)		(g kg⁻¹)				(g kg⁻¹)		
AP_1	Omi Adio (*Fluvaquent*)	5.9	7.3	54.1	1.05	0.89	0.49	0.21	2.76	0.91	251.0	10.1	0.3	2.3	0.3	159	144	697
AP_2	Ojoo Barrack (*Fluvaquent*)	5.8	7.4	60.1	1.07	0.78	0.48	0.20	2.67	1.10	275.0	8.5	0.4	4.5	0.4	177	123	700
ID	University of Ibadan (*Tropaquent*)	6.1	5.3	81.3	2.82	1.16	0.50	0.23	4.81	3.28	412.1	6.0	0.4	6.4	0.4	289	141	570
LSD		ns	1.3**	18.5*	1.16**	0.25*	ns	ns	1.77*	0.86**	23.4*	1.3*	ns	1.4*	ns	ns	ns	ns

Notes: AP: alluvial plain wetland; ID: inland depression wetland; AC: active carbon; OM: organic matter; ECEC: effective cation exchange capacity; *n*: number of samples; ns: statistically not significant.

[a]Crystalline Fe and Al extracted by dithionite-citrate-bicarbonate.

[b]Amorphous Fe and Al extracted by 0.2 M oxalic acid + ammonium oxalate solution; LSD: least significant difference value between wetlands.

*Significant at $p \leq 0.05$.

**Significant at $p \leq 0.01$.

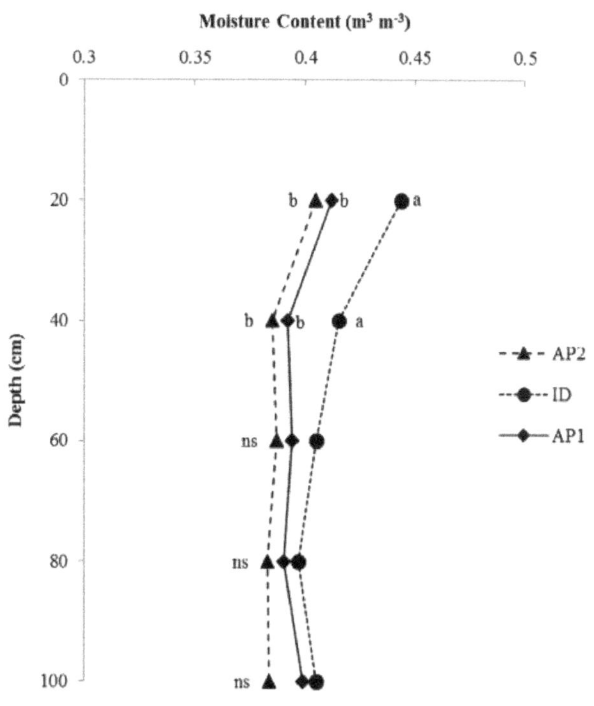

Figure 2. Variation in moisture content with soil depth. Depths with the same letter amongst the wetlands are not significantly different at $p \leq 0.05$, ns: not significantly different. AP: alluvial plain wetland; ID: inland depression wetland.

The higher amount of Fe_d and Al_d in these soils suggested that more amorphous Fe and Al had been transformed to crystalline form. The high tropical soil temperature, prolonged dry season coupled with high precipitation and age of Fe and Al released during weathering may have resulted in more Fed and Al_d in the wetland soils (Olaleye, Ogunkunle, & Sahrawat, 2000). High temperature causes Al and Fe oxides to dehydrate and subsequently shift from an form to a more crystalline state. In comparison, there were no significant differences in Fe and Al forms among the wetlands, except Fe_d (Table 3). Although, a low value of AC was obtained from ID, the significantly lower values of Fed can be attributed to poor drainage which led to accumulation of organic material (Table 3). It has been reported that organic matter hinders the rate of crystallization of iron oxides (Schwertmann, 1964). Furthermore, the higher amount of crystalline-Fe (Fe_d) in the alluvial plain wetlands could also have contributed to the significantly lower values of ECEC in these soils. Hendershot and Lavkulich (1983) reported that iron oxides coatings on clay minerals may decrease the cation exchange capacity of soils.

The mean concentration of Pb and Cd in 0–20 cm soil depth was used to compare the variation in heavy metal distribution in the wetlands. Within the wetlands, there were significant differences ($p \leq 0.05$) in the surface distribution of Pb and Cd (Table 3) with Pb having the higher concentration in each of the soils. The statistically higher values of Pb and Cd reported under ID could be as a result of the accumulation of organic material in the wetland and higher input of heavy metals resulting from human activities. Retained organic materials in the soil surface enhanced the concentrations of heavy metals by adsorption, complexation, and chelation. A high positive correlation between SOM and concentrations of Pb and Cd ($p \leq 0.05$) (Table 4) partly explained the metal concentration distribution order of ID > AP_2 > AP_1 with ascending SOM content of AP_1 < AP_2 < ID. Du Laing et al. (2007) also established that increasing organic matter enhances metal concentration of floodplain soils. Meanwhile, the relatively higher amount of clay fraction also accounted for the metal distribution in the inland depression wetlands.

Table 4. Pearson correlation matrix among Pb and Cd concentrations and soil properties in wetlands[a]

	Clay	Sand	pH	AC	OM
Pb	0.571	−0.172	0.288	−0.633*	0.766*
Cd	0.510	−0.067	0.312	−0.614*	0.636*
ECEC	0.423	−0.611*	0.436	0.230	0.698*
pH	−0.268	0.306	-	0.431	0.241
AC	−0.831**	0.411	0.399	-	-
OM	0.443	−0.532	0.296	−0.532*	-

[a]Correlation analysis of all wetland samples (0–20 cm).

*Significant at $p \leq 0.05$.

**Significant at $p \leq 0.01$.

AC: active carbon; OM: organic matter; ECEC: effective cation exchange capacity.

Profile Distribution of Metals in Wetlands

The linear relationships between metal concentrations and soil depth are identified by the high values of the correlation coefficient (R^2 ranges from 0.69 to 0.91, Figures 3 and 4). Results showed that the distribution of Pb and Cd varied significantly with sampling depth. Similar reports by Du Laing, Meers, et al. (2009) indicated that the concentrations of some metals (Cd, Ni, and Pb) are significantly affected by depth. Generally, Pb concentrations in the top layers (0–20 cm) of the wetlands were much higher than those in the deeper layers except for ID, where Pb concentration (383.2 mg kg^{-1}) in the 0–20 cm depth was lower than the 20–40 cm (391.1 mg kg^{-1}) and 40–60 cm (384.7 mg kg^{-1}) depths (Figure 3). This distribution pattern suggests a middle depth enrichment of Pb under ID resulting probably from metal enrichment during deposition. The reduction in Pb concentration with depth can be ascribed to the decrease in SOM with depth in the wetlands (Ferraz & Lourençlo, 2000). Although, Pb showed a positive correlation with clay content in the 0–20 cm depth (Table 4), there was a reduction in Pb concentration with depth despite the slight increase in clay content with depth. This indicates that the mobility of Pb in these wetlands was strongly controlled by the presence of organic complexes which was lower at deeper depths. In contrast to the reduction of Pb down the profile under alluvial plain wetlands (AP_1 and AP_2), the distribution of Cd increased with depth under AP_1 and AP_2 but showed a similar trend with Pb under ID (Figure 4). The mixed distribution pattern of Cd amongst the wetlands may be attributed to redox-induced changes, resulting from the improved drainage in the alluvial plain wetlands. In these wetlands, processes such as metal binding capacity of humic materials, insoluble metal sulfide formation, and changes in Fe/Mn-oxyhydroxides are inhibited (Du Laing, Rinklebe, et al., 2009) thereby enhancing the mobility of Cd in the alluvial plain wetlands. This view is corroborated by Vandecasteele, Du Laing, and Tack (2007), who reported that Cd is more mobile and readily available for plant uptake in oxidized soils with good drainage system when compared to reduced conditions.

Chemical Fractions of Pb and Cd in the Wetlands

Geochemical fractions of Pb and Cd in the wetlands from the sequential extraction are shown in Figures 5 and 6. The concentration of Pb associated with different fractions in the alluvial plain wetlands was in the order of organic > metal–organic complex > carbonate > crystalline Fe–Mn oxide > amorphous Fe–Mn oxide > exchangeable > detrital. In contrast, the order under inland depression wetland was organic > carbonate > metal–organic complex > amorphous Fe–Mn oxide > crystalline Fe–Mn oxide > exchangeable > detrital. In all the wetlands, Pb was mostly concentrated in the organic bound fraction (Figure 5). The concentration of organic bound Pb (Figure 5) accounted for over 50% of the total Pb content in all the wetland soils. The affinity of Pb to organic material explains the high values of organic bound Pb observed in the wetlands (Du Laing et al., 2007). Lead tends to undergo complexation reaction with strong binding sites which are probably offered in organic matter. Apart from the organic bound Pb, lead had specific binding sites in the different solid phases within the soils of the wetlands. The Pb loading might have contributed to the exchangeable fraction (2.2 to 3.3% of the total Pb) observed in all the wetlands since Pb occupied the weaker binding sites with its increasing concentration in the soils (Figure 5). The least concentration of Pb form in all the wetlands was observed under detrital (residual) fractions (Figure 5). The observed concentration of the inert (residual) fraction of Pb (detrital, amorphous Fe–Mn and crystalline Fe–Mn oxide fractions) in all the wetlands is close to the concentrations of Pb in the continental crust (12.5 mg kg^{-1}) and worldwide shale (20.0 mg kg^{-1}). Therefore, it is possible to suggest that the high concentration of Pb in the non-residual (readily available) fractions in the studied wetlands were probably as a result of anthropogenic activities.

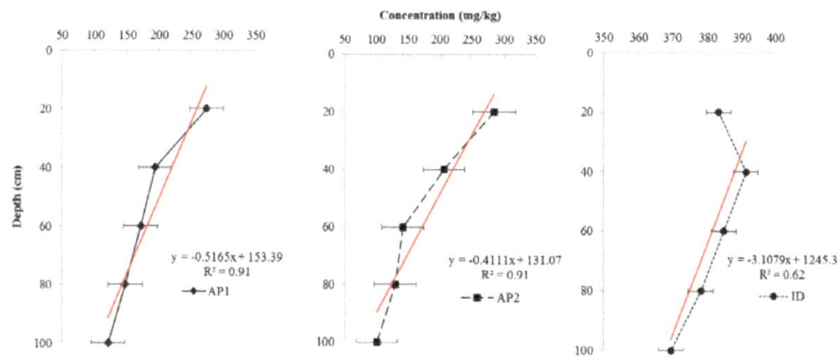

Figure 3. Variation in Pb concentrations with soil depth and regression models in different wetlands. Error bars = standard error. The red line is the regression line and in the regression equation, y: metal concentration and x: soil depth. AP: alluvial plain wetland; ID: inland depression wetland.

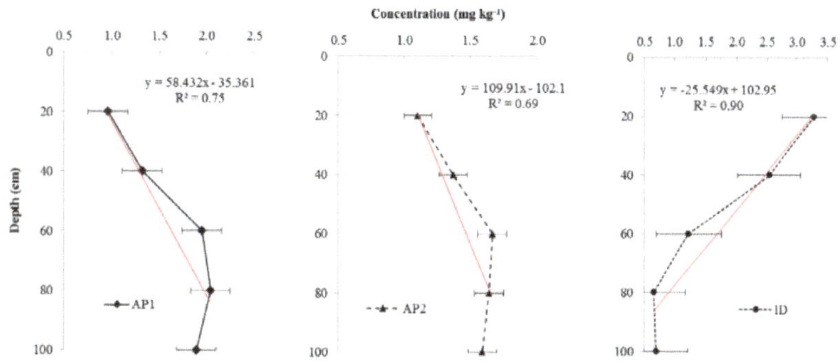

Figure 4. Variation in Cd concentrations with soil depth and regression models in different wetlands. Error bars = standard error. The red line is the regression line and in the regression equation, y: metal concentration and x: soil depth. AP: alluvial plain wetland; ID: inland depression wetland.

Figure 6 shows the geochemical fractions of Cd in the different wetlands. The concentration of water-soluble exchangeable fraction differ significantly ($p \leq 0.05$) between the wetlands, and it varied from 0.11 mg kg^{-1} (ID) to 0.42 mg kg−1 (AP2). Approximately 38–52% of the total Cd was associated with organic matter. The values of Cd obtained from sequential extraction suggest that it prefers to form complexes with organic matter in the wetland soils. The apparent stability and lability of Cd complexes in the soils is found to be dependent on Cd loading. It was observed that Cd prefers to form thermodynamically strong complexes (especially with organic matter) at low Cd loading. However, the concentration of thermodynamically weak complexes (exchangeable and carbonate bound) gradually increases with the increasing concentrations of Cd in the soils (Ferraz & Lourençlo, 2000). The detrital fraction of Cd in the studied wetlands was less than 1% of the total metal contents. The high concentration of Cd in the readily available fraction may lead to huge potential risk for environmental sustainability and human health. With the amount of Cd in these fractions, there is enough Cd for accumulation from the soil matrix into plant tissues.

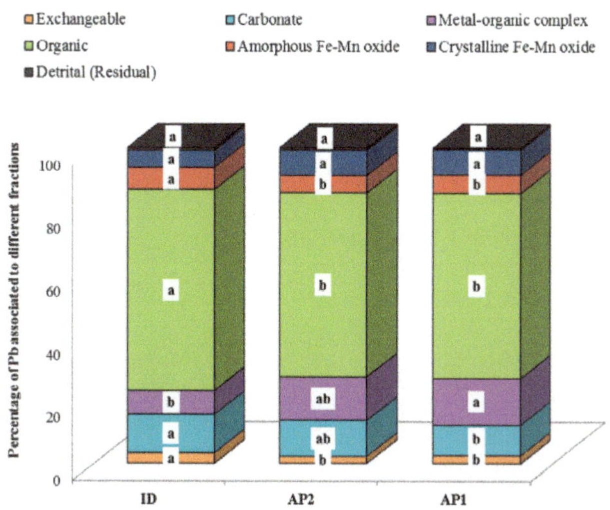

Figure 5. Lead concentration in each geochemical fraction of the wetland soils (0–20 cm). AP: alluvial plain wetland; ID: inland depression wetland; Concentrations with the same letter within the same geochemical fraction are not significantly different at $p \leq 0.05$.

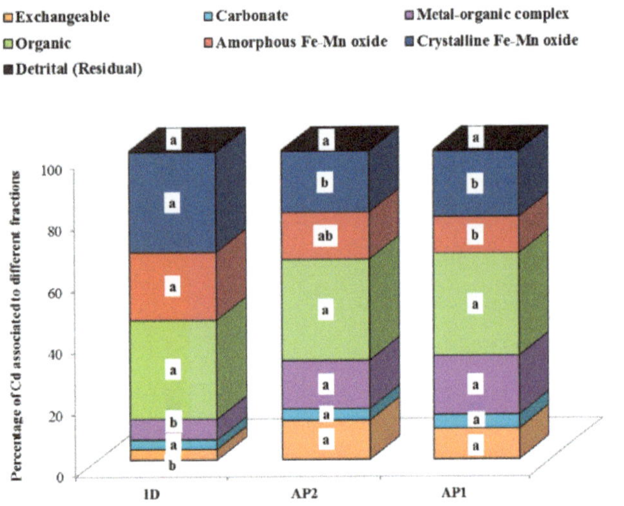

Figure 6. Cadmium concentration in each geochemical fraction of the wetland soils (0–20 cm). AP: alluvial plain wetland; ID: inland depression wetland; Concentrations with the same letter within the same geochemical fraction are not significantly different at $p \leq 0.05$.

CONCLUSIONS

Surface and profile distributions of Cd and Pb in the three wetlands showed that there were accumulation and mobility of these metals, which vary with soil depths. The surface distribution was influenced by anthropogenic activities, hydraulic conditions, AC, SOM, and particle size distribution. The observed profile distribution patterns in the wetlands were linear distribution and middle enrichment. The linear distribution showed an increase in Cd and decrease in Pb concentration under AP wetland with increasing depth, while the middle enrichment distribution had concentration of metals at a depth higher than values obtained from layers above and below it under ID wetland. These patterns are primarily related to SOM, hydraulic conditions, and depositional history. Concentrations of non-residual fraction of Cd and Pb (which are good indicators of their bioavailability in a system) may increase with increasing Cd and Pb input in these soils due to their affinity to complex with organic matter. These Cd and Pb complexes can easily be leached out into water bodies or taken up into plant tissues of crops that are cultivated in these wetlands. Therefore, the need to monitor the uptake of these metals by plants in order to reduce the potential risk to human health becomes important.

REFERENCES

1. Aiyelari, E. A., & Oshunsanya, S. O. (2008). Preliminary studies of soil erosion in a valley bottom in Ibadan under some tillage practices. *Global Journal of Agricultural Sciences, 7*, 221–228.
2. Binns, J. A., Maconachie, R. A., & Tanko, A. I. (2003). Water, land and health in urban and peri-urban food production: The case of Kano, Nigeria. *Land Degradation and Development, 14*, 431–444.10.1002/(ISSN)1099-145X,
3. Christine, G., Sylvaine, T., & Michel, A. (2002). Fractionation studies of trace elements in contaminated soils and sediments: A review of sequential extraction procedures.*Trends in Analytical Chemistry, 21*, 451–467.,
4. Clozel, B., Ruban, V., Durand, C., & Conil, P. (2006). Origin and mobility of heavy metals in contaminated sediments from retention and infiltration ponds. *Applied Geochemistry, 21*, 1781–1798.10.1016/j.apgeochem.2006.06.017,
5. Coffin, D. E. (1963). A method for determination of free iron oxides in soils and clays.*Canadian Journal of Soil Science, 43*, 9–17.
6. Cui, B., Yang, Q., Yang, Z., & Zhang, K. (2009). Evaluating the ecological performance of wetland restoration in the Yellow River Delta, China. *Ecological Engineering, 35*,1090–1103.10.1016/j.ecoleng.2009.03.022,
7. Du Laing, G., Vandecasteele, B., De Grauwe, P., Moors, W., Lesage, E., Meers, E., …,Verloo, M. G. (2007). Factors affecting metal concentrations in the upper sediment layer of intertidal reedbeds along the river Scheldt. *Journal of Environmental Monitoring, 9*, 449–455.10.1039/b618772b,
8. Laing, G., Meers, E., Dewispelaere, M., Rinklebe, J., Vandecasteele, B., Verloo, M. G., & Tack, F. M. G. (2009). Effect of water table level on metal mobility at different depths in wetland soils of the Scheldt estuary (Belgium). *Water Air Soil Pollution, 202*,353–367.10.1007/s11270-009-9982-2,
9. Du Laing, G., Rinklebe, J., Vandecasteele, B., Meers, E., & Tack, F. M. G. (2009). Trace metal behaviour in estuarine and riverine floodplain soils and sediments: A review.*Science of The Total Environment, 407*,3972–3985.10.1016/j.scitotenv.2008.07.025, ,
10. Ferraz, M. C. M., & Lourençlo, J. C. N. (2000). The influence of organic matter content of contaminated soils on the leaching rate of heavy metals. *Environmental Progress,19*, 53–58.10.1002/(ISSN)1547-5921
11. Gee, G. W., & Or, D. (2002). Particle size analysis. In J. H. Dane & G. C. Topp (Eds.),*Methods of soil analysis, Part 4, physical methods* (pp. 255–294). Madison, WI: Soil Science Society of America.
12. Giani, L., Ahrens, V., Duntze, O., & Irmer, S. K. (2003). Geo-pedogenesis of salic fluvisols on the North Sea Island of Spiekeroog. *Journal of Plant Nutrition and Soil Science, 166*, 370–378.10.1002/jpln.200390057,
13. Grossman, R. B., & Reinsch, T. G. (2002). Bulk density and linear extensibility: Core method. In J. H. Dane & G. C. Topp (Eds.), *Methods of Soil Analysis, Part 4, Physical Methods* (pp. 208–228). Madison, WI: Soil Science Society of America.
14. Hendershot, W. H., & Lavkulich, L. M. (1983). Effect of sesquioxide coating on surface charge of standard minerals and soil samples. *Soil Science Society of America Journal,47*, 1252–1260.10.2136/sssaj1983.0361599500470006037x,
15. Ip, C. C. M., Li, X., Zhang, G., Wai, O. W. H., & Li, Y. (2007). Trace metal distribution in sediments of the Pearl River Estuary and the surrounding coastal area, South China.*Environmental Pollution, 147*, 311–323.10.1016/j.envpol.2006.06.028,,

16. Krishnamurti, G. S. R., & Naidu, R. (2000). Speciation and phytoavailability of cadmium in selected surface soils of South Australia. *Australian Journal of Soil Research, 38*,991–1004.10.1071/SR99129,

17. Lowery, B., Arshad, M.A., Lal, R., & Hickey, W.J. 1996. Soil water parameters and soil quality. In: J.W. Doran & A.J. Jones (Eds.), *Methods for assessing soil quality*, (pp.143–157). Madison, WI: Soil Science Society of America Special Publication 49.

18. May, P. A., & Edwards, G. S. (2001). Comparison of heavy metal accumulation in a natural wetland and constructed wetlands receiving acid mine drainage. *Ecological Engineering, 16*, 487–500.,

19. Mbabazi, J., Bakyayita, G., Wasswa, J., Muwanga, A., Twinomuhwezi, H., & Kwetegyeka,J. (2010). Variations in the contents of heavy metals in arable soils of a major urban wetland inlet drainage system of Lake Victoria, Uganda. *Lakes & reservoirs: Research and management, 15*, 89–99.

20. Ngole, V. M. (2011). Using soil heavy metal enrichment and mobility factors to determine potential uptake by vegetables. *Plant Soil Environment, 57*, 75–80.

21. Ogban, P. I., & Babalola, O. (2009). Characteristics, classification and management of inland valley bottom soils for crop production in sub-humid southwestern Nigeria.*Journal of Tropical Agriculture, Food, Environment and Extension, 8*, 1–13.

22. Okalebo, J. R., Gathua, K. W., & Woomer, P. L. (1993). *Laboratory methods of soil and plant analysis: A working manual*. Nairobi: TSBF-UNESCO, EPZ.

23. Okusami, T. A., & Rust, R. H. (1992). Occurrence, characteristics and classification of some hydromorphic soils from south Nigeria. In J. M. Kimble (Ed.), *Characterization, classification and utilization of wet soils* (pp. 185–197). USDA, Soil Conservation Service:Proceedings VIII ISCOM.

24. Olaleye, A. O., Ogunkunle, A. O., & Sahrawat, K. L. (2000). Forms and pedogenic distribution of extractable iron in selected wetland soils in Nigeria. *Communications in Soil Science and Plant Analysis, 31*, 923–941.10.1080/00103620009370488Taylor & Francis Online,

25. Oluwatosin, G. A., Adeyolanu, O. D., Dauda, O. T., & Akinbola, G. E. (2008). Level and geochemical fractions of Pb, Cd and Zn in valley bottom soils of some urban cities in south west Nigeria. *African Journal of Biotechnology, 7*, 3455–3465.

26. Oluwatosin, G. A., Adeoyolanu, O. D., Ojo, A. O., Are, K. S., Dauda, T. O., &Aduramigba-Modupe, V. O. (2010). Heavy metal uptake and accumulation by edible leafy vegetable (*Amaranthus hybridus* L.) grown on urban valley bottom soils in southwestern Nigeria. *Soil and Sediment Contamination: An International Journal, 19*,1–20.Taylor & Francis Online,

27. Pagnanelli, F., Moscardini, E., Giuliano, V., & Toro, L. (2004). Sequential extraction of heavy metals in river sediments of an abandoned pyrite mining area: Pollution detection and affinity series. *Environmental Pollution, 132*,189–201.10.1016/j.envpol.2004.05.002, ,

28. SAS Institute. 2002. *SAS/STAT user's guide, version 8.2.* Cary, NC: SAS Institute .

29. Schwertmann, U. (1964). The differentiation of iron oxide in soils by a photochemical extraction with acid ammonium oxalate. *Zeitschrift für Pflanzenern. Bodenkunde, 105*,194–202.10.1002/(ISSN)1522-2624

30. Vandecasteele, B., Du Laing, G., & Tack, F. M. G. (2007). Effect of submergence-emergence sequence and organic matter or aluminosilicate amendment on metal uptake by woody wetland plant species from contaminated sediments. *Environmental Pollution, 145*, 329–338.10.1016/j.envpol.2006.03.003, ,

31. Wan, Z., Song, C., Guo, Y., Wang, L., & Huang, J. (2008). Effects of water gradients on soil enzyme activity and active organic carbon composition under *Carex lasiocarpa*marsh. *Acta Ecologica Sinica, 28*, 5980–5986.

32. Weil, R. R., Islam, K. R., Stine, M. A., Gruver, J. B., & Samson-Liebig, S. E. (2003).Estimating active carbon for soil quality assessment: A simplified method for laboratory and field use. *American Journal of Alternative Agriculture, 18*, 1–15.

CHAPTER 5 EFFECTS OF LAND COVER CHANGE ON EVAPOTRANSPIRATION AND STREAM FLOW OF SMALL CATCHMENTS

Lívia Cristina Pinto Dias, Márcia N. Macedo, Marcos Heil Costa, Michael T. Coe, Christopher Neill

Keywords: Evapotranspiration; Streamflow; Modeling; Xingu Basin; Amazon; Cerrado Gadget timed out while loading

INTRODUCTION

The Amazon is the Earth's largest remaining tropical forest and the location of the largest absolute extent of tropical forest clearing each year. In the eastern Amazon, the Xingu River Basin has a drainage area of approximately 500,000 km^2, and annual mean discharge of about 8000 m^3 s^{-1}, making it the 5th largest tributary of the Amazon. The Xingu headwaters region is located in Mato Grosso state (Upper Xingu) and the river flows northward for roughly 2000 km before reaching the Amazon River in Pará state (Fig. 1a). In addition to its large extent, the Xingu Basin is economically and environmentally important. The third-largest hydroelectric dam complex in the world, the Belo Monte Dam, is currently under construction at Altamira, near the confluence of the Xingu and the Amazon River. At the same time, the Xingu Basin contains one of the world's largest mosaics of protected areas and indigenous reserves, which protect indigenous cultures and 280,000 km^2 of tropical forest.

High deforestation rates and rapid changes in the agricultural landscape have occurred in the Upper Xingu outside of the indigenous reserve, with about 35% of the original forest converted to pasture and croplands (especially soybeans) by 2010 (Leite et al., 2011, Macedo et al., 2012 and Morton et al., 2006). These transitions from native forests and savannas to pastures or croplands drive widespread changes in vegetation that have the potential to modify the hydroclimatology of the Upper Xingu (Coe et al., 2013 and Panday et al., 2015).

The classic study by Bosch and Hewlett (1982) quantified how reduction of vegetation cover increases average annual discharge. This has been confirmed for various ecosystems by Andréassian (2004), Brown et al. (2005), Bruijnzeel (1990) and Sahin and Hall (1996). Recently, Hayhoe et al. (2011) conducted a paired catchment experiment at the Tanguro Farm in the Upper Xingu (Fig. 1a) to examine how the conversion of tropical forests to soybeans influences the seasonality of streamflow and to quantify the discharge of small catchments under these two types of vegetation cover. The authors found that water yield in soybean catchments was approximately fourfold that of forest catchments and that stormflows contributed less than 13% of the annual discharge, while baseflow accounted for nearly 90% of annual streamflow in both forest and soybean catchments.

Citation: Lívia Cristina Pinto Dias, Márcia N. Macedo, Marcos Heil Costa, Michael T. Coe, Christopher Neill, Effects of land cover change on evapotranspiration and stream flow of small catchments, Journal of Hydrology: Regional Studies, Volume 4, Part B, September 2015, Pages 108-122, ISSN 2214-5818, http://dx.doi.org/10.1016/j.ejrh.2015.05.010.

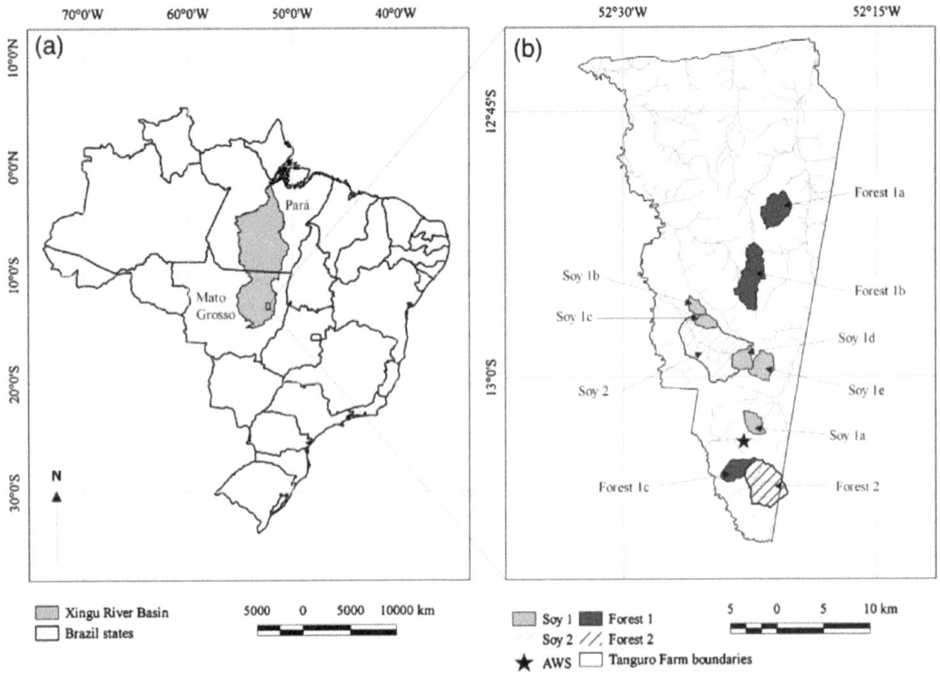

Figure 1. Map of the study area. (a) Location of the Xingu River Basin and Tanguro Farm within the Amazon. (b) Inset depicting the location of the study catchments within Tanguro Farm. First-order catchments are labeled with a 1; second-order catchments are labeled with 2. The star indicates the location of the automatic weather station (AWS).

The Upper Xingu is an ideal study area for examining the effects of land cover change on hydrology because it spans the transition between evergreen tropical forest in the north and cerrado (Brazilian savanna) in the south. This transition zone provides a unique opportunity to study four types of land cover (tropical forest, cerrado, pasture and soybean) under similar climatological and edaphic conditions.

Land surface models (LSM) have been used to simulate the behavior of the hydrological system at various scales in central Brazil (Coe et al., 2011, Panday et al., 2015 and Pongratz et al., 2006) and Amazonia (Coe et al., 2009 and Costa and Foley, 1997). These models describe the flow of water between soil, vegetation, and atmosphere and allow researchers to investigate land surface processes (evapotranspiration, runoff and drainage), as well as the historical and potential future consequences of climate and deforestation for the regional water balance.

Since the first LSM was developed in the late 1960s, LSMs have been iteratively refined and they can be classified into generations according to the processes included, as described by Sellers et al. (1997). After the second generation, several important hydrological processes were included, such as biophysical control of evapotranspiration, precipitation interception, and soil moistures availability. Third- and fourth-generation models include improvements in carbon assimilation and dynamic vegetation, respectively, and are the most common models used in previous land use change studies in Brazil.

Costa and Foley (1997) used a modified version of the Land Surface Transfer Scheme (LSX; third-generation model; Pollard and Thompson, 1995) to analyze the water balance in the Amazon Basin. This study demonstrated that forest clearing decreases annual evapotranspiration by 12% in the Amazon Basin. Pongratz et al. (2006) used the Simple Biosphere Model (SiB2; third-generation model; Sellers et al., 1996a and Sellers et al., 1996b), to study the surface energy and water balance in Mato Grosso. Their results indicated that forest clearing increases the daily temperature range and that conversion of evergreen forest to C3 croplands decreases the latent heat flux by 21% during the wet season. The Integrated Biosphere Simulator (IBIS; Foley et al., 1996 and Kucharik et al., 2000), a fourth-generation model, has been extensively used to study the influence of deforestation on the discharge of the Amazon River and its tributaries (Coe et al., 2011, Coe et al., 2009, Lima et al., 2014, Panday et al., 2015 and Stickler et al., 2013). The authors found that large-scale historical deforestation has increased discharge by as much as 20% in large watersheds of the southeastern Amazon.

Independent of the model chosen, the accuracy of simulated results is fundamentally dependent on the parameters chosen. Parameters representing soil hydrological properties such as porosity, field capacity, wilting point and saturated hydraulic conductivity are particularly important because they influence soil water retention, the amount of water available for evapotranspiration (Cuenca et al., 1996, Delire et al., 1997 and Marthews et al., 2014), and the resulting water cycle.

This study presents a combination of field measurements and model simulation results for the Upper Xingu focused around three main goals: (1) analysis of water yield data collected in tropical forest and soybean catchments to understand the effects of land cover change on regional water dynamics; (2) determination of the best set of soil hydrological properties for simulating the partitioning of annual precipitation into evapotranspiration and discharge in forest and soybean catchments; and (3) investigation of the differences in annual evapotranspiration and streamflow in tropical forest, cerrado, pasture and soybean catchments in the Upper Xingu.

MATERIALS AND METHODS

Site Description

The study is centered on the 80,000 ha Tanguro Farm, a large soybean farm located in the Upper Xingu Basin, in Mato Grosso, Brazil (Fig. 1). This region is located in the Brazilian Central Plateau, characterized by a generally flat topography and dominated by oxisols with a mean soil texture of 55% sand, 2% silt, and 43% clay (Hayhoe et al., 2011 and Scheffler et al., 2011). Tanguro was originally covered by transitional tropical forests, with an average canopy height of 20 m and relatively low species diversity compared to the wetter tropical forests occurring further north and west (Balch et al., 2008). During the 1980s, about 32,000 ha were cleared and converted to pasture grasses, consisting mostly of *Brachiaria* sp. (Scheffler et al., 2011). Between 2004 and 2008, these pasturelands were converted to mechanized soy production. This land use history is typical of much of the Upper Xingu region.

During the study period (2007–2010), soybean was planted near the onset of the rainy season (usually in November). In general, in the first year after conversion, only a single crop of soybeans was planted in the field (without crop rotation), and the soil remained exposed between growing seasons. After the first year, typically a single soybean crop was planted in November (using a no-till planting system) and harvested around March, when a cover crop (usually millet) was planted to protect the soils between growing seasons.

Discharge and Meteorological Data

Discharge was measured in twelve headwater catchments located entirely on the Tanguro Farm from September 2007 to December 2010. The study catchments included seven soybean (six first-order and one second-order) and five forested (four first-order and one second-order) streams. Because ten of these catchments were also studied byHayhoe et al. (2011), we named them following Hayhoe et al. (2011) to facilitate comparison of results between these studies (Fig. 1b). Our study differs from Hayhoe et al. (2011) because it includes two additional catchments (one in forest and one in soybean), an 18-month longer time series, and updated rating curves.

Discharge was estimated based on hourly water level data, following the methods described in Hayhoe et al. (2011). Briefly, each basin was gauged with a staff gauge and Onset HOBO U20 Water Level data logger, which monitored hourly barometric pressure. Pressure data was converted to water level (depth) using reference air pressure data (logged with an identical unit in a nearby field laboratory) and calibrated using the permanent staff gauges at each site. Finally, the entire water level time series, including the catchments and time periods reported by Hayhoe et al. (2011), was converted to discharge using updated rating curves. The depth-to-discharge relationships were developed for each catchment based on periodic discharge measurements using a Marsh McBirney Flo-Mate 2000 electromagnetic flow meter. Catchment boundaries were delineated based on vegetation-corrected Shuttle Radar Topography Mission (SRTM) data, using standard hydrology tools in ESRI ArcGIS 10.2.

Hourly temperature, humidity, wind speed, rainfall, and incoming solar radiation data were measured by an automatic weather station (AWS, Fig. 1b) located inside the Tanguro Farm (Latitude: 13°37′16″ S; Longitude: 52°22′55,68″ W; altitude: 358 m), between January 2007 and December 2011. Where necessary (typically before September 2008) we used the PERSIANN (*Precipitation Estimation from Remotely Sensed Information using Artifical Neural Networks*; Hsu et al., 1997 and Sorooshian and Hsu, 2000) 0.25° global precipitation dataset to fill gaps in the rainfall time series. Missing values for the other variables were filled following the same behavior of the previous adjacent data points.

Incident longwave radiation (L_{in}) is an input variable required to run the INLAND and Agro-IBIS models (described in Section 2.3). Although it was not measured directly by the Tanguro AWS, L_{in} can be estimated from other measured meteorological variables using previously published equations (Brutsaert, 1975, Idso and Jackson, 1969, Idso, 1981, Prata, 1996 and Satterlund, 1979), which are only valid for clear sky conditions. Here, we estimated L_{in} for all days of the simulation period according to the empirical equation proposed by Idso (1981), which uses the Stefan–Boltzmann constant (σ), air temperature (T) in Kelvin and water vapor pressure (e) in kPa:

$$L_{in}=[0.7+5.95\times10^{-5}(e\exp(1500/T))]\sigma T^4$$

We assume that the errors associated with L_{in} estimation influenced all simulations equally, enabling analysis of the relative differences in the behavior of the simulated water balance. We chose the Idso (1981) equation because it showed the best agreement with measurements of incident longwave radiation in pasture and forest sites in southwestern Amazonia (Aguiar et al., 2011).

Numerical Experiment

The Integrated Model of Land Surface Processes (INLAND; Costa et al., manuscript in preparation) is a fifth-generation model that has been developed by Brazilian researchers as part of the Brazilian Earth System Model project. This model is an evolution of the IBIS. One of the main goals of the INLAND project is to better represent biomes and important processes in Brazil (such as agriculture). To include agriculture, the Dynamic Model of Agroecosystems (AgroIBIS) is being incorporated into INLAND. AgroIBIS is a modified version of IBIS capable of simulating phenological and physiological characteristics of crops (Kucharik and Twine, 2007 and Kucharik, 2003). Because the integration of these two models was underway when we performed the simulations in this study, we elected to use INLAND version 1.0 and AgroIBIS separately. Because both models are modifications of IBIS, they share most water balance equations and their results are directly comparable in the context applied here.

These models simulate the exchanges of energy, water, carbon and momentum in the soil–vegetation–atmosphere system. They represent two layers of vegetation with 12 plant functional types, as well as six soil layers of variable thickness, representing a total of 8 m in depth. Water infiltration in soil is modeled using Darcy's equation and soil moisture is based on the flow of Richards's equation. Plant transpiration is governed by stomatal conductance coupled to photosynthesis (Ball et al., 1987). Total evapotranspiration is the sum of bare soil evaporation, plant transpiration and evaporation from precipitation intercepted by the canopy (Pollard and Thompson, 1995). Water uptake by plants is a function of atmospheric demand, soil physical properties, root distribution and water in the soil profile (Kucharik et al., 2000). Finally, surface and subsurface runoff are explicitly simulated as a function of the water not evapotranspired to the atmosphere (i.e. the difference between rainfall and total evapotranspiration), soil characteristics, vegetation and climate.

We performed single point simulations for tropical forest, cerrado, and pasture land uses in the INLAND model and soybean cropland in the AgroIBIS model. Although we ran the models for the period from January 1, 2007 to December 31, 2011, we analyzed simulation results only for two hydrological years (September 1, 2008 to August, 31 2010). Hourly meteorological data from the AWS at Tanguro Farm served as model input data and we considered it representative for all catchments in the study. We used fixed vegetation for the natural ecosystems (cerrado and forest) and pastureland simulations. We specified November 1st as the planting date for single-cropped soybeans and assumed that soils remained exposed in the periods between plantings. We used fixed concentrations of CO_2 at 390 ppm in all simulations.

Soil Hydrological Properties

Soil properties in INLAND and AgroIBIS are based on soil texture, and the soil at Tanguro Farm is classified as sandy clay according to the USDA soil texture triangle. Here, we considered three potential data sources to describe the hydrological properties of sandy clay soils: the mean values described by Campbell and Norman (1998), the pedotransfer functions proposed by Cosby et al. (1984) and observed data for Tanguro Farm described by Scheffler et al. (2011).

INLAND and AgroIBIS currently use values for porosity (Θ_s), field capacity (Θ_{fc}), wilting point (Θ_{wp}), Campbell's "b" exponent (b), air entry potential (Ψ_{ad}) and saturated hydraulic conductivity (K_S), as recommended by Campbell and Norman (1998). Recently, however, hydrologic properties of soils estimated using pedotransfer functions based on soil texture, proposed by Cosby et al. (1984), have been used by the Large-Scale Biosphere-Atmosphere Experiment in Amazonia Data Model Intercomparison Project (LBA-DMIP; De Gonçalves et al., 2013). In addition, Scheffler et al. (2011) measured in situ values of K_S under forest, pasture and soybean covers on Tanguro Farm.

For each land cover (forest, cerrado, pasture, and soybean), we assessed four sets of soil hydraulic parameters to identify the combination that provides the smallest difference between simulated data and mean observed values for forest and soybeans (Table 1). The four resulting simulation scenarios were:

A. All soil hydraulic parameters estimated using Cosby et al. (1984) equations;

B. K_S measured at the Tanguro Farm (Scheffler et al., 2011) and other parameters estimated using Cosby et al. (1984) equations;

C. All soil hydraulic parameters recommended by Campbell and Norman (1998);

D. K_S measured at the Tanguro Farm (Scheffler et al., 2011) and other parameters recommended by Campbell and Norman (1998).

Table 1. Soil sand-clay parameters used for simulations A (all parameters from Cosby et al., 1984), B (K_S fromScheffler et al., 2011; other parameters from Cosby et al., 1984), C (all parameters from Campbell and Norman, 1998) and *D* (K_S from Scheffler et al., 2011; other parameters from Campbell and Norman, 1998).Θ_s, porosity (m³ m⁻³); Θ_{fc}, field capacity (m³ m⁻³); Θ_{wp}, wilting point (m³ m⁻³); b, Campbell's "b" exponent;Ψ_{ad}, air entry potential (mH_2O); and K_S, saturated hydraulic conductivity ($m\ s^{-1}$).

Simulation	Land Cover	Θ_s	Θ_{fc}	Θ_{wp}	b	Ψ_{ad}	K_S
A	All	0.411	0.284	0.194	9.7	0.18	4.638×10^{-6}
B	Forest/Cerrado	0.411	0.242	0.194	9.7	0.18	1.565×10^{-4}
	Pasture	0.411	0.262	0.194	9.7	0.18	2.781×10^{-5}
	Soy	0.411	0.260	0.194	9.7	0.18	3.189×10^{-5}
C	All	0.430	0.339	0.239	6.0	0.29	3.333×10^{-7}
D	Forest/Cerrado	0.430	0.339	0.239	6.0	0.29	1.565×10^{-4}
	Pasture	0.430	0.339	0.239	6.0	0.29	2.781×10^{-5}
	Soy	0.430	0.339	0.239	6.0	0.29	3.189×10^{-5}

We assume identical K_S for forest and cerrado (Table 1), since both are natural ecosystems.

RESULTS

Observed Data

Observed mean annual streamflow from September 2008 to August 2010 was approximately three times higher in soybean catchments than in forest catchments (Table 2). Average total observed runoff was 534.8 ± 214.7 mm yr^{-1} for soybean catchments (95% confidence interval (CI), 320.1–749.5 mm yr^{-1}) and 194.6 ± 144.1 mm yr^{-1} for forest catchments (CI, 50.5–338.7 mm yr^{-1}).

Table 2. Average observed total runoff (R_{obs}), runoff coefficient (C) and area of six soy catchments and four forest catchments in Tanguro Farm. Soy 1 and Forest 1 are first order catchments; Soy 2 and Forest 2 are second order catchments. Catchments that overlap with Hayhoe et al. (2011) are indicated with an asterisk.

Land cover	R_{obs} (mm yr^{-1})	C	Drainage area (km^2)
Soy 1a*	470.9	0.36	3.7
Soy 1b*	994.6	0.76	2.3
Soy 1c*	457.3	0.35	2.5
Soy 1d*	441.8	0.34	4.1
Soy 1e	125.1	0.10	5.6
Soy 2*	719.0	0.55	27.1
Mean	534.8	0.41	–
Forest 1a*	238.1	0.18	8.4
Forest 1b*	45.8	0.04	12.7
Forest 1c*	78.5	0.06	5.2
Forest 2	415.9	0.32	15.2
Mean	194.6	0.15	–

During this period, the average rainfall was 1301 mm yr^{-1}. The ratio between the average annual total runoff and average annual rainfall (runoff coefficient, C) was 0.41 (standard deviation, SD = 0.20) in soybean catchments and 0.15 (SD = 0.11) in forest catchments (Table 2).

The dry and wet seasons were well separated, with the wet season occurring between September and April and the dry season between May and August (Fig. 2). Mean precipitation during the wet and dry months was 5.9 and 0.1 mm day^{-1}, respectively. The average maximum streamflows in the two hydrological years was 1.8 mm day^{-1} for soybean catchments and 0.6 mm day^{-1} for forest catchments. The mean annual amplitude of the flow (the mean difference between maximum and minimum streamflows) in soybean catchments (0.7 mm day^{-1}) was more than twice that of forest catchments (0.3 mm day^{-1}).

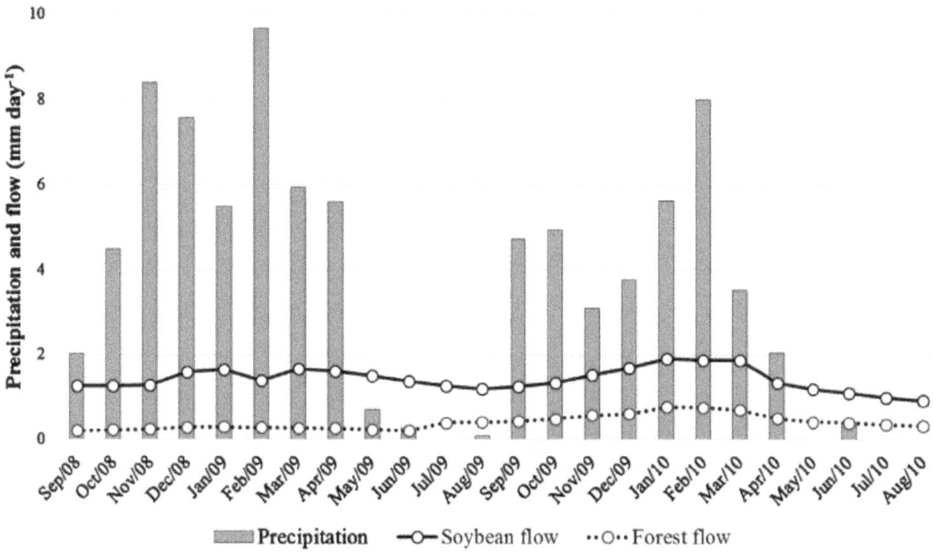

Figure 2. Precipitation and mean daily total runoff for forest and soybean catchments in the Upper Xingu (time period September 2008 and August 2010).

Simulated Results

The difference between mean total precipitation (Fig. 3 dashed line) and the sum of the three simulated components of the water balance (evapotranspiration, surface runoff and subsurface runoff) was the mean net change in groundwater storage (Fig. 3). The mean annual groundwater storage term was negative for all simulated sets of parameters and land covers, except for soybeans in simulation C. Negative groundwater storage meant that there was a net removal of water stored in the soil during the analyzed period. The simulated annual discharge (total runoff) was the sum of simulated annual surface and subsurface runoffs.

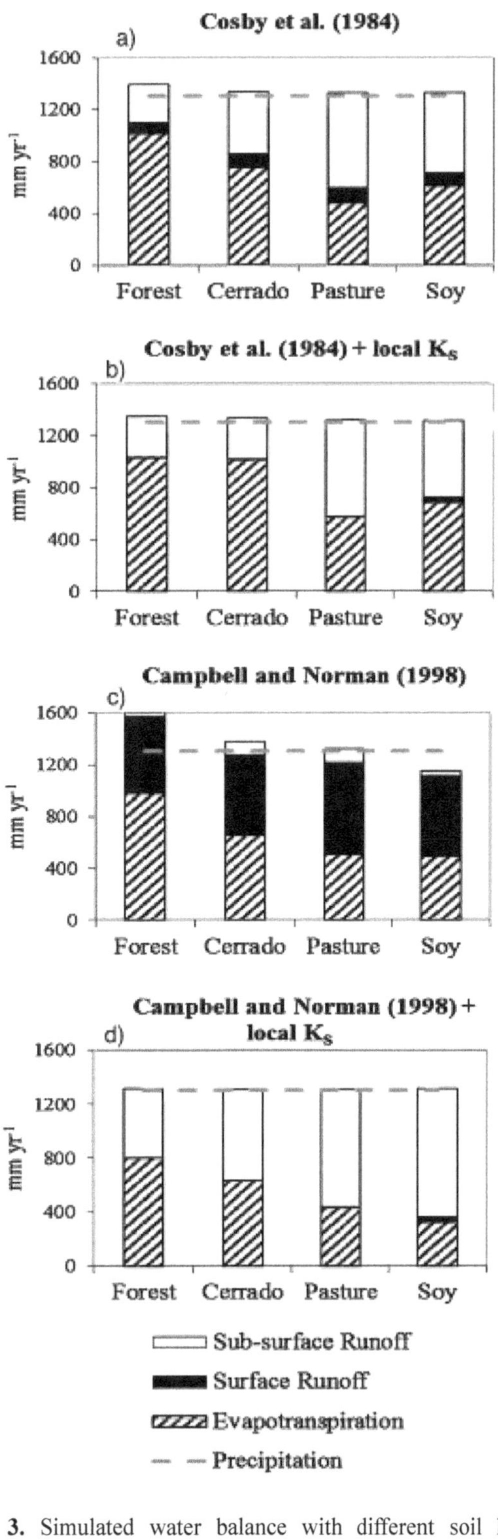

Figure 3. Simulated water balance with different soil hydraulic parameters. The difference between mean total precipitation (1301 mm yr^{-1}, dashed line) and the sum of the three simulated components of the water balance is the mean net change in groundwater storage. Simulations with (A) all parameters estimated using Cosby et al. (1984); (B) K_S measured at Tanguro Farm by Scheffler et al. (2011) and the other parameters estimated using Cosby et al. (1984); (C) the values of Campbell and Norman (1998) were used for all parameters; and (D) K_S measured at Tanguro Farm by Scheffler et al. (2011) and Campbell and Norman (1998) values in the other parameters.

Simulation *B* predicted the smallest difference between simulated data and the mean observed values for both forest and soybeans (Fig. 4). Moreover, for soybean catchments, simulations A, B, and C predict simulated total runoff within the range of observed discharge, but only simulation B predicts total runoff within the observed range for forest catchments. Therefore, simulation B was used to analyze the effect of land cover change on evapotranspiration and streamflow in the Upper Xingu Basin. Unless otherwise noted, henceforth all simulated results refer to simulation B.

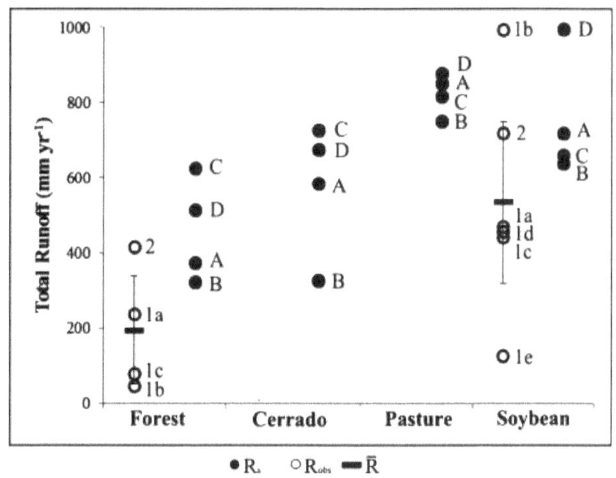

Figure 4. Mean annual discharges observed (R_{obs}), mean of R_{obs} ($\bar{R}$) for each land cover, and mean annual discharge simulated (R_s) for each land cover and four sets of soil hydraulic parameters for sandy clay soil. Small letters define catchments with forest or soy (Fig. 1 and Table 1) and capital letters define the different set of soil hydraulic parameters. Error bars indicate the confidence interval at 95%.

Simulated evapotranspiration was 39% lower (Table 3) in agricultural ecosystems (pasture and soybean) compared with natural ecosystems (tropical forest and cerrado). Simulated mean annual total runoff for soybean cover was twice that of forest cover. Considering the four types of land use, the mean annual simulated streamflow in agricultural ecosystems (pasture and soybeans) was 114% higher than in natural ecosystems (forest and cerrado).

Table 3. Water balance predicted by simulation B (K_S from Scheffler et al., 2011; other parameters from Cosby et al., 1984) for the period from September 2008 and August 2010 (mm yr^{-1}). The mean precipitation in this period was 1301 mm yr^{-1}.

	Evapotranspira-tion	Surface runoff	Subsurface runoff	Total run-off	Net change in groundwater storage
Forest	1024.7	0.2	322.3	322.5	−46.7
Cer-rado	1009.5	0.2	324.9	325.1	−34.1
Mean	1017.1	0.2	323.6	323.8	−40.4
Pasture	567.3	5.5	744.2	749.7	−16.6
Soy	678.5	43.3	594.3	637.6	−15.6
Mean	622.9	24.4	669.3	693.7	−16.1

Simulated surface runoff was greater in agricultural ecosystems than in natural ecosystems (Table 3). Subsurface runoff represented 99% of the total runoff in the simulation with forest cover, compared to 93% for soybean cover. Considering all simulated land uses, subsurface runoff represented 96% of total runoff in agricultural ecosystems and approximately 100% of total runoff in natural ecosystems. Although the simulated surface runoff was slightly greater in agricultural ecosystems than in natural ecosystems, most of the increase in mean annual discharge was due to an increase in groundwater flow. The runoff coefficients for forest, cerrado, pasture and soybeans catchments were 0.25, 0.25, 0.58 and 0.49, respectively.

DISCUSSION

By combining field measurements in paired catchments with numerical modeling, our results provide insights into how the conversion of tropical forest or cerrado to pasture or soybean alters the partitioning of rainfall into evapotranspiration and streamflow. We found that replacement of native vegetation with other land covers substantially increased water yield, which supports the results of several previous studies (Andréassian, 2004, Bosch and Hewlett, 1982, Brown et al., 2005, Bruijnzeel, 1990,Hayhoe et al., 2011, Sahin and Hall, 1996 and Tomasella et al., 2009). Mean annual observed streamflow was three times larger in soybean catchments than in forest catchments, primarily because of decreased evapotranspiration. The lower leaf area index (LAI), shallower root depth, higher albedo, lower surface roughness and lower transfer of energy and momentum all contribute to reducing evapotranspiration in soybean areas (Bonan et al., 1992, Costa and Foley, 2000, Costa et al., 2007, Pongratz et al., 2006 and Sampaio et al., 2007).

The variability in observed discharge among Upper Xingu catchments of comparable land use cannot be fully explained by stream slope or catchment area (Hayhoe et al., 2011). Based on our simulated results, which examined the impact of different K_S on forest discharge, and field measurements of discharge in the Forest 1a and 1b catchments, we conclude that soil properties alone cannot explain the discharge variability between sites. Total simulated runoff decreased exponentially with increasingK_S in forest sites, converging to roughly 300 mm yr^{-1}. A simulation using K_S 100-times smaller than that measured in situ resulted in a simulated discharge only twice as high that under simulation B conditions. Although these results suggest that flow should vary significantly with changes in K_S, Scheffler et al. (2011) reported no significant difference in K_S at the Forest 1a and 1b catchments. Despite similar soil hydraulic characteristics, Forest 1a discharge was fivefold that of Forest 1b (Table 2).

Having eliminated K_S, stream slope, and catchment area as major contributing factors, we conclude that other factors account for the large difference in discharge. Possible causes of this variability include measurement errors, underflow beneath the monitored streams and flow between the basins (Hayhoe et al., 2011). This phenomenon is possible in flat landscapes with deep well-drained soils (Bruijnzeel, 1990) like those found in the Upper Xingu, but we were unable to refute or confirm it in this study.

The partitioning of total runoff into surface and sub-surface components was considerably altered with measured K_S. The Campbell and Norman (1998) and Cosby et al. (1984) pedotransfer equations underestimate K_S compared to the field measurements reported by Scheffler et al. (2011). This is likely because these equations were developed to predict temperate soil properties and poorly represent the behavior of tropical soils. Unlike temperate soils, tropical oxisols are highly leached, are predominantly 1:1 clay minerals and have high concentrations of aluminum and iron oxides (Barros and van Lier, 2014). Aluminum and iron oxides are the primary stabilizing agents for soil aggregates and well-aggregated soils are well-drained. Because the two indirect sources of soil properties data used in this study did not represent the well-drained sandy clay soils typical of our study area, the simulations relying exclusively on using only Campbell and Norman (1998) or Cosby et al. (1984) resulted in a greater difference between simulated data and observed runoff, compared with simulations that partially included field measurements of soil characteristics. Simulation B included the measured K_S and better represented the water balance components for the Upper Xingu.

In simulation B, evapotranspiration was 1025 mm yr^{-1} for forest and 1010 mm yr^{-1}cerrado. The forests at Tanguro Ranch occur at the transition between the Cerrado and Amazon biomes and are characterized by shorter stature and lower LAI than the dense forests found in northern Amazonia (Balch et al., 2008). Studies of similar transitional forests in Sinop (Costa et al., 2010, Vourlitis et al., 2008 and Vourlitis et al., 2002) estimated evapotranspiration rates ranging from 822 to 1300 mm yr^{-1}. For dense cerrado physiognomies (*cerradão*) and cerrado *sensu stricto*, tree-dominant savannahs expected along the Amazon-Cerrado ecotone, previous studies suggest evapotranspiration ranging from approximately 800 to 1442 mm yr^{-1} (Giambelluca et al., 2009 and Oliveira et al., 2005). Our simulated evapotranspiration for both forest and cerrado was well within the range of those previously reported in the literature.

Our simulation results indicate that evapotranspiration was similar in forest and cerrado catchments. This is consistent with previous findings, which suggest that transitional forests and cerrados can have similar evapotranspiration if surface soil water availability is sufficient (Rodrigues et al., 2014). Tropical forests and cerrados thus exhibit similar evapotranspiration in transitional areas, given that both were constrained by the same climate and water availability during our simulations.

Our simulated pasture evapotranspiration (567 mm yr^{-1}) was lower than values previously reported in the literature for this region, which range from 822 to 982 mm $yr^{-1}$for Mato Grosso (Lathuillière et al., 2012 and Priante-Filho et al., 2004). Measured evapotranspiration for soybean areas in Mato Grosso were unavailable, but Lathuillière et al. (2012) conducted an indirect measurement experiment with remote sensing products and crop modeling. Their estimates suggest that mean soybean evapotranspiration in Mato Grosso (540 mm yr^{-1} or 4.4 mm day^{-1}, over the growing season) was lower than our simulated evapotranspiration for soybean (679 mm yr^{-1}). Our simulated soybean evapotranspiration may be higher than that of Lathuillière et al. (2012) because we included transpiration by soybeans during the growing season and evaporation throughout the year. Our simulated pasture and soybean evapotranspiration corresponded to 44% and 52%, respectively, of the mean precipitation.

Simulated evapotranspiration in agricultural ecosystems (pasture and soybean catchments) was about 39% lower than in natural ecosystems (forest and cerrado catchments). This is comparable to observed evapotranspiration reductions of about 36% following deforestation and pasture establishment in an area of Maranhão state (Oliveira et al., 2014).

Between April and August each year, the amount of precipitation was lower than the sum of evapotranspiration and total runoff (Fig. 5), indicating that vegetation was under water stress during this period. Unlike wet equatorial forests, where solar radiation availability is an important factor controlling evapotranspiration, seasonally dry tropical forests are largely controlled by seasonal trends in rainfall and the biotic response of plants to water stress (Costa et al., 2010 and Vourlitis et al., 2008). The dry season was an equally important control on evapotranspiration in forest and cerrado ecosystems, consistent with previous studies (Biudes et al., 2015, Da Rocha et al., 2009, Rodrigues et al., 2014 and Vourlitis et al., 2014). Evapotranspiration in perennial vegetation (forest, cerrado, and pasture) was somewhat greater during the dry season than the wet season, in response to the larger vapor pressure deficit, as previously reported by Costa et al. (2010). For soybeans, the growing season from November to February had greater evapotranspiration than the dry season.

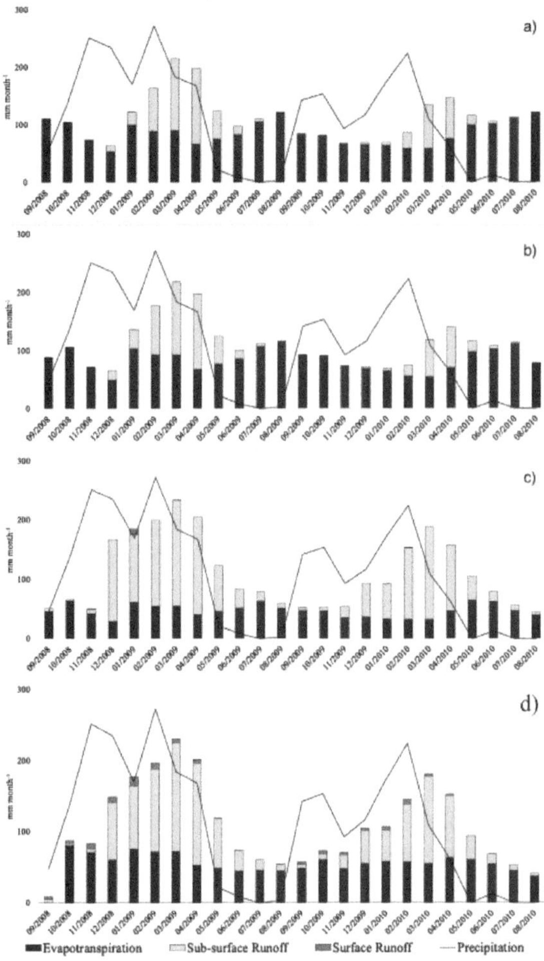

Figure 5. Observed precipitation and simulated evapotranspiration, surface runoff and sub-surface runoff seasonality for (a) tropical forest, (b) cerrado, (c) pasture, and (d) soybeans in mm month-1. The difference between monthly precipitation (solid line) and the sum of the three simulated components of the water balance (vertical bars) is equal to the net change in soil water storage (positive numbers represent soil water recharge; negative numbers represent soil water depletion).

Both observed and simulated results agree that the replacement of natural vegetation by agriculture increases streamflow by more than 100% in the Upper Xingu. Simulated mean annual total runoff for soybeans was twice that of forests (638 and 323 mm yr^{-1}).

Simulated subsurface runoff was 96% of annual total runoff in agricultural ecosystems and nearly 100% of annual total runoff in the natural ecosystems. These findings were consistent with the results reported by Hayhoe et al. (2011), who found that baseflow represented 96% of the discharge in forest and 94% of the discharge in soybean catchments in the Upper Xingu. However, the simulated subsurface runoff was highly seasonal, and reduced to near zero during the dry season (Fig. 5).

Agricultural ecosystems are generally expected to have larger surface runoff than natural ecosystems, which may cause environmental problems such as erosion and the loss of soil nutrients. However, the soil infiltrability and saturated hydraulic conductivity in this region were very high even after conversion of the landscape to agriculture. As a result, large soil infiltration and groundwater recharge – not increases in surface runoff – were the primary factor accounting for observed increases in total runoff in agricultural ecosystems. It is unclear whether other environmental problems may result from these increases in surface runoff. As noted by Neill et al. (2013), we do not yet know the full extent of areas with erosion nor the potential for increased erosion if a prolonged period of agriculture were to increase soil compaction in this region.

CONCLUSIONS

Observed and simulated results show that converting natural vegetation to agriculture substantially modifies the water balance components in small catchments in the Upper Xingu. Field observations indicated that mean streamflow in soy catchments was about three times greater than that of forest catchments, while the mean annual amplitude of observed flow was more than two times larger in soy than in forest areas.

Modeled results showed about 40% less evapotranspiration in agricultural ecosystems (pasture and soybean catchments) than in natural ecosystems (forest and cerrado catchments), whereas average total discharge was about 100% higher in agricultural ecosystems than in natural ecosystems in the Upper Xingu.

Our study is a first step toward understanding the evapotranspiration and streamflow for four types of land cover in the study region. Future studies can investigate whether the consequences of increased water yield in headwater catchments can be detected in higher-order watersheds and how these cumulative disturbances might compromise the economic and environmental health of the Xingu River Basin. In situ data on soil parameters was essential to simulating the hydrologic behavior of this region. Therefore, future research should also focus on improving estimates of soil hydraulic properties in tropical soils, given the importance of these soil parameters for accurately simulating the water balance.

ACKNOWLEDGMENTS

We would like to acknowledge the Amazon Environmental Research Institute (IPAM) for weather station data and assistance with field data collection, as well as Grupo A. Maggi for providing access to the study site. We thank Shelby Riskin (nee Hayhoe), Richard McHorney, and Suzanne Spitzer for contributing and compiling streamflow data from Tanguro. This study was supported by the US National Science Foundation (DEB-0949996, DEB-0743703), the Gordon and Betty Moore Foundation, and the Conselho Nacional de Desenvolvimento Científico e Tecnológico (CNPq, process 135648/2011-4).

REFERENCES

1. Aguiar, L.J.A., Costa, J.M.N., Fischer, G.R., Aguiar, R.G., Costa, A.C.L., Ferreira, W.P.M., 2011. Estimativa da radiac¸ ão de onda longa atmosférica em áreas de floresta e de pastagem no sudoeste da Amazônia. Rev. Bras. Meteorol. 26, 215–224.
2. Andréassian, V., 2004. Waters and forests: from historical controversy to scientific debate. J. Hydrol. 291, 1–27, http://dx.doi.org/10.1016/j.jhydrol.2003.12.015.
3. Balch, J.K., Nepstad, D.C., Brando, P.M., Curran, L.M., Portela, O., de Carvalho, O., Lefebvre, P., 2008. Negative fire feedback in a transitional forest of southeastern Amazonia. Glob. Chang. Biol. 14, 2276–2287, http://dx.doi.org/10.1111/j.1365-2486.2008.01655.x.
4. Ball, J.T., Woodrow, I.E., Berry, J.A., 1987. A model predicting stomatal conductance and its contribution to the control of photosynthesis under different light conditions. In: Biggins, J. (Ed.), Progress in Photosynthetic Research. Springer, Netherlands, Dordrecht, pp. 221–224.
5. Barros, A.H.C., van Lier, Q.J., 2014. Pedotransfer functions for Brazilian soils. In: Teixeira, W.G., Ceddia, M.B., Ottoni, M.V., Donnagema, G.K. (Eds.), Application of Soil Physics in Environmental Analyses. Springer International Publishing, pp. 131–162, http://dx.doi.org/10.1007/978-3-319-06013-2 6.
6. Biudes, M.S., Vourlitis, G.L., Machado, N.G., de Arruda, P.H.Z., Neves, G.A.R., de Almeida Lobo, F., Neale, C.M.U., de Souza Nogueira, J., 2015. Patterns of energy exchange for tropical ecosystems across a climate gradient in Mato Grosso. Brazil. Agric. For. Meteorol. 202, 112–124, http://dx.doi.org/10.1016/j.agrformet.2014.12.008.
7. Bonan, G.B., Polard, D., Tompson, S.L., 1992. Effects of boreal forest vegetation on global climate. Nature 359, 716–718.
8. Bosch, J.M., Hewlett, J.D., 1982. A review of catchment experiments to determine the effect of vegetation changes on water yield and evapotranspiration. J. Hydrol. 55, 3–23.
9. Brown, A.E., Zhang, L., McMahon, T.A., Western, A.W., Vertessy, R.A., 2005. A review of paired catchment studies for determining changes in water yield resulting from alterations in vegetation. J. Hydrol. 310, 28–61, http://dx.doi.org/10.1016/j.jhydrol.2004.12.010.
10. Bruijnzeel, L.A., 1990. Hydrology of Moist Forest and the Effects of Conversion: A State of Knowledge Review. UNESCO, Paris and Amsterdam.
11. Brutsaert, W., 1975. On a derivable formula for long-wave radiation from clear skies. Water Resour. Res. 11, 742–744.
12. Campbell, G.S., Norman, J.M., 1998. An Introduction to Environmental Biophysics, 2nd. ed. Springer, New York, NY, http://dx.doi.org/10.1007/978-1-4612-1626-1.
13. Coe, M.T., Costa, M.H., Soares-Filho, B.S., 2009. The influence of historical and potential future deforestation on the stream flow of the Amazon River – Land surface processes and atmospheric feedbacks. J. Hydrol. 369, 165–174, http://dx.doi.org/10.1016/j.jhydrol.2009.02.043.
14. Da Rocha, H.R., Manzi, A.O., Cabral, O.M., Miller, S.D., Goulden, M.L., Saleska, S.R., R.-Coupe, N., Wofsy, S.C., Borma, L.S., Artaxo, P., Vourlitis, G., Nogueira, J.S., Cardoso, F.L., Nobre, A.D., Kruijt, B., Freitas, H.C., von Randow, C., Aguiar, R.G., Maia, J.F., 2009. Patterns of water and heat flux across a biome gradient from tropical forest to savanna in Brazil. J. Geophys. Res. 114, G00B12, http://dx.doi.org/10.1029/2007JG000640.
15. De Gonc¸ alves, L.G.G., Borak, J.S., Costa, M.H., Saleska, S.R., Baker, I., Restrepo-Coupe, N., Muza,

M.N., Poulter, B., Verbeeck, H., Fisher, J.B., Arain, M.A., Arkin, P., Cestaro, B.P., Christoffersen, B., Galbraith, D., Guan, X., van den Hurk, B.J.J.M., Ichii, K., Imbuzeiro, H.M.A., Jain, A.K., Levine, N., Lu, C., Miguez-Macho, G., Roberti, D.R., Sahoo, A., Sakaguchi, K., Schaefer, K., Shi, M., Shuttleworth, W.J., Tian, H., Yang, Z.-L., Zeng, X., 2013. Overview of the Large-Scale Biosphere–Atmosphere Experiment in Amazonia Data Model Intercomparison Project (LBA-DMIP). Agric. For. Meteorol. 182–183, 111–127, http://dx.doi.org/10.1016/j.agrformet.2013.04.030.

16. Foley, J.A., Prentice, I.C., Ramankutty, N., Levis, S., Pollard, D., Sitch, S., Haxeltine, A., 1996. An integrated biosphere model of land surface processes, terrestrial carbon balance, and vegetation dynamics. Global Biogeochem. Cycles 10, 603–628.

17. Giambelluca, T.W., Scholz, F.G., Bucci, S.J., Meinzer, F.C., Goldstein, G., Hoffmann, W.a., Franco, A.C., Buchert, M.P., 2009. Evapotranspiration and energy balance of Brazilian savannas with contrasting tree density. Agric. For. Meteorol. 149, 1365–1376, http://dx.doi.org/10.1016/j.agrformet.2009.03.006

18. Idso, S.B., 1981. A set of equations for full spectrum and 8 to 14 m and 10.5 to 12.5 m thermal radiation from cloudless skies. Water Resour. Res. 17, 295–304.

19. Kucharik, C.J., 2003. Evaluation of a Process-Based Agro-Ecosystem Model (Agro-IBIS) across the U.S. Corn Belt: Simulations of the Interannual Variability in Maize Yield. Earth Interact. 7, 1–33, http://dx.doi.org/10.1175/ 1087-3562(2003)0072.0.CO;2.

20. Lathuillière, M.J., Johnson, M.S., Donner, S.D., 2012. Water use by terrestrial ecosystems: temporal variability in rainforest and agricultural contributions to evapotranspiration in Mato Grosso. Brazil. Environ. Res. Lett. 7, 1–12, http://dx.doi.org/10.1088/1748-9326/7/2/024024.

21. Marthews, T.R., Quesada, C.A., Galbraith, D.R., Malhi, Y., Mullins, C.E., Hodnett, M.G., Dharssi, I., 2014. Highresolution hydraulic parameter maps for surface soils in tropical South America. Geosci. Model Dev. 7, 711–723, http://dx.doi.org/10.5194/gmd-7-711-2014.

22. Neill, C., Coe, M.T., Riskin, S.H., Krusche, A.V., Elsenbeer, H., Macedo, M.N., McHorney, R., Lefebvre, P., Davidson, E.A., Scheffler, R., Figueira, A.M.S., Porder, S., Deegan, L.A., 2013. Watershed responses to Amazon soya bean cropland expansion and intensification. Philos. Trans. R. Soc. Lond. B. Biol. Sci. 368, 20120425, http://dx.doi.org/10.1098/rstb.2012.0425.

23. Oliveira, R.S., Bezerra, L., Davidson, E.A., Pinto, F., Klink, C.A., Nepstad, D.C., Moreira, A., 2005. Deep root function in soil water dynamics in cerrado savannas of central Brazil. Funct. Ecol. 19, 574–581, http://dx.doi.org/ 10.1111/j.1365-2435.2005.01003.x.

24. Pollard, D., Thompson, S.L., 1995. Use of a land-surface-transfer scheme (LSX)in a global climate model:the response to doubling stomatal resistance. Glob. Planet. Change 10, 129–161, http://dx.doi.org/10.1016/0921-8181(94)00023-7.

25. Priante-Filho, N., Vourlitis, G.L., Hayashi, M.M.S., Nogueira, J.D.S., Campelo, J.H., Nunes, P.C., Souza, L.S.E., Couto, E.G., Hoeger, W., Raiter, F., Trienweiler, J.L., Miranda, E.J., Priante, P.C., Fritzen, C.L., Lacerda, M., Pereira, L.C., Biudes, M.S., Suli, G.S., Shiraiwa, S., Paulo, S.R.Do, Silveira, M., 2004. Comparison of the mass and energy exchange of a pasture and a mature transitional tropical forest of the southern Amazon Basin during a seasonal transition. Glob. Chang. Biol. 10, 863–876, http://dx.doi.org/10.1111/j.1529-8817.2003.00775.x.

26. Sampaio, G., Nobre, C., Costa, M.H., Satyamurty, P., Soares-Filho, B.S., Cardoso, M., 2007. Regional climate change over eastern Amazonia caused by pasture and soybean cropland expansion. Geophys. Res. Lett. 34, L17709, http://dx.doi.org/10.1029/2007GL030612.

27. Tomasella, J., Neill, C., Figueiredo, R., Nobre, A.D., 2009. Water and chemical budgets at the catchment scale including nutrient exports from intact forests and disturbed landscapes. In: Keller, M., Bustamante, M., Gash, J., Silva Dias, P. (Eds.), Amazonia and Global Change. American Geophysical Union, Washington, pp. 505–524.

28. Vourlitis, G.L., Priante-Filho, N., Hayashi, M.M.S., Nogueira, J.S., Caseiro, F.T., Campelo Jr., J.H., 2002. Seasonal variations in the evapotranspiration of a transitional tropical forest of Mato Grosso. Brazil. Water Resour. Res. 38, 1094.

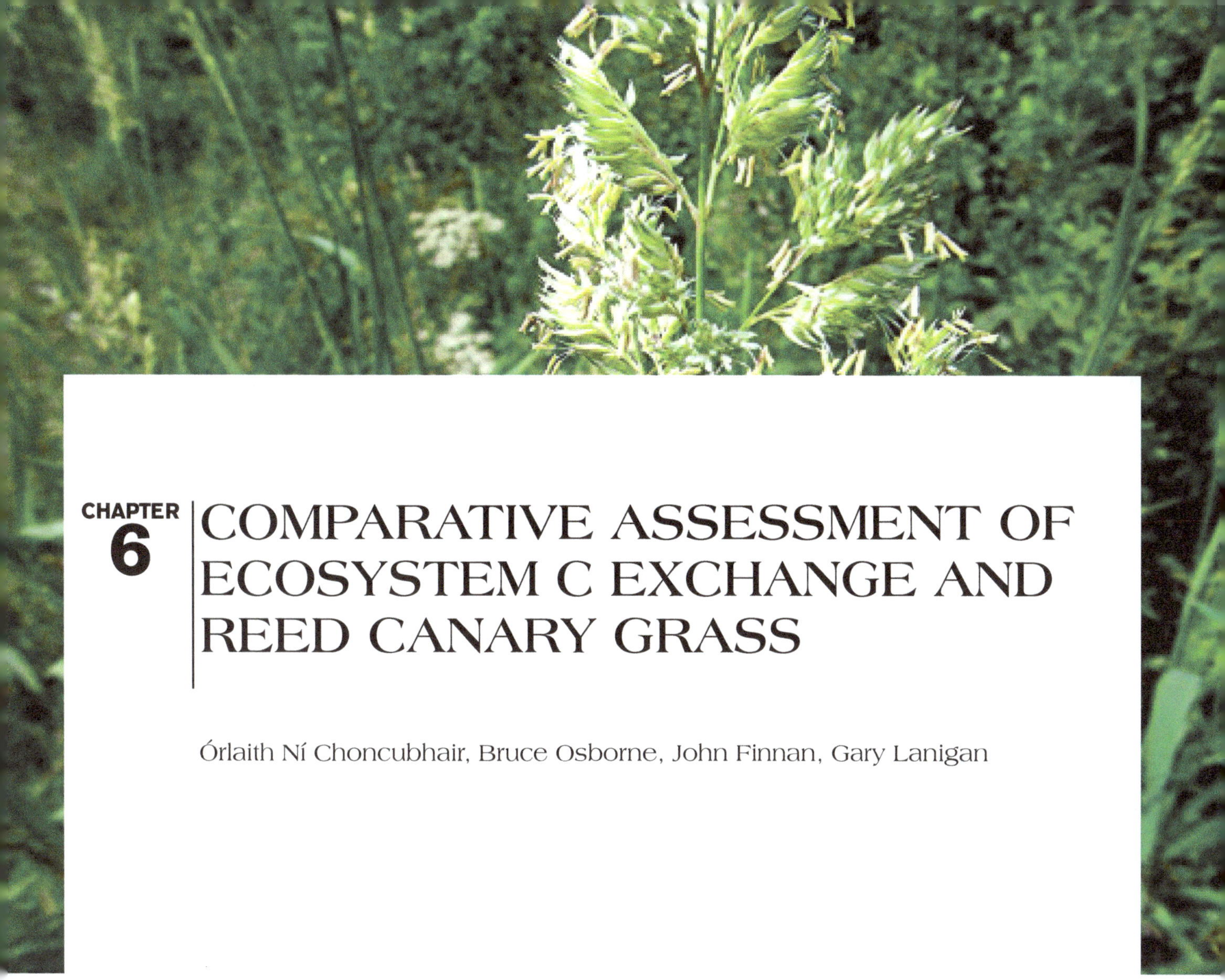

CHAPTER 6

COMPARATIVE ASSESSMENT OF ECOSYSTEM C EXCHANGE AND REED CANARY GRASS

Órlaith Ní Choncubhair, Bruce Osborne, John Finnan, Gary Lanigan

INTRODUCTION

Global concerns surrounding the impact of anthropogenic greenhouse gas (GHG) emissions on our climate allied with challenges to global energy security are driving interest in renewable energy, including bioenergy. In the European Union, member states have committed to increasing the contribution of renewable energy to 20% of total energy consumption by 2020 (EU, 2009). Furthermore, limitations identified in first-generation liquid biofuels have engendered an increased focus on second-generation alternatives produced from ligno-cellulosic plant materials (Sims *et al*., 2010). Realisation of the potential of bioenergy in the EU will, however, require significant agricultural land area, estimated at between 18 and 21 million hectares (Özdemir *et al*., 2009).

Perennial rhizomatous grasses (PRGs), such as *Miscanthus* × *giganteus* and reed canary grass (RCG), confer many advantages as potential nonfood bioenergy crops and have received increasing attention in Europe and the USA in recent decades (Landström *et al*., 1996; Lewandowski *et al*., 2003). *Miscanthus*, a C_4 plant originating from East Asia, can be highly productive, yielding 10–25 t dry matter ha^{-1} y^{-1} in central and northern Europe (Lewandowski*et al*., 2000; Finnan & Burke, 2014), while yields of RCG, a C_3 perennial grass indigenous to temperate regions of Europe, Asia and North America, ranged from 5 to 12 t dry matter ha^{-1} y^{-1} in trials in northern Europe (Landström *et al*., 1996; Saijonkari-Pahkala, 2001; Lewandowski*et al*., 2003; Kandel *et al*., 2013). As perennial species, both *Miscanthus* and RCG invest significant resources below ground, thus building reserves for more rapid canopy development in the spring compared with annual crops (Beale & Long, 1995; McLaughlin & Walsh, 1998).

High productivity in *Miscanthus* has historically been attributed to the superior light-, water- and nitrogen-use efficiency afforded by the C_4 photosynthetic pathway (Long, 1983). For example, the maximum efficiency of solar energy conversion in C_4 crops has been estimated to be 40% higher than that of C_3 species (Monteith, 1978). Furthermore, studies have highlighted the crop's exceptional ability to maintain high photosynthetic productivity even in cool temperate climates (Beale & Long, 1995). This is most likely achieved through a combination of reduced susceptibility to photoinhibition and decreased sensitivity to chilling temperatures (Beale *et al*., 1996; Naidu & Long, 2004; Wang *et al*., 2008).

More recent studies have, however, identified extended leaf longevity and high leaf area as the driving factors contributing to greater productivity in *Miscanthus*. In a side-

Citation: Órlaith Ní Choncubhair, Bruce Osborne, John Finnan, Gary Lanigan, Comparative assessment of ecosystem C exchange and reed canary grass, DOI: 10.1111/gcbb.12343.

by-side comparison of field-scale stands of *Miscanthus* and C_4 maize (*Zea mays*) in the USA, Dohleman & Long (2009) showed that the efficiency of captured sunlight-to-biomass conversion was almost identical in both crops averaged over two growing seasons. However, light interception efficiency was 61% higher in *Miscanthus*, which developed a closed canopy a month earlier than maize and maintained it a month longer in the autumn. This resulted in substantial net carbon gains for the cold-tolerant *Miscanthus* crop due to enhanced leaf area duration (Dohleman & Long, 2009).

The question remains, however, as to whether *Miscanthus* can out-perform native C_3bioenergy crop candidates in temperate regions in terms of productivity, leaf longevity and net C sequestration. RCG, in particular, grows vigorously after seed establishment and easily out-competes weeds after the first year of establishment (Lewandowski *et al.*, 2003). Indeed, its competitive advantages have allowed RCG to become an invasive species in certain wetland areas of the mid-western and north-western USA (Wrobel *et al.*, 2009). To date, no direct comparisons of *Miscanthus* and RCG under the same environmental conditions have been made to understand the relative dynamics of crop development and C assimilation in these ecosystems. This information would: (1) provide focus for exploiting favourable plant traits and developing superior genotypes for bioenergy production, (2) highlight the relative merits of*Miscanthus* and RCG in different climatic zones with varying growing seasons and (3) reveal the C balance implications of land-use change (LUC) to these crops.

Additional factors, such as the previous land use and the magnitude of LUC-related emissions, will significantly impact the long-term C balance of established bioenergy crops. The initial transition phase of LUC can be associated with substantially increased GHG emissions (Guo & Gifford, **2002**; Fargione *et al.*, **2008**; Donnelly *et al.*, **2011**; Poeplau *et al.*, **2011**; Houghton*et al.*, **2012**), particularly if bioenergy crop yields are low (Don *et al.*, **2012**). The time taken for the crop to achieve maximum productivity is also likely to be a significant determinant of the early establishment C balance.

In mature *Miscanthus* crops, soil C sequestration rates of 0.4–0.66 Mg ha^{-1} y^{-1} have been reported for plantations established on former croplands (Don *et al.*, **2012**; Zimmermann*et al.*, **2012**; Poeplau & Don, **2014**). Less information is available on grasslands converted to*Miscanthus*, but limited meta-analysis data has shown an annual carbon sequestration rate close to zero (no positive or negative effect) (Don *et al.*, **2012**; McCalmont *et al.*, **2016**; Qin*et al.*, **2016**). Measurements of carbon exchange in RCG plantations have provided evidence of net C uptake in these ecosystems (Shurpali *et al.*, **2009**; Mander *et al.*, **2012**; Lind *et al.*, **2015**); however, these studies were largely confined to drained peat extraction areas and there is little information on the carbon balance of RCG crop plantations established on mineral soils.

This study presents the first side-by-side field-scale comparison of *Miscanthus* and RCG to investigate comparative differences in crop development, leaf longevity and ecosystem-scale C fluxes from initial establishment to near maturity. The bioenergy crops were established on land previously under permanent grass. Conversion of grassland to bioenergy crops is of particular relevance in Ireland and in the wider European continent as over 90% and 35% of utilised agricultural area in Ireland and the EU-27, respectively, is currently used for grass production (Central Statistics Office, **2014**; Huyghe *et al.*, **2014**). The specific aim of the study was to address the following questions: (1) what are the C emissions associated with the initial LUC from grassland, which we define as the 'transitional phase'? (2) What are the longer term C balance implications of establishing *Miscanthus* and RCG crops on permanent grasslands (the 'post-establishment phase')? (3) How does leaf longevity in *Miscanthus* compare to that of an indigenous C_3 bioenergy crop and does it result in higher net C uptake?

MATERIALS & METHODS

Site Description & Management

The study was carried out from late April 2009 to the end of December 2011 at the Teagasc Environmental Research Centre, Johnstown Castle, Co. Wexford in the south-east of Ireland (52.3°N, 6.5°W, 67 m above sea level). This region has a maritime temperate climate with a mean annual rainfall of 1038 mm distributed evenly across the year and a mean annual air temperature of 10.4 °C. The seasonal range in temperature is narrow (Fig. 1), with an average summer air temperature of 14.9 °C and an average winter air temperature of 6.3 °C. The prevailing wind direction is south-westerly.

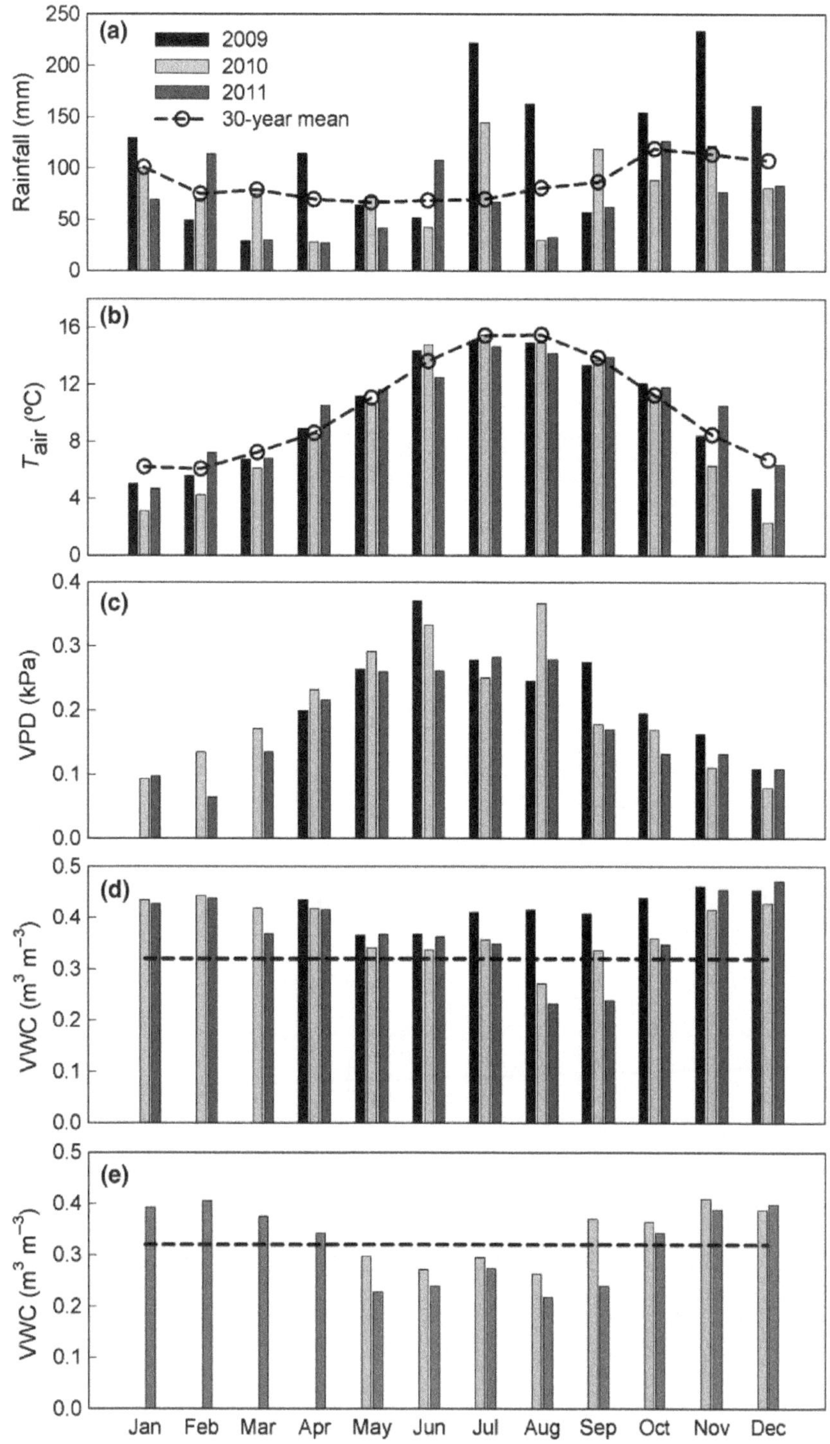

Figure 1. Mean monthly values for total rainfall (a), air temperature (b), vapour pressure deficit (VPD) (c) and soil volumetric water content (VWC) at the *Miscanthus* (d) and RCG (e) sites for 2009, 2010 and 2011. The 30-year (1981–2010) mean monthly rainfall and air temperature values for Johnstown Castle are also shown. The soil VWC at field capacity is included in (d) and (e) as dashed black lines.

The *Miscanthus* and RCG crops were established on two former grassland sites (2 ha and 1 ha in area, respectively) (Fig. 2). Most of the experimental area had been maintained as grassland for at least 37 years and was managed organically for beef production since 2006. The*Miscanthus* site had been conventionally tilled and reseeded with perennial ryegrass in 2000 and surface seeded with white clover in 2005. The RCG site had been conventionally tilled and reseeded with perennial ryegrass and red clover in 2005. Otherwise, historical land management activities were comparable at both sites. The swards received organic fertiliser in the form of cattle slurry and farmyard manure. Grazing took place every 3–4 weeks until October 2008. Soils in this area are variable and are classified as imperfectly drained Gleys (FAO classification: Gleyic Cambisol) or moderately to well-drained Brown Earth soils (Cambisol). Selected soil physical and chemical properties of the *Miscanthus*, RCG and adjacent reference grassland sites are summarised in Table 1. The wilting point, field capacity and water content at saturation were calculated using a hydraulic properties model (http://hydrolab.arsusda.gov/soilwater/Index.htm) as 0.17, 0.32 and 0.59, respectively, for all sites.

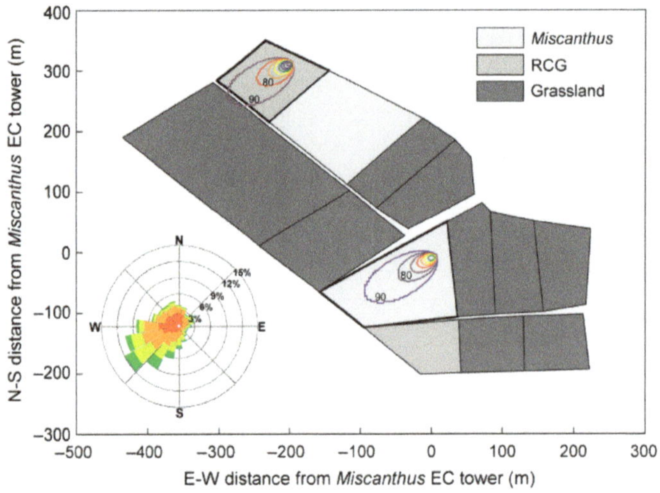

Figure 2. Experimental site map showing the location and dimensions of the *Miscanthus* and RCG field sites and surrounding grassland fields, as well as a wind rose displaying the distribution of mean wind direction for the site. The wind rose was generated using the freeware WRPLOT View (http://weblakes.com). Contributions from wind speed classes 0.5–2.1 (red), 2.1–3.6 (orange), 3.6–5.7 (yellow), 5.7–8.8 (green) and ≥ 8.8 m s^{-1} (dark green) are depicted. Sample footprints are shown for the*Miscanthus* and RCG eddy covariance towers at their maximum height under unstable conditions ($z/L = -0.06$, where z is the height and L is the Monin-Obukhov length). Contour lines represent the crosswind integrated cumulative contribution (%) to the estimated footprint.

Table 1. Physical and chemical characteristics of the soils at the *Miscanthus*, RCG and adjacent reference grassland sites in autumn 2011. Bulk density (BD), texture, pH and total nitrogen (TN) are reported for 0–15 cm soil depth. The soils at all sites are classified as loam. Total carbon (TC) and total organic carbon (TOC) are shown for three depths: 0–15 cm (A), 15–30 cm (B) and 30–45 cm (C)

	BD (g cm^{-3})	Sand (%)	Silt (%)	Clay (%)	pH	TN (%)	TC (%)			TOC (%)		
							A	B	C	A	B	C
Miscanthus	0.98	50	32	18	6.4	0.3	3.2	2.9	1.5	2.4	2.2	1.1
Grassland reference	0.99	51	32	18	6.5	0.3	2.9	2.9	2.1	2.2	2.2	1.5
RCG	0.97	48	33	19	6.7	0.3	2.8	2.5	1.9	2.2	1.9	1.4
Grassland reference	0.88	52	30	19	6.2	0.3	2.7	2.5	2.1	2.0	1.8	1.5

On the 1st April 2009, both sites were sprayed with glyphosate to eradicate the extant vegetation. The soil was conventionally tilled using a mouldboard plough to a depth of 20 cm on the 27th April 2009 (approximately one tenth of the *Miscanthus* site) and completed on the 29th April 2009, followed by power-harrowing on the 1st and 5th June 2009. *Miscanthus*rhizomes were planted on one site on the 9th and 10th June 2009 and the soil was consolidated using a heavy roller 1 week later. Additional herbicides were applied to the *Miscanthus* site in the early establishment phase to reduce competition from grass and broad-leaf weeds. The selective herbicide MCPA was sprayed to control broad-leaf species (13th August 2009 and 27th July 2010) while glyphosate was applied soon after harvesting (5th March 2010 and 8th March 2011) to control grass weeds. The crop was cut with a conditioner mower on the 4th March 2010 and 4th March 2011 but the limited biomass material that was cut was left on the ground.

The second site lay fallow until April 2010 when RCG was established. For this, the site was power-harrowed and seeded with RCG (*Phalaris arundinacea* L.) at a rate of 30 kg ha^{-1} on the 15th April 2010, and then consolidated with a heavy roller 1 day later. MCPA was applied on the 29th June 2010 to reduce competition from broad-leaf species. The RCG crop was harvested once during the study period, on the 12th October 2010, while harvesting of the 2011 crop was delayed until spring 2012, which is the preferred time for harvesting. No fertilisers were applied over the duration of the study.

Micrometeorological Measurements

Ecosystem-scale CO_2 fluxes were measured using an open-path eddy covariance (EC) system commencing on the 28th April 2009 and 15th April 2010 at the *Miscanthus* and RCG sites, respectively. The instrumentation was identical at both sites and consisted of an open-path infrared gas analyser (IRGA) (LI-7500, LI-COR Biosciences, Lincoln, NE, USA) coupled with a 3D sonic anemometer (CSAT3, Campbell Scientific, Logan, UT, USA). EC data were collected at a frequency of 10 Hz and averaged over 30-minute intervals. The flux tower was located in the north-eastern corner of the fields to maximise the fetch in the direction of the prevailing south-westerly wind (Fig. 2). Tower height was increased during periods of active growth to maintain its position above the canopy while restricting the flux footprint to the experimental area for as much time as possible. The maximum tower height was 4 m and 3 m at the *Miscanthus* and RCG sites, respectively, while the minimum sensor to canopy height ratios were 1.47 (*Miscanthus*) and 1.52 (RCG).

Ancillary biometric sensors included an air temperature and relative humidity probe (HMP45C, Campbell Scientific, Logan, UT, USA), a net radiation sensor (NR-Lite, Kipp & Zonen, Delft, The Netherlands) and a down-welling quantum sensor (SKP 215, Skye Instruments Ltd., Llandrindod Wells, UK). Two self-calibrating soil heat flux plates were installed at 8 cm soil depth (HFP01SC, Hukseflux, Delft, The Netherlands) and averaging soil temperature probes were installed at 2 cm and 6 cm depth above the soil heat flux plates. Time domain reflectometers (CS616, Campbell Scientific, Logan, UT, USA) measured soil volumetric water content (VWC) in the upper 15 cm of soil. Daily meteorological data (mean air temperature, total rainfall and global solar radiation) were obtained from a Met Éireann synoptic weather station located 1.7 km from the field site.

Quality Assurance and Flux Analysis

Data quality control procedures included spike removal (Vickers & Mahrt, **1997**), time lag compensation using a covariance maximisation procedure and compensation for air density fluctuations using the WPL term (Webb *et al.*, **1980**). The double rotation method (Kaimal & Finnigan, **1994**) was used to correct for sonic anemometer tilt as the alternative planar fit method requires several weeks of measurement with constant instrumental set-up and is often not recommended for measurements over canopies with dynamic height variation (Moureaux*et al.*, **2012**). Spectral attenuation effects were corrected following the analytical methods of Moncrieff *et al.* (**1997**). Tests on developed turbulence and stationarity were applied to the calculated fluxes (Table 13, Mauder & Foken, **2004**) and data of questionable quality (QC-flag = 2) were removed, while data of moderate quality (QC-flag = 1) were retained but not included in the regression analysis performed in the gap-filling procedure. Data screening based on the results of these tests has been shown to result in a less systematic distribution of data gaps compared to the removal of fluxes below a derived friction velocity (u_*) threshold (Ruppert *et al.*, **2006**). Plausible limits were also applied to net ecosystem exchange (NEE) ($-50 < \text{NEE} < 30$ μmol CO_2 m^{-2} s^{-1}), latent heat ($-20 < \text{LE} < 600$ W m^{-2}) and sensible heat ($-100 < \text{H} < 300$ W m^{-2}) fluxes.

The peak location of the flux footprint, *xmax*, and the distance from the flux tower which includes 90% of the source area contributing to the measured flux, *xR90*, were estimated for the prevailing south-westerly wind direction using the Kljun *et al.* (**2004**) model as 35 m and 95 m, respectively, for the *Miscanthus* site and 32 m and 87 m, respectively, for the RCG site. Owing to the relatively constrained field sizes, detailed footprint analysis was conducted on half-hourly fluxes based on the analytical Kormann & Meixner (**2001**) footprint model. Since analytical models tend to overestimate flux footprints in comparison to more complex Lagrangian stochastic models due to the neglect of along-wind velocity fluctuation (Kljun *et al.*, **2003**), this approach is likely to be conservative. Sample flux footprints for the two towers at their maximum height under unstable conditions are shown in Fig. 2. At the maximum height, the footprint of the *Miscanthus* tower was well conserved within the *Miscanthus* field while close to 90% of the flux footprint from the RCG tower was derived from the RCG field. At lower heights (April 2009 to August 2011 for the *Miscanthus* tower and April 2010 to June 2011 for the RCG tower), the flux footprints were better confined to the experimental areas. Half-hourly data were rejected if less than 70% of the derived footprint originated from the respective experimental areas, similar to previous studies (Ammann *et al.*, **2007**; Baum *et al.*, **2008**; Davis*et al.*, **2010**; Vanderborght *et al.*, **2010**). This resulted in 17.4% and 14.9% of fluxes being rejected overall for the *Miscanthus* and RCG sites, respectively.

The positioning of the EC sensors in relation to the inertial boundary layer was also monitored due to the relatively low measurement heights. The thickness of the inertial boundary layer, calculated according to equation 5 of Munro & Oke (**1975**), reached maximum values of 4.9 m, 4.9 m and 7.0 m in 2009, 2010 and 2011, respectively, for *Miscanthus* and values of 2.8 m and 3.7 m in 2010 and 2011, respectively, for RCG. The lower limit of the inertial boundary layer was estimated to occur between $(z\text{–}d) = 5z0$ and $(z\text{–}d) = 10z0$ (Garratt, **1992**), where z is the height above ground, d is the zero-plane displacement height and $z0$ is the roughness length of the crop surface. This yielded maximum values in the range 1.1–1.6 m, 1.1–1.5 m and 3.0–4.2 m in 2009, 2010 and 2011, respectively, for *Miscanthus* and values of 0.8–1.1 m and 1.7–2.5 m in 2010 and 2011, respectively, for RCG. These estimates suggested that the sensors were within the appropriate inertial boundary layer for the vast majority of the experiment.

The quality of the CO_2 flux estimates was assessed by examining energy balance closure (EBC) at the site. This routine provides an independent check of the degree to which turbulent fluxes are captured in the boundary layer and may highlight significant bias in measurements (Twine*et al.*, **2000**). Energy balance closure was tested by comparing half-hourly and daily sums of net radiation with the sum of good quality (QC-flag = 0, 1) *LE* and *H* fluxes and energy storage terms using the equation

$$R_n = LE + H + G + S_s + S_p \tag{1}$$

where *Rn* is net radiation, *LE* is the latent heat flux, *H* is the sensible heat flux, *G* is the soil heat flux, *Ss* is soil heat storage above the heat flux plates and *Sp* is the energy stored in photosynthate. Heat storage in the soil surface layer (of depth Δz) was calculated as

$$S_s = \frac{\Delta T(\theta_v \rho_w c_w + \rho_s c_s)\Delta z}{\Delta t} \tag{2}$$

where ΔT is the change in average soil temperature above the heat flux plates over the time interval Δt, θv is the soil volumetric water content, ρw is the density of water, *cw* is the specific heat capacity of water, ρs is the soil bulk density and *cs* is the specific heat capacity of soil [a value of 837 J kg^{-1} K^{-1} was used for *cs* (Scott, **2000**)]. The energy captured during photosynthesis and stored in biomass, *Sp*, was computed by equating photosynthetic fixation of 2.5 mg CO_2 m^{-2} s^{-1} to an energy flux of 28 W m^{-2} (Meyers & Hollinger, **2004**). For this calculation, photosynthetic rates were based on gross primary productivity (GPP) estimates calculated by subtracting modelled total ecosystem respiration (TER) from measured NEE. Heat storage in plant biomass and canopy air was not measured and considered minor relative to the other terms in equation 1, particularly in the case of *Miscanthus* which had a sparse, open canopy for much of the study.

The regression analysis of half-hourly energy fluxes yielded a slope of 0.945 ± 0.004 [95% confidence interval (CI)] and intercept of 5.2 W m^{-2} for the *Miscanthus* site and a slope of 0.905 ± 0.004 and intercept of 6.3 W m^{-2} for the RCG site (Fig. 3). Calculating daily sums of energy fluxes resulted in a small increase in the slope for *Miscanthus* to 0.963 ± 0.013 and a larger increase in the slope for RCG to 0.964 ± 0.016 (Fig. 3). This suggests that there was a small contribution from additional nonestimated storage terms, such as heat storage in plant biomass or canopy air, to the energy balance of these ecosystems. These contributions tend to be negligible on a daily scale (Oncley *et al.*, **2007**). Although these slopes were significantly different from one, they compare well with reported ranges of EBC of 0.53–0.99 (Wilson *et al.*, **2002**) and 0.70–0.94 (Stoy *et al.*, **2013**) and suggest that the vast majority of the available energy was accounted for. Consequently, turbulent energy exchange was satisfactorily resolved by the EC measurement system.

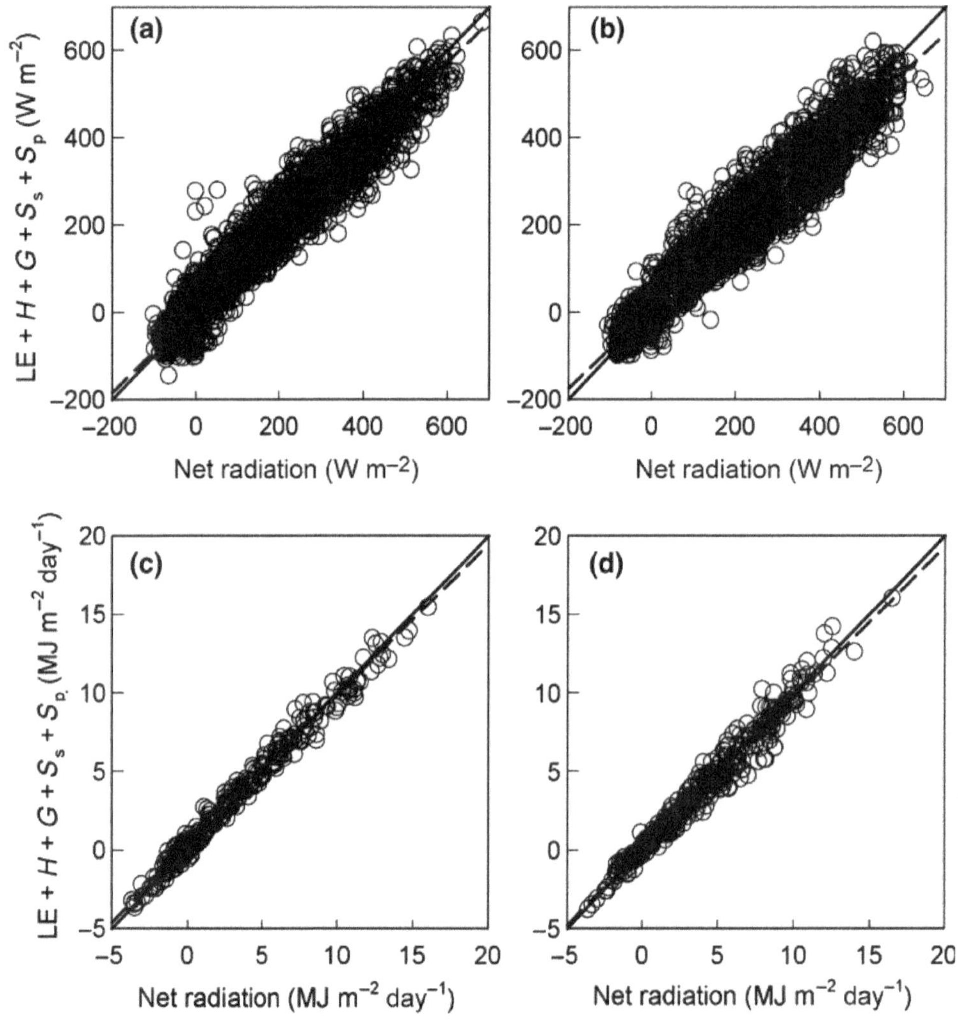

Figure 3. Energy balance closure on a half-hourly timescale for *Miscanthus* (a) and RCG (b) and a daily timescale for *Miscanthus* (c) and RCG (d) in 2011. The dashed black lines represent the regression lines [$y = 0.95x + 5.24$, $R^2 = 0.96$ for *Miscanthus* (half-hourly); $y = 0.91x + 6.3$, $R^2 = 0.96$ for RCG (half-hourly); $y = 0.96x + 0.2$, $R^2 = 0.99$ for *Miscanthus* (daily); $y = 0.96x + 0.04$, $R^2 = 0.97$ for RCG (daily)]. The solid black lines indicate a 1 : 1 relationship. Daily sums were calculated using the available data in a 24-h period.

In addition, cospectral analysis of the high frequency data was performed to assess the frequency response of the EC system. Fast Fourier transforms were applied to the 10 Hz data to yield full cospectra, which were subsequently reduced into exponentially spaced frequency bins and ensemble-averaged. Normalised, ensemble-averaged and frequency-weighted cospectra for both sites are presented in Fig. 4 for unstable conditions and compared with the universal theoretical Kaimal cospectral function (Moncrieff *et al.*, **1997**). The shapes of the measured cospectra are generally consistent with the Kaimal curve, except for small deviations from the theoretical −4/3 slope in the inertial subrange. This result is indicative of some small-scale dampening of the high frequency signal, however, the flux postprocessing routines employed included a correction for high frequency losses. The CO_2 spectral correction factor (SCF) for *Miscanthus* had a median of 1.17 and mean ± 95% CI of 1.19 ± 0.001. For RCG, the median was 1.15 and the mean was 1.16 ± 0.002. These values fall in the expected range of CO_2 correction factors (1.04–1.25; Aubinet *et al.*, **2001**) and compare well with an SCF of 1.12 for a similar open-path system (Haslwanter *et al.*, **2009**).

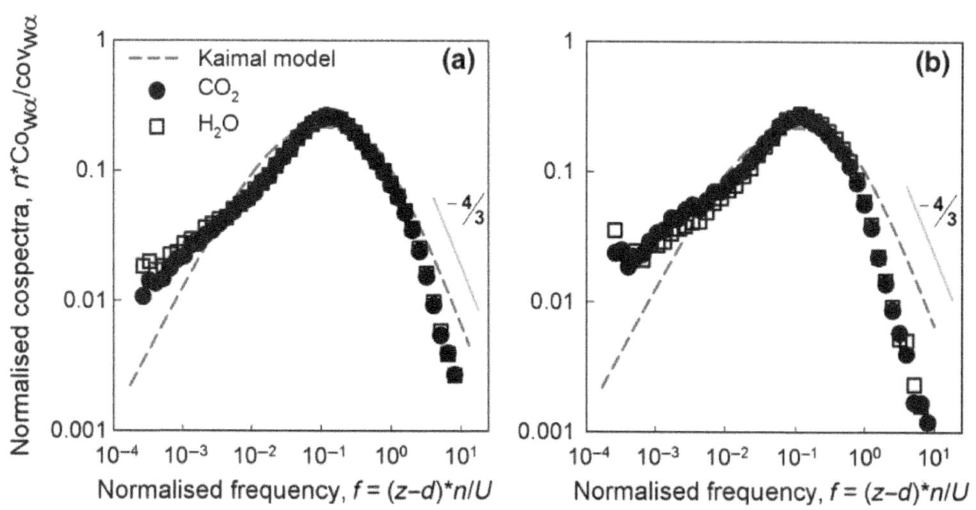

Figure 4. Normalised ensemble-averaged cospectra of CO_2 and H_2O in unstable conditions plotted against normalised frequency, *f*, for the *Miscanthus* (a) and RCG (b) sites. The dashed grey curve and solid grey line represent the theoretical universal Kaimal cospectra and the −4/3 slope predicted by Kolmogorov's Law, respectively (Kaimal*et al.*, **1972**).

Gaps in the flux data created by measurement failures and data quality analysis procedures were filled using semiempirical gap-filling techniques. TER was modelled by relating soil temperature to nocturnal measured NEE using the exponential Lloyd & Taylor (**1994**) equation

$$\mathrm{TER} = R_{10}e^{E_0(1/(283.15-T_0)-1/(T-T_0))} \tag{3}$$

where R_{10} is ecosystem base respiration at a reference temperature of 10 °C, E_0 is an activity energy parameter, T_0 was set to 227.13 K as in the original study and *T* is soil temperature (K). Significant seasonality has been demonstrated in the E_0 parameter, particularly in summer active ecosystems (Reichstein *et al.*, **2005**), and therefore bimonthly best fit estimates of E_0were obtained in each year using equation 3. These short-term bimonthly estimates were then averaged over each year to derive more realistic long-term E_0 values. Temporal variability in the R_{10} parameter was also considered. For each available nocturnal NEE value and adopting the relevant long-term E_0 value, equation 3 was rearranged to provide an estimate of R_{10} in that half-hourly period. The weekly running mean, R_{10Mean}, was then computed and employed instead of R_{10} in equation 3 above to incorporate time-varying influences on R_{10} such as soil moisture status, plant phenology etc. The dependence of nocturnal TER on temperature was extended to daytime and estimates of TER were obtained for all 30-minute periods in this way.

GPP was initially computed by subtracting the estimated TER from measured daytime half-hourly NEE. Daytime GPP values were then pooled for bimonthly periods and subdivided into fixed 4 °C temperature bins within each bimonthly period. GPP model parameters were then derived for each subset using a rectangular hyperbola function (Falge *et al.*, **2001**)

$$\mathrm{GPP} = \frac{\alpha Q_{\mathrm{PPFD}} A_{\mathrm{max}}}{\alpha Q_{\mathrm{PPFD}} + A_{\mathrm{max}}} \tag{4}$$

where α is the ecosystem quantum (photon) yield (mol CO_2 [mol photon]$^{-1}$), Q_{PPFD} is the photosynthetic photon flux density (μmol [photon] m^{-2} s^{-1}) and A_{max} is the maximum assimilation rate (μmol CO_2 m^{-2} s^{-1}).

The net ecosystem carbon balance (NECB) is defined as the net rate of C accumulation or loss from an ecosystem, taking account of all physical, biological and anthropogenic sources and sinks of carbon in an ecosystem (Chapin *et al.*, **2006**). NECB was calculated as $-\Sigma$NEE − $C_{harvest}$, where $C_{harvest}$ is the carbon removed from the ecosystem at harvest, assuming that contributions from carbon monoxide, methane, volatile organic carbon (VOC), dissolved carbon and particulate carbon were negligible.

An assessment of the uncertainty associated with annual CO_2 flux estimates was performed in a similar way to Black *et al.* (**2007**). Firstly, the contribution of uniform systematic errors to the measured fluxes was evaluated following Goulden *et al.* (**1996**). Assuming spectral similarity between latent heat, sensible heat and CO_2 flux, the imbalance between available energy and measured latent and sensible heat, as calculated in the long-term energy balance, was used as an approximation of the underestimation of ecosystem exchange by the EC measurement system. Secondly, sampling uncertainty errors, associated with the imputation of half-hourly missing data during gap-filling procedures, were assessed in a similar way to Falge *et al.* (**2001**). Artificial datasets were created with 10, 25, 35 and 45% of the data replaced by gaps and gap-filling procedures were followed to fill the artificial gaps. The mean bias error was calculated as the mean difference between measured and calculated values for the complete data series. The final uncertainty estimate was computed for each measured or gap-filled half-hourly flux based on either the uniform systematic error derived from the energy balance closure, expressed as a percentage, or the mean bias error arising from sampling uncertainty (gap-filling).

Soil Respiration

Soil respiration was monitored using a closed dynamic chamber coupled with a portable infrared gas analyser (Environmental Gas Monitor EGM-4, PP Systems, Hitchin, UK). Chamber measurements were made on bare soil at 10 sampling points in the *Miscanthus* and RCG sites, yielding an estimate of the flux of CO_2 from the soil surface which combines contributions from autotrophic (plant root) and heterotrophic (microbes and soil fauna) components. Measurements were carried out from early June 2010 to early September 2011 at approximately weekly intervals during the growing season and monthly intervals during the winter. Soil temperature and VWC (0–7 cm) were recorded at each sampling point using a WET sensor (Delta-T Devices, Cambridge, UK).

Crop Analysis

Above-ground *Miscanthus* biomass was calculated as the product of average shoot density and total dry biomass (leaf + stem) per shoot sampled prior to harvesting and at least monthly during the growing seasons of 2010 and 2011. Shoot density estimates (shoots m^{-2}) were based on the number of shoots counted in a quadrat of area 2.1 m^2 at 10 random locations in the *Miscanthus* plantation. Total dry biomass per shoot was measured by cutting a minimum of 35 random shoots at ground level during each sampling event and drying the leaf and stem biomass at 70 °C to constant weight. Below-ground biomass was sampled on three occasions (February 2010, January 2011 and September 2011) by excavating all below-ground plant material associated with three randomly located *Miscanthus* plants to a depth of 40 cm. Roots, live rhizomes and dead rhizomes were separated, washed free of soil over a 2 mm sieve and dried to constant weight at 70 °C. Below-ground biomass was up-scaled from the plant scale to a per unit area (m^{-2}) basis using calculated above- to below-ground ratios and average above-ground biomass on each individual sampling date. The contribution of understory vegetation was assessed by clipping all above-ground material in quadrats of area 0.25 m^2 placed inside the large *Miscanthus* quadrats with subsequent determination of dry matter yields after oven drying at 70 °C.

Above-ground RCG biomass was assessed by cutting all vegetation at ground level in quadrats of area 0.25 m^2 randomly positioned at five locations within the crop. Sampling was carried out in September 2010, prior to harvest in October 2010 and in June and September 2011. A root auger of volume 750 cm^3 (Eijkelkamp Agrisearch Equipment, Giesbeek, The Netherlands) was used to sample below-ground biomass to a depth of 45 cm in January and October 2011 at 4 random locations in the crop. The intact soil cores were mixed with water in the laboratory, washed over a 0.2 mm sieve and dried at 70 °C to constant weight.

Miscanthus green leaf area index (GLAI) was determined by firstly establishing an allometric relationship between leaf area and the product of leaf length and width, following Clifton-Brown *et al.* (**2000**). The length and width of leaves harvested on three different occasions were measured and images of the leaves were captured with a flatbed scanner (CanoScan LiDE 35, Canon, Tokyo, Japan). Leaf area estimates were obtained using the open-source softwareimagej (http://imagej.nih.gov/ij/) to generate the linear regression equation

$$\text{leaf area} = 0.76\,(\text{length} \times \text{width}) \qquad (5)$$

where both measured quantities are in cm^2, $R^2 = 0.93$ and $n = 295$. This equation was applied to leaf length × width measurements made during above-ground biomass sampling. *Miscanthus* GLAI was computed as the product of green leaf area per shoot and shoot density. The GLAI of understory vegetation was estimated based on an allometric relationship between dry biomass weight and calculated leaf area.

RCG leaf area index (LAI) was monitored with a Sunscan Canopy Analysis System (Delta-T Devices, Cambridge, UK) at approximately monthly intervals, providing a measure of the total leaf area per unit ground area. Individual measurements made at 40 locations within the crop were averaged on each sampling date. These measurements involved a non-destructive sampling technique based on the fraction of photosynthetically active radiation (PAR) intercepted by the canopy (Campbell & Norman, **1989**). Therefore, RCG LAI values included contributions from both live and senescent leaves as both influence the transmission of light through the canopy.

RESULTS

Environmental Conditions

Mean annual air temperatures were close to the 30-year mean temperature in 2009 (10.0 °C) and 2011 (10.4 °C) but dipped in 2010 to 9.3 °C. Two prolonged periods of cold weather were experienced during these years (December 2009 to February 2010 and November 2010 to January 2011) (Fig. 1). In addition, summer mean temperature was 7% lower than the long-term mean in 2011. Rainfall totals were highly variable, with higher than average rainfall in 2009 (1427 mm), close to the average in 2010 (972 mm) and below average in 2011 (839 mm). Exceptionally wet summer months characterised 2009, with atypically high rainfall also occurring in November. The mean monthly soil VWC at the *Miscanthus* site was high for much of the study, only reducing below field capacity (0.32 m^3 m^{-3}) in August 2010 and August and September 2011. Soil VWC in the RCG site was below field capacity from May to August of each year but remained above field capacity (0.32 m^3 m^{-3}) outside of the peak growth period. In general, soil VWC was higher in *Miscanthus* than RCG, most obviously from May to July 2011. Vapour pressure deficit (VPD) was always low, ranging from a monthly average of 0.06 kPa in February 2011 to a maximum monthly average of 0.37 kPa in June 2009 and August 2010 (Fig. 1).

Crop Development

Establishment and growth of the *Miscanthus* crop was slow in the first 2 years after planting. Prior to the first harvest in March 2010, above-ground dry biomass totalled just 10 ± 4.0 (standard error of the mean, SEM) g m^{-2} while below-ground biomass stocks were estimated at 13.8 ± 7.1 g m^{-2} (Table 2). Following the growing season of 2010, above-ground biomass reached a seasonal maximum of over 100 ± 20.2 g m^{-2} in September and a maximum GLAI of 0.55 ± 0.05 m^2 m^{-2} was recorded in October (Fig. 5). The contribution of understory vegetation to ecosystem productivity was substantial at this time. Above-ground biomass stocks of understory vegetation, dominated mostly by grass species (*Agrostis stolonifera, Poa trivialis* and*Alopecurus geniculatus*), totalled 110 ± 21.7 g m^{-2} in September 2010, with an associated GLAI of 3.56 ± 0.70 $m^2 m^{-2}$.

Table 2. Above- and below-ground biomass stocks in the *Miscanthus* and RCG crops in g (dry matter) m^{-2}. Total above- and below-ground stocks in *Miscanthus* are divided into their constituent pools. Values in parentheses represent the standard error of the mean

Date	Above-ground (g m^{-2})			Live Rhizome	Below-ground (g m^{-2})		
	Stem	Leaf	Total		Dead Rhizome	Root	Total
Miscanthus							
February 2010	8.2 (3.3)	1.6 (0.7)	9.9 (4.0)	10.7 (6.2)	0	3.0 (0.9)	13.8 (7.1)
September 2010	63.9 (17.3)	36.6 (5.2)	100.5 (20.2)				
January 2011	33.1 (13.9)	6.5 (2.2)	39.6 (16.1)	73.6 (3.8)	9.3 (2.1)	22.3 (2.7)	105.2 (8.5)
March 2011	53.3 (15.2)	4.9 (1.4)	58.2 (16.4)				
June 2011	52.3 (23.3)	31.5 (9.2)	83.8 (32.5)				
September 2011	1001.6 (362.3)	308.9 (102.6)	1310.5 (464.6)	331.1 (138.4)	16.0 (6.3)	63.7 (21.8)	410.8 (166.4)
Reed Canary Grass							
September 2010			653.0 (76.4)				
October 2010			589.3 (28.5)				
January 2011							489.0 (46.3)
June 2011			675.4 (93.9)				
September 2011			706.4 (61.2)				834.5 (72.5)

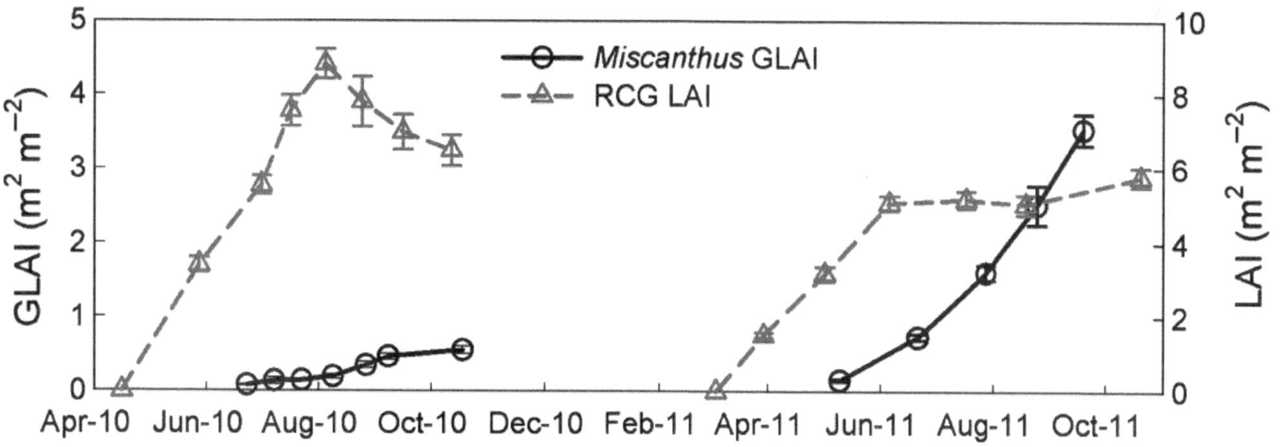

Figure 5. Temporal pattern of green leaf area index (GLAI) in *Miscanthus* and leaf area index (LAI) in RCG during the 2010 and 2011 growing seasons. Vertical bars represent the standard error of the mean.

Following initial slow development in 2011, growth rates increased exponentially towards the end of the summer, yielding 1311 ± 465 g m^{-2} in late September 2011. GLAI increased steadily over this time (Fig. 5, Table 2). Above-ground crop expansion was facilitated by a four-fold increase in below-ground biomass between January and late September 2011 when total below-ground *Miscanthus* biomass was 410.8 ± 166.4 g m^{-2}, comprising 80.6% live rhizome, 3.9% dead rhizome and 15.5% root (Table 2). Correspondingly, stocks of understory vegetation declined, associated with herbicide application in spring 2011 and increased competition from*Miscanthus* plants. Estimated above-ground biomass from weeds was 5.8 ± 2.2 g m^{-2} and 14.5 ± 6.3 g m^{-2} in late June and late September 2011, respectively, while their GLAI was calculated as 0.19 ± 0.07 m^2 m^{-2} and 0.47 ± 0.20 m^2 m^{-2} at the same times, representing significant reductions from 2010 values.

In contrast with the slow establishment of *Miscanthus*, rapid growth was observed in the new RCG plantation after sowing in mid-April 2010. A dense green canopy developed early in the first growing season resulting in a seasonal maximum LAI of 8.8 ± 0.4 m^2 m^{-2} at the start of August 2010 (Fig. 5). A reduction in LAI was then observed as the crop senesced. At the time of the first harvest in October 2010, 589 ± 29 g m^{-2} had accumulated in above-ground biomass and the LAI was 6.5 ± 0.4 m^2 m^{-2}. In the second year of establishment, above-ground growth began earlier than in the conversion year and, by early June 2011, a growing season maximum LAI of 5.1 ± 0.2 m^2 m^{-2} was recorded and above-ground biomass totalled 675 ± 94 g m^{-2}. Flowering and senescence occurred earlier in 2011 than in 2010 with the result that, by August 2011, the crop had completely senesced. Significant below-ground biomass stocks were recorded in the RCG crop in the second year of establishment (Table 2), with the maximum below-ground stock in September 2011 (835 ± 73 g m^{-2}) exceeding above-ground biomass estimates. Understory vegetation made a minor contribution to above-ground biomass in the RCG ecosystem, accounting for just 3.4% and 0.5% of total above-ground biomass in June and late September 2011, respectively.

Carbon Fluxes

Flux measurements at the *Miscanthus* site began 1 day before the majority of the grassland field was tilled. Ecosystem exchange of carbon was dominated by respiratory fluxes in the first month post-disturbance. Half-hourly NEE fluxes ± SEM averaged 4.0 ± 0.1, 3.3 ± 0.1, 3.3 ± 0.1 and 2.7 ± 0.1 μmol CO_2 m^{-2} s^{-1} in the first 4 weeks following tillage, respectively, with positive fluxes indicating a net release of C to the atmosphere. Monthly cumulative NEE ± uncertainty amounted to 104.7 ± 3.1, 62.7 ± 2.7 and −12.1 ± 3.1 g C m^{-2} in the first 3 months, respectively, post-tillage, representing significant initial net losses of carbon from the ecosystem.

A small recovery in photosynthetic activity was apparent in subsequent months, concurrent with the slow development of the newly-established *Miscanthus* crop and the emergence of understory vegetative species prompted by a flush of germination after the initial soil disturbance. However, seasonal maximum rates of instantaneous net C assimilation, recorded in mid-October 2009, were no greater than −15 μmol CO_2 m^{-2} s^{-1} (Fig. 6). The magnitude of photosynthetic uptake was similarly modest in 2010, peaking at −17 μmol CO_2 m^{-2} s^{-1} in late September. The seasonal pattern of NEE changed dramatically in 2011, however, concomitant with significant increases in above-ground *Miscanthus* biomass, large-scale leaf expansion and higher GLAI. Maximum instantaneous rates of net C assimilation were −30 to −35 μmol CO_2 m^{-2} s^{-1} between mid-August and late September 2011.

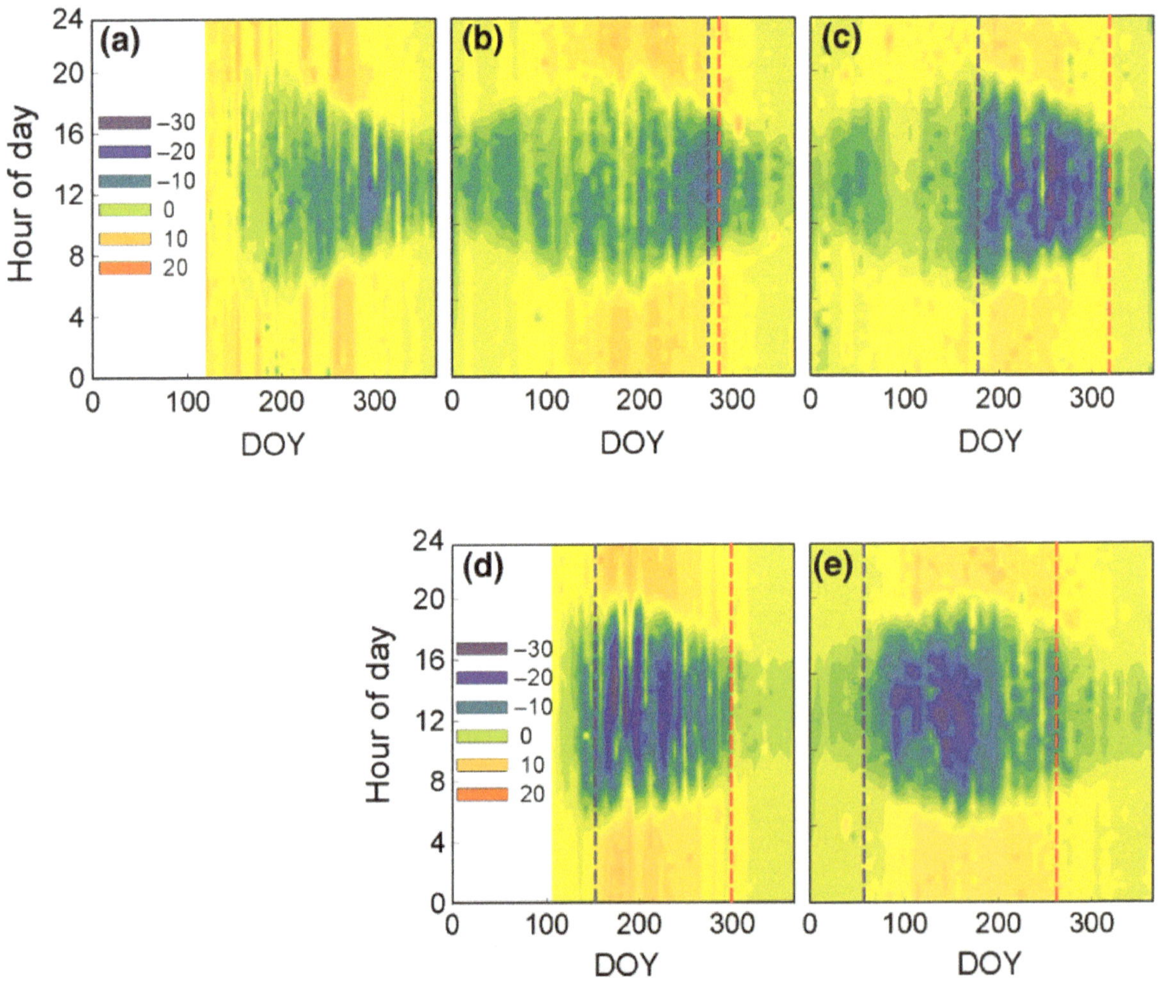

Figure 6. Diurnal course of half-hourly fluxes of net ecosystem carbon exchange in μmol CO_2 m^{-2} s^{-1} in the *Miscanthus* ecosystem during 2009 (beginning in late April) (a), 2010 (b) and 2011 (c) and in the RCG ecosystem during 2010 (beginning in mid-April) (d) and 2011 (e). Dashed vertical dark blue lines mark the 'break-even point' at which cumulative NEE becomes negative (net C uptake). Dashed vertical red lines indicate the timing of the end of net C uptake.

Measurements of NEE in the RCG crop began the same day the site was harrowed and seeded in mid-April 2010. Early half-hourly fluxes of NEE were dominated by ecosystem respiration but C uptake became evident from mid-May onwards (Fig. 6). Weekly-averaged NEE was 2.6 ± 0.1, 1.7 ± 0.1, 0.8 ± 0.1 and −0.2 ± 0.1 μmol CO_2 m^{-2} s^{-1} in the first 4 weeks following planting, respectively. As a result, the ecosystem represented a net source of carbon over the first month (cumulative NEE was 33.3 ± 2.4 g C m^{-2}) but switched to a net sink during the second (−88.0 ± 4.8 g C m^{-2}) and third month (−127.3 ± 13.9 g C m^{-2}).

Rapid establishment and extensive early canopy development in the RCG crop accompanied strong photosynthetic activity early in the first growing season. Just over 2 months after planting (late June 2010), a seasonal maximum NEE of −35 μmol CO_2 m^{-2} s^{-1} was recorded. Net C uptake continued until late September 2010 at a decreasing rate (Fig. 6). Winter NEE fluxes were small as a result of the exceptionally cold temperatures recorded from November 2010 to January 2011. However, photosynthetic activity resumed when daily mean temperatures began to exceed about 5 °C in February 2011. Net ecosystem C accumulation continued until a seasonal maximum NEE of −38 μmol CO_2 m^{-2} s^{-1} was recorded in mid-May 2011, after which rates of C uptake declined.

Daily fluxes of TER and GPP are shown in Fig. 7. Maximal daily values of GPP in the *Miscanthus*crop increased from −7.3 and −9.2 g C m^{-2} d^{-1} in September 2009 and June 2010, respectively, to −14.1 g C m^{-2} d^{-1} in August 2011. Maximal daily TER was 10.0, 9.5 and 7.4 g C m^{-2} d^{-1} in September 2009, June 2010 and July 2011, respectively for *Miscanthus*. In the RCG crop, maximal daily GPP values of −17.6 and −16.3 g C m^{-2} d^{-1} were recorded in late June 2010 and mid-May 2011, respectively, while maximal daily TER values of 9.2 and 7.6 g C m^{-2} d^{-1} were observed in mid-July 2010 and 2011, respectively. Higher TER values were generally associated with warmer soil temperatures and higher ecosystem productivity.

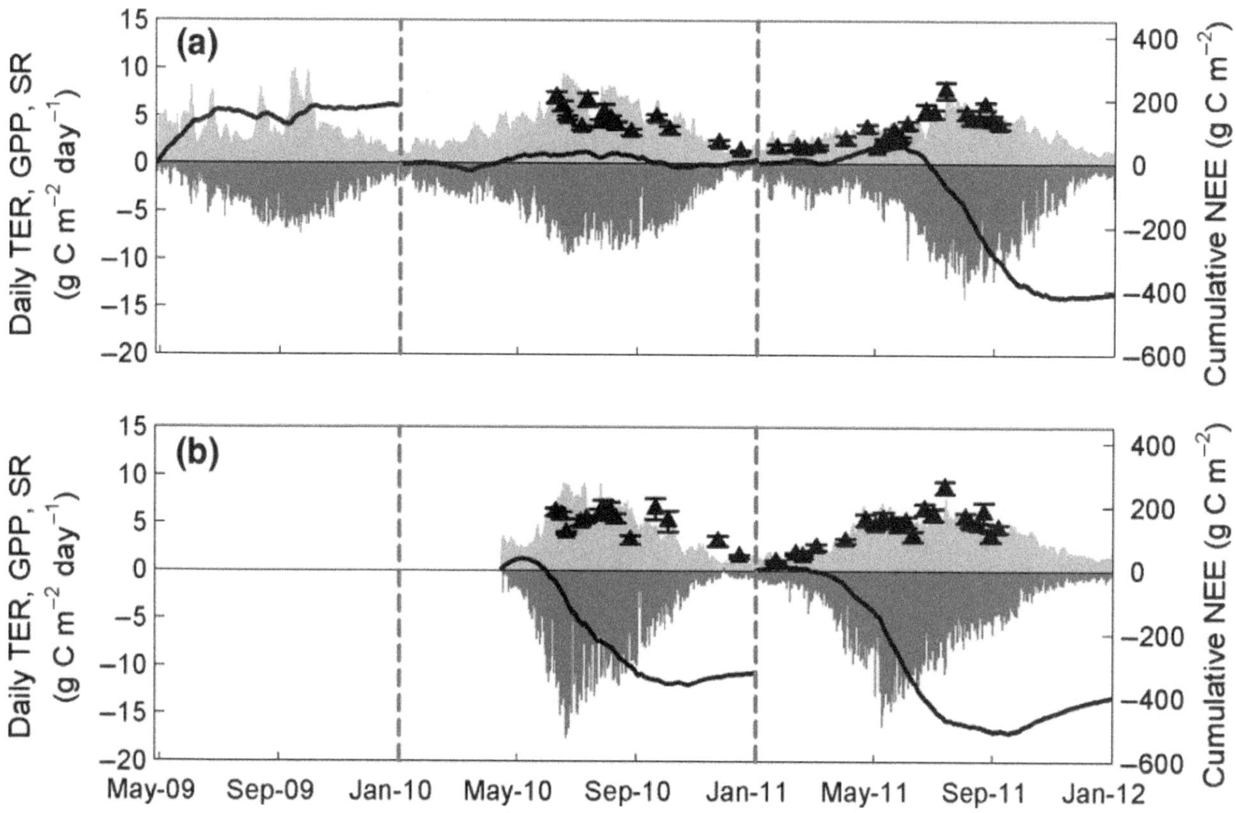

Figure 7. Daily total ecosystem respiration (TER) (light grey area), daily gross primary productivity (GPP) (dark grey area) and cumulative net ecosystem exchange (NEE) (solid black line) over the duration of the *Miscanthus* (a) and RCG (b) studies. Cumulative NEE was reset to zero at the end of each calendar year (indicated by a dashed vertical grey line). Measured soil respiratory (SR) fluxes are also shown (black triangles), representing below-ground respiration from both autotrophic and heterotrophic sources. Vertical bars represent the standard error of the measurement means.

The cumulative effect of the measured NEE fluxes over the course of the study is also shown in Fig. 7. In the *Miscanthus* crop, cumulative NEE was 183 ± 28 g C m^{-2} from late April to the end of December 2009, 13 ± 45 g C m^{-2} in 2010 and −411 ± 63 g C m^{-2} in 2011 (Table 3). GPP increased from −797 g C m^{-2} in 2009 to −1684 g C m^{-2} in 2011. Cumulative TER peaked in 2010 (1514 g C m^{-2}) and decreased in 2011 to 1273 g C m^{-2}, concomitant with below average summer temperatures. In the RCG crop, cumulative NEE values were −319 ± 57 g C m^{-2} from mid-April to the end of December 2010 and −397 ± 114 g C m^{-2} during 2011. An increase in GPP was observed from the conversion year (−1430 g C m^{-2}, 8½-month period) to the second year of establishment (−1708 g C m^{-2}), while TER also increased over this period from 1112 to 1311 g C m^{-2}.

Table 3. Annual cumulative fluxes of gross primary productivity (GPP), total ecosystem respiration (TER), net ecosystem exchange (NEE) and net ecosystem carbon balance (NECB) for the *Miscanthus* and RCG ecosystems

	GPP	TER	NEE* ± uncertainty	NECB† ± uncertainty
Miscanthus				
28th Apr 2009 – 31st Dec 2009 (Ploughing – year end)	−797	979	183 ± 28	−183 ± 28
2010	−1501	1514	13 ± 45	−13 ± 45
2011	−1684	1273	−411 ± 63	411 ± 63
RCG				
15th Apr 2010 – 31st Dec 2010 (Harrowing/planting – year end)	−1430	1112	−319 ± 57	66 ± 58
2011	−1708	1311	−397 ± 114	397 ± 114

*Positive NEE values represent a net release of C to the atmosphere, negative values a net uptake of C by the ecosystem.
†NECB = $-\Sigma$NEE − $C_{harvest}$ where $C_{harvest}$ is the carbon removed from the ecosystem at harvest.

The 'break-even' point at which cumulative NEE became negative (net C uptake) occurred in early October in 2010 and late June in 2011 for *Miscanthus* (Fig. 6). For RCG, the 'break-even' point occurred much earlier in the year than for *Miscanthus*: in early June in the conversion year (2010) and late February in 2011. With regard to winter fluxes, net C uptake did not cease in *Miscanthus* until mid-October and mid-November in 2010 and 2011, respectively, and the*Miscanthus* ecosystem subsequently remained relatively C-neutral until year end. In contrast, net C loss was observed from late October onwards in the conversion year for RCG and at a higher rate in 2011 from late September onwards.

The net rate of long-term C accumulation in an ecosystem is better described by the NECB. Since no removal of C through harvesting occurred during the *Miscanthus* experiment, the cumulative NEE values quoted for 2009, 2010 and 2011 are representative of the NECB of this ecosystem. For the RCG crop, the NECB was calculated as 66 ± 58 g C m^{-2} (a net C sink), while the NECB for the second year of RCG establishment was 397 ± 114 g C m^{-2}, as no biomass was harvested in this year. These values assume that contributions from leaching, methane, VOC, carbon monoxide and particulate carbon were negligible. Leaching of dissolved organic carbon (DOC) and dissolved inorganic carbon (DIC) are likely to be the most significant of these nonestimated fluxes (Osborne *et al.*, **2010**; Smith *et al.*, **2010**). Kindler *et al.* (**2011**) reported average DOC and biogenic DIC losses from European croplands of 4.1 ± 1.3 and 14.6 ± 4.8 g C m^{-2} y^{-1}, respectively. This study included an Irish cropland (Carlow) with well-drained, sandy loam soil which lost 2.6 ± 0.5 and 15.2 ± 4.1 g C m^{-2} y^{-1} as DOC and biogenic DIC, respectively. Following tillage disturbance of a grassland lysimeter at Johnstown Castle with comparable soil to the *Miscanthus* and RCG sites (low to medium drainage capacity), DOC leaching amounted to 1.6 g C m^{-2} over a 33-week period (Ó. Ní Choncubhair, B. Osborne, K. Richards and G. Lanigan, unpublished results). While such C losses comprise a small fraction of the annual NEE estimates for *Miscanthus* and RCG in 2011, their significance for the long-term NECB is far greater.

Based on chamber measurements, soil respiration (below-ground autotrophic and heterotrophic respiration) accounted for the majority of TER, with small contributions from leaf and stem respiration (Fig. 7). This was particularly evident in 2011 when significant below-ground biomass had developed in both crops. Soil respiration in *Miscanthus* varied from 1.3 ± 0.1 (SEM) g C m^{-2} d^{-1} in December 2010 to 7.7 ± 0.9 g C m^{-2} d^{-1} in July 2011. In RCG, fluxes ranged from 0.9 ± 0.1 in January 2011 to 8.7 ± 1.0 g C m^{-2} d^{-1} in July 2011. During intervals of active RCG growth and slow *Miscanthus* development, soil respiration in RCG exceeded that of *Miscanthus*, most notably in autumn 2010 and in April and May 2011. Outside of these periods, soil respiration fluxes were comparable in both ecosystems.

Controls on Ecosystem Carbon Fluxes

Net assimilation of carbon by the *Miscanthus* ecosystem was closely related to crop development during the growing season, as shown in Fig. 8(a). When *Miscanthus* GLAI values were 0.5 m^2 m^{-2} or lower, gross photosynthesis was cancelled out by a comparable ecosystem respiration rate and monthly sums of NEE fluctuated around zero. As leaf expansion progressed and a greater fraction of incident light was intercepted by the *Miscanthus* crop, large increases in monthly NEE were recorded, reaching a peak of −161.8 g C m^{-2} during August 2011.

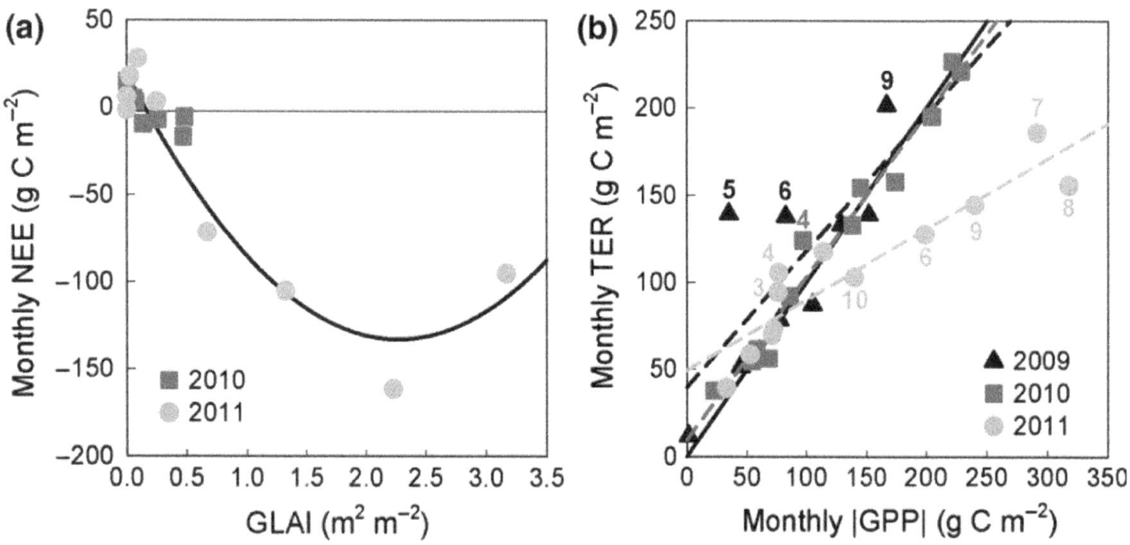

Figure 8. Monthly sums of net ecosystem exchange (NEE) in the *Miscanthus* ecosystem are plotted against green leaf area index (GLAI) for 2010 and 2011 in (a). Panel (b) shows the correlation for the *Miscanthus* ecosystem of monthly sums of total ecosystem respiration (TER) with monthly sums of gross primary productivity (GPP), plotted in absolute values (|GPP|) to facilitate the comparison. The black solid line indicates a 1 : 1 relationship while the dashed lines represent regression lines: 2009 (black line)$y = 0.78x + 39.6$, $R^2 = 0.58$; 2010 (dark grey line) $y = 0.93x + 9.6$, $R^2 = 0.97$; 2011 (light grey line) $y = 0.40x + 49.3$, $R^2 = 0.85$. Months showing deviations from the 1 : 1 line are marked with their respective month number in the calendar year.

A strong positive relationship was observed between monthly sums of TER and GPP for much of 2009 and 2010 and the winter months of 2011 (Fig. 8b). Notable deviations from a 1 : 1 linear relationship in May and June 2009 correspond to significant reductions in photosynthetic capacity after conventional inversion tillage. Monthly TER also exceeded monthly GPP in September 2009. This month was characterised by a 3-week period when rainfall was significantly reduced (total 1.2 mm) after a summer of exceptionally wet weather. In addition, respiration outweighed photosynthesis in March and April 2010 and in April 2011 following systemic herbicide application. However, the largest departures from slopes close to unity were observed in the growing season of 2011, when GPP was approximately 1.5 times greater than TER.

The relationship between monthly sums of TER and GPP in the RCG ecosystem is shown in Fig. 9(a) for 2010 and 2011. A strong positive association was observed between the two variables in both years, with respiratory fluxes representing roughly 50% of GPP across the full measurement period. The only months when TER exceeded GPP were April 2010 (after planting), November – December 2010 and October – December 2011. The impact of soil VWC on monthly TER fluxes in the RCG crop can be seen in Fig. 9(b). A negative relationship was evident in 2010 as VWC increased and reached values in excess of 0.4 $m^3\ m^{-3}$, presumably due to oxygen-limitation. Monthly sums of TER in 2011 (drier year) were comparable to 2010 values at high moisture contents, but when soil VWC decreased below a mid-range level (~ 0.26 $m^3\ m^{-3}$), a positive relationship was observed between TER and VWC.

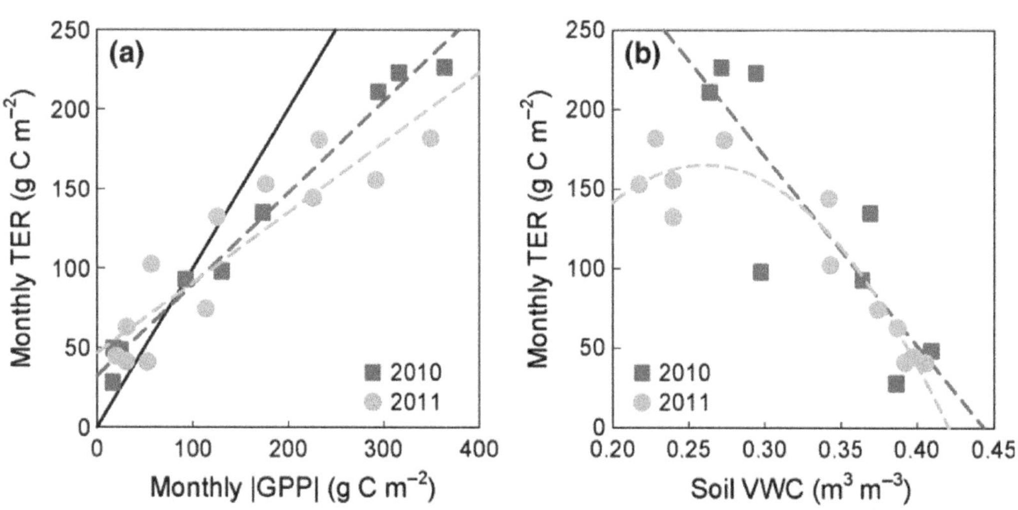

Figure 9. Correlation of monthly sums of total ecosystem respiration (TER) with monthly sums of gross primary productivity (GPP) for 2010 and 2011 in the RCG ecosystem (a). GPP is plotted in absolute values (|GPP|). The black solid line indicates a 1 : 1 relationship while the dotted lines represent regression lines: 2010 (dark grey line)$y = 0.57x + 32.1$, $R^2 = 0.98$; 2011 (light grey line) $y = 0.44x + 46.5$, $R^2 = 0.81$. Panel (b) shows the relationship between TER and soil volumetric water content (VWC) during 2010 and 2011 in the RCG ecosystem. The dashed lines represent regression lines: 2010 (dark grey line) $y = -1194.2x + 529.4$, $R^2 = 0.72$; 2011 (light grey line) $y = -6451.4x^2 + 3362.2x - 273.1$, $R^2 = 0.91$.

DISCUSSION

In the discussion below, we address the three questions posed, relating to (1) the transitional-phase C fluxes, (2) the postestablishment phase C fluxes associated with the *Miscanthus* and RCG crops and (3) the role of leaf longevity in regulating net C exchange. We discuss how our results compare with previous studies involving land-use transitions and examine the implications of our findings for bioenergy crop selection.

Carbon Fluxes during Land-Use Transition To Bioenergy Crops

Significant carbon emissions and changes in SOC have been attributed to land-use transitions such as deforestation and conversion to cropland (Davidson & Ackerman, **1993**; Houghton*et al.*, **2012**; Poeplau & Don, **2013**). Net C losses in the early stage of LUC can arise as a result of diminished photosynthetic capacity following herbicide application, enhanced mineralisation after tillage as well as reduced C uptake while the new crop develops to full maturity.

Table 4 summarises findings from previous research on C fluxes during land-use transitions. In a reseeding experiment in Johnstown Castle, cumulative TER from a grass sward with similar soil type to the bioenergy crop sites was 278 ± 29 g C m^{-2} in the month following herbicide application (Ó. Ní Choncubhair, B. Osborne, K. Richards and G. Lanigan, unpublished results). Including photosynthetic uptake (albeit at a diminishing rate) would reduce this value significantly. For example, Zenone *et al.* (**2011**) recorded a cumulative NEE of 19–46 g C m^{-2} in the month after herbicide application to long-term grassland. However, further variable net C losses occur in the posttillage period, dependent on the former land use, environmental conditions and tillage intensity. Cumulative NEE in the first 3 months posttillage in the*Miscanthus* ecosystem was 155.4 ± 8.9 g C m^{-2} and falls in the mid-range of estimated short-term tillage-induced emissions. These losses were primarily driven by minimal photosynthetic activity and slow above-ground development of the rhizome-propagated plants. This slow establishment is typical of Irish and many northern European trials where 3–5 years is generally required for full establishment (Lewandowski *et al.*, **2000**; Clifton-Brown *et al.*, **2007**).

Table 4. Transitional-phase and post-establishment phase C fluxes associated with different land-use transitions. Annual fluxes from permanent grasslands are included for comparison. Cumulative fluxes refer to net ecosystem exchange (NEE), total ecosystem respiration (TER) or soil respiration (SR). For transitional-phase fluxes, the time period refers to the sampling duration in months (M). For long-term fluxes, the length of establishment is given in years (Y), where Y1 is the conversion year

Original land use	New land use	Location	Time period	Cumulative NEE^A, TER^B or SR^C (g C m^{-2})	Research note	Reference
Transitional-phase fluxes						
Post-herbicide						
Grass	Grass	Ireland	1 M	278 ± 29^B	Summer reseeding	Unpublished results*
Grass	Soybean	USA	1 M	19–46^A		Zenone *et al.* (2011)
Post-tillage						
Grass	*Miscanthus*	Ireland	3 M	155 ± 9^A		This study
Wheat	Wheat	USA	2 M	29–41^C		Dao, 1998
Grass	Spring barley	Denmark	3 M	260^C		Eriksen & Jensen (2001)
Conversion year (Y1)						
Grass	*Miscanthus*	Ireland	8 M	183 ± 28^A	From tillage	This study
Arable	*Miscanthus*	USA	7 M	−58^A	From planting	Zeri *et al.* (2011)
Grass	RCG	Ireland	8.5 M	−319 ± 57^A	From planting	This study
Grass/Arable	RCG	Finland	5.3 M	−57^A	From 1.5M post-planting	Lind *et al.* (2015)
Grass/Arable	Poplar	Belgium	7 M	75 ± 4.4^A	From 2M post-planting	Zenone *et al.* (2015)
Arable	Poplar	Canada	12 M	312^A	From 4M post-tillage	Cai *et al.* (2011)
Grass/Wheat	Switchgrass	USA	12 M	−31^A	From 4M post-planting	Skinner & Adler (2010)
Grass	Soybean	USA	12 M	205–262^A	No-till	Zenone *et al.* (2013)
Post-establishment (long-term) fluxes						
Grass	*Miscanthus*	Ireland	Y2	13 ± 45^A		This study
Grass	*Miscanthus*	Ireland	Y3	−411 ± 63^A		This study
Arable	*Miscanthus*	USA	Y3	−554 ± 20^A		Zeri *et al.* (2011)
Grass	RCG	Ireland	Y2	−397 ± 114^A		This study
Grass/Arable	RCG	Finland	Y2, 3†	−259^A	Fertilised	Lind *et al.* (2015)
Drained peat	RCG	Finland	Y4–7	−9 to −211^A		Shurpali *et al.* (2009)
Drained peat	RCG	Estonia	Y4	−91 to −163^A‡	± Fertiliser	Mander *et al.* (2012)
Arable	Switchgrass	USA	Y3	−485 ± 20^A		Zeri *et al.* (2011)
NA	Switchgrass	USA	Y3, 4†	−448^A	Growing season only	Wagle *et al.*, 2015;
Grass/Arable	Poplar	Belgium	Y2	−96 ± 15^A		Zona *et al.* (2013)
Arable	Poplar	Canada	Y5	−17^A		Cai *et al.* (2011)
Permanent grassland						
Grass		Ireland	Annual	−200 to −385^A	Johnstown Castle; similar soil	Peichl *et al.* (2012)
Grass		Ireland	Annual	−193 to −258^A	South-west region	Jaksic *et al.* (2006)
Grass		Europe	Annual	−240 ± 70^A	9 sites	Soussana *et al.* (2007)

NA, not available.
*Órlaith Ní Choncubhair, Bruce Osborne, Karl Richards, Gary Lanigan.
†Mean value given for these years.
‡Chamber measurements.

Following RCG planting in April, net C losses of 33.3 ± 2.4 g C m^{-2} were recorded in the first month. However, the ecosystem became a strong net C sink after this. A number of factors may have contributed to this result. Firstly, the RCG site had been ploughed 1 year previous to crop establishment. Therefore, any increased mineralisation of SOC made available by tillage is most likely to have occurred in the months following the initial soil disturbance (Vellinga *et al.*, **2004**; Willems *et al.*, **2011**). Secondly, a high seeding rate was employed to guarantee successful establishment. Thirdly, RCG is considered a highly competitive plant that grows rapidly and pre-empts the development of other vegetation early in the growing season (Lavergne & Molofsky, **2004**). For example, Adams & Galatowitsch (**2005**) highlighted a shift from a low root:shoot ratio (< 1) in the first 4 months of development to a higher ratio (> 2) in the remainder of the 2-year study, which enabled RCG to initially monopolise above-ground space and later to spread vegetatively below ground.

Rapid early colonisation and crop development can therefore reduce the early transitional C losses associated with land-use change. However, our results suggest that significant net ecosystem C losses can be generated as a result of reduced photosynthetic assimilation and enhanced decomposition rates, particularly in C-rich grasslands. This loss represents a short-term land conversion carbon debt that must be overcome before net C sequestration is attained in the newly established bioenergy crop.

C Balance Implications Postestablishment

The *Miscanthus* ecosystem switched from a net C source of 183 ± 28 g C m^{-2} in the conversion year (8-month period) to C-neutral in the second year and a strong C sink in the third year (−411 ± 63 g C m^{-2}). In contrast, rapid and early development of the RCG crop resulted in the ecosystem being a net C sink both during the conversion year (−319 ± 57 g C m^{-2}, 8½-month period post-planting) and in the second year of establishment (−397 ± 114 g C m^{-2}). Previous assessments of C fluxes in the conversion year show highly variable results (Table 4), with cumulative NEE ranging from −58 to +312 g C m^{-2} depending on crop type and establishment technique, location and duration of measurements. Only a limited number of studies revealed small C sinks in the conversion year, similar to RCG in this study. Even after harvested biomass losses were incorporated, the NECB of this RCG ecosystem postplanting was still positive, indicating a net C sink of 66 ± 58 g C m^{-2} in the conversion year.

Net C fluxes in 2011 compare favourably with long-term fluxes from previous studies (Table 4), which range from close to C-neutral for poplar and RCG in certain years to in excess of −450 g C m^{-2} y^{-1} for *Miscanthus* and switchgrass. Good agreement exists between our results and the studies on *Miscanthus* in the USA (−554 ± 20 g C m^{-2} y^{-1}, Zeri *et al.*, **2011**) and RCG on mineral soil in Finland (−259 g C m^{-2} y^{-1}, Lind *et al.*, **2015**), taking account of location and comparative differences in season length and environmental controls. Studies focussed on the LUC transition from permanent grassland to *Miscanthus* or RCG are severely lacking, however, despite the fact that almost one-third of utilised agricultural land in Europe is grassland (Fischer*et al.*, **2010**).

In this study, the key driver of ecosystem C fluxes in *Miscanthus* was crop development, constraining monthly-cumulated NEE in the first 2 years close to zero as primary production was counterbalanced by ecosystem respiration. In addition to the inherent limitations of slow establishment in northern Europe, various other factors may have contributed to the poor agronomic performance observed including relatively late planting, possible rhizome failure and competition from understory grass species. However, in the third growing season, exponential above-ground growth and leaf expansion was accompanied by a 4-fold increase in below-ground biomass, resulting in strong net C uptake from June to October of that year.

In the case of RCG, environmental conditions and crop phenology were the main drivers of ecosystem C exchange. Since crop establishment was rapid and early canopy development was uninhibited by competition from other species, physiological activity was dictated by soil temperature, moisture and incident radiation. Net C uptake of −35 μmol CO_2 m^{-2} s^{-1} was observed just over two months after planting, culminating in a strong net C sink of −319 ± 57 g C m^{-2} in the conversion year and an increased net sink in the second year (−397 ± 114 g C m^{-2}). Furthermore, below-ground biomass represented a significant C stock of 3.4 t C ha^{-1}, similar to the results of Xiong & Kätterer (**2010**).

During periods when RCG productivity exceeded that of *Miscanthus* (autumn of the conversion year and early in its second year), higher soil respiration (below-ground autotrophic and heterotrophic) was recorded in RCG than *Miscanthus*. This may indicate that there was a larger contribution from the autotrophic component to total below-ground respiration in RCG at these times. Soil respiration fluxes were similar in both crops outside of these periods. In a Finnish RCG crop which had 70% of total plant biomass below ground during the peak growth period, autotrophic respiration was the dominant component (about 55%) of TER (Shurpali*et al.*, **2008**). Soil respiration constituted the majority of TER in this study, particularly in the final year. While recognising the methodological differences between chamber and EC estimates, this result highlights the significant and increasing contribution of below-ground processes to C cycling in these developing perennial crops. Enhanced below-ground C allocation and root-associated C cycling may favour long-term soil carbon sequestration (Xiong & Kätterer, **2010**; Anderson-Teixeira *et al.*, **2013**), boosted also by the absence of regular soil disturbance through tillage (Freibauer *et al.*, **2004**).

For the early postestablishment phase, our findings suggest that RCG provides a more favourable C balance than *Miscanthus*. However, the outperformance of RCG by *Miscanthus* in the third growth year, both in terms of yield and net C accumulation, indicates that the C sink potential of *Miscanthus* will improve further as the crop matures and above-ground productivity approaches the maximum yield.

Leaf longevity in *Miscanthus* and RCG

Light-use efficiency is a critical determinant of primary production and crop yield (Monteith, **1977**) and is closely related to the crop's ability to maintain a closed canopy during the growing season (Beale & Long, **1995**). Greater leaf longevity and higher leaf area have been demonstrated in *Miscanthus* compared with C_4 maize (Dohleman & Long, **2009**). However, little information exists on the relative performance of *Miscanthus* when compared to native C_3 plants well-adapted to the temperate conditions of northern Europe.

In this study, strong coupling of GPP and crop phenology was highlighted by a striking 3-month difference in the timing of peak C assimilation in RCG (May) and *Miscanthus* (August) in 2011. RCG emerged early in the spring and achieved net C uptake (the 'break-even' or compensation point) by late February. This is consistent with the temperate climatic conditions of the current study and the associated long growing season. However, the subsequent occurrence of peak RCG productivity in mid-May resulted in the early onset of senescence in August. Crop phenology, therefore, introduced an asynchrony between the timing of maximum leaf area and maximum solar radiation for RCG, which could have implications for the ability of RCG to fully exploit peak irradiances and to maximise yields under some environmental conditions. Furthermore, RCG TER fluxes tended to exceed GPP in the winter months and reduced the cumulative C sink by 113 g C m^{-2} in the second year of establishment.

In contrast with RCG which exhibited early season leaf development, significant increases in leaf area did not occur until June for *Miscanthus*. As a result, the 'break-even' point of net C uptake did not occur until late June in 2011. However, *Miscanthus* maintained growth and substantial leaf area late into the autumn and sustained net ecosystem C accumulation until mid-November. After this, the crop remained C-neutral until the end of the year, similar to the findings of Zeri *et al.* (**2011**). Indeed, strong coupling of TER and GPP was observed during much of the current study, which meant that significant C losses did not occur.

Dohleman & Long (**2009**) demonstrated a 59% longer growing season in *Miscanthus* (199 days on average) than C_4 maize (126 days on average). Our results show that the longevity of photosynthetically active leaves in *Miscanthus* is comparable even to native C_3 crops. Indeed, the number of recorded days with net C uptake in 2011 was almost identical for both crops (215 and 212 days for *Miscanthus* and RCG, respectively). This result compares very favourably with an average cropping season length of 149 days in Irish spring barley (Davis *et al.*, **2010**) and 212 days of net C uptake (six-year average) in an Irish grassland (Peichl *et al.*, **2011**). This further highlights the exceptional performance of C_4 *Miscanthus* in cool temperate climates where C_4 metabolism should be temperature-limited and implies that *Miscanthus* may confer an advantage as a bioenergy crop in the long term. If leaf photosynthetic capacity and longevity is comparable to native C_3 crops and C emissions outside of the growing season are close to zero, the net C balance of *Miscanthus* is likely to be favourable.

Implications of the study

Bioenergy crop cultivation in northern Europe may focus more on grassland conversion to avoid a reduction in the area of croplands dedicated to food and feed production. However, permanent grasslands in northern Europe show strong annual net C uptake, with NEE values ranging from −193 to −385 g C m^{-2} y^{-1} (Table 4). This highlights the potential negative impacts associated with disturbing grasslands that are highly productive and supply substantial amounts of stabilised C to the soil (Jackson *et al.*, **1996**; Jones & Donnelly, **2004**; Poeplau *et al.*,**2011**). Our study demonstrated that a significant C debt can be associated with the early-establishment phase of these bioenergy crops but highlighted the future potential of*Miscanthus* to surpass RCG and possibly long-term grasslands in terms of its C sink strength. Additional measurements of other associated GHGs, such as N_2O and CH_4, will be necessary, however, to assess the full GHG implications of land-use change to this crop.

Furthermore, the duration of the full crop production cycle must be considered. In the case of*Miscanthus*, productive yields may be achievable for up to 15 years or more (Clifton-Brown*et al.*, **2007**; Christian *et al.*, **2008**; Arundale *et al.*, **2014**) but this is likely to be much shorter for RCG (approximately 7–10 years; Saijonkari-Pahkala, **2001**; Finnan, **2007**). Therefore, more cultivation and replanting will be required in long-term RCG plantations and this has associated C balance implications due to more regular soil disturbance and concomitant reductions in productivity.

Although the theoretical light-use efficiency benefits afforded by the C_4 photosynthetic pathway are often highlighted, the empirical results of this work suggest that high biomass productivity will be controlled more by leaf- or canopy-related factors, both genetic and environmentally derived, rather than the photosynthetic characteristics of individual leaves. Further to this, the distinct difference in the timing of peak C uptake between the two crops is significant and provides information on the suitability of these crops in different climatic zones. While the late season performance of *Miscanthus* may constrain its productivity in regions with a short growing season, RCG may be a better candidate in these regions due to its early emergence in spring and subsequent rapid development. Growing *Miscanthus* under a clear, plastic film, as is common practice for maize production in parts of northern Europe, may also be a valuable tool to encourage earlier emergence in spring and enhance further the duration of net C uptake in the crop (Clifton-Brown *et al.*, **2011**).

A final point worth noting is the relative allocation of biomass above and below ground in these two bioenergy crops. At the time of peak biomass yield in 2011, more than 50% of total RCG biomass was below ground compared with 24% in *Miscanthus* and above-ground biomass in*Miscanthus* was almost double that recorded in RCG. Greater investment of resources below ground could enhance long-term C sequestration; however, lower above-ground yields have significant implications for the economic viability and C-offsetting potential of the RCG crop.

ACKNOWLEDGEMENTS

This research was funded by the Department of Agriculture, Food and the Marine Research Stimulus Fund (Project Ref. 07 527). The authors also gratefully acknowledge the technical assistance of Brendan Swan, Kevin McNamara, Vincent Staples, Carmel O' Connor and Teresa Cowman in Johnstown Castle.

REFERENCES

1. Adams CR, Galatowitsch SM (2005) *Phalaris arundinacea* (reed canary grass): rapid growth and growth pattern in conditions approximating newly restored wetlands. *Ecoscience*, **12**, 569–573.
2. Ammann C, Flechard C, Leifeld J, Neftel A, Fuhrer J (2007) The carbon budget of newly established temperate grassland depends on management intensity. *Agriculture Ecosystems & Environment*, **121**, 5–20.
3. Anderson-Teixeira K, Masters M, Black C, Zeri M, Hussain M, Bernacchi C, Delucia E (2013) Altered belowground carbon cycling following land-use change to perennial bioenergy crops. *Ecosystems*, **16**,508–520.
4. Arundale R, Dohleman F, Heaton E, McGrath J, Voigt T, Long S (2014) Yields of *Miscanthus × giganteus* and*Panicum virgatum* decline with stand age in the Midwestern USA. *Global Change Biology Bioenergy*, **6**, 1–13.
5. Aubinet M, Chermanne B, Vandenhaute M, Longdoz B, Yernaux M, Laitat E (2001) Long term carbon dioxide exchange above a mixed forest in the Belgian Ardennes. *Agricultural and Forest Meteorology*, **108**,293–315.
6. Baum K, Ham J, Brunsell N, Coyne P (2008) Surface boundary layer of cattle feedlots: implications for air emissions measurement. *Agricultural and Forest Meteorology*, **148**, 1882–1893.
7. Beale C, Long S (1995) Can perennial C_4 grasses attain high efficiencies of radiant energy conversion in cool climates? *Plant, Cell and Environment*, **18**, 641–650.
8. Beale CV, Bint DA, Long SP (1996) Leaf photosynthesis in the C_4-grass *Miscanthus × giganteus*, growing in the cool temperate climate of southern England. *Journal of Experimental Botany*, **47**, 267–273.
9. Black K, Bolger T, Davis P *et al.* (2007) Inventory and eddy covariance-based estimates of annual carbon sequestration in a Sitka spruce (*Picea sitchensis* (Bong.) Carr.) forest ecosystem. *European Journal of Forest Research*, **126**, 167–178.
10. Cai T, Price D, Orchansky A, Thomas B (2011) Carbon, water, and energy exchanges of a hybrid poplar plantation during the first five years following planting. *Ecosystems*, **14**, 658–671.
11. Campbell G, Norman J (1989) The description and measurement of plant canopy structure. In: Plant Canopies: Their Growth, Form and Function (eds Russell G, Marshall B, Jarvis P), pp. 1–19. Cambridge University Press, Cambridge, UK.
12. Central Statistics Office (2014) Crops and Livestock Survey June 2013 - Final Results. Central Statistics Office, Cork, Ireland.
13. Chapin F, Woodwell G, Randerson J *et al.* (2006) Reconciling carbon-cycle concepts, terminology, and methods. *Ecosystems*, **9**, 1041–1050.
14. Christian D, Riche A, Yates N (2008) Growth, yield and mineral content of *Miscanthus × giganteus* grown as a biofuel for 14 successive harvests. *Industrial Crops and Products*, **28**, 320–327.
15. Clifton-Brown J, Neilson B, Lewandowski I, Jones M (2000) The modelled productivity of *Miscanthus ×giganteus* (GREEF et DEU) in Ireland. *Industrial Crops and Products*, **12**, 97–109.
16. Clifton-Brown J, Breuer J, Jones M (2007) Carbon mitigation by the energy crop, *Miscanthus*. *Global Change Biology*, **13**, 2296–2307.
17. Clifton-Brown J, Robson P, Sanderson R, Hastings A, Valentine J, Donnison I (2011) Thermal requirements for seed germination in *Miscanthus* compared with switchgrass (*Panicum virgatum*), reed canary grass (*Phalaris arundinaceae*), maize (*Zea mays*) and perennial ryegrass (*Lolium perenne*). *GCB Bioenergy*, **3**,375–386.
18. Dao T (1998) Tillage and crop residue effects on carbon dioxide evolution and carbon storage in a Paleustoll. *Soil Science Society of America Journal*, **62**, 250–256.
19. Davidson E, Ackerman I (1993) Changes in soil carbon inventories following cultivation of previously untilled soils. *Biogeochemistry*, **20**, 161–193.
20. Davis P, Brown J, Saunders M *et al.* (2010) Assessing the effects of agricultural management practices on carbon fluxes: spatial variation and the need for replicated estimates of net ecosystem exchange.*Agricultural and Forest Meteorology*, **150**, 564–574.
21. Dohleman F, Long S (2009) More productive than maize in the Midwest: how does *Miscanthus* do it? *Plant Physiology*, **150**, 2104–2115.
22. Don A, Osborne B, Hastings A *et al.* (2012) Land-use change to bioenergy production in Europe: implications for the greenhouse gas balance and soil carbon. *Global Change Biology Bioenergy*, **4**, 372–391.
23. Donnelly A, Styles D, Fitzgerald J, Finnan J (2011) A proposed framework for determining the environmental impact of replacing agricultural grassland with *Miscanthus* in Ireland. *Global Change Biology Bioenergy*, **3**, 247–263.
24. Eriksen J, Jensen L (2001) Soil respiration, nitrogen mineralization and uptake in barley following cultivation of grazed grasslands. *Biology and Fertility of Soils*, **33**, 139–145.
25. EU (2009) Directive 2009/28/EC of the European Parliament and of the Council of 23 April 2009 on the promotion of the use of energy from renewable sources and amending and subsequently repealing Directives 2001/77/EC and 2003/30/EC. EU, Brussels.
26. Falge E, Baldocchi D, Olson R *et al.* (2001) Gap filling strategies for defensible annual sums of net ecosystem exchange. *Agricultural and Forest Meteorology*, **107**, 43–69.
27. Fargione J, Hill J, Tilman D, Polasky S, Hawthorne P (2008) Land clearing and the biofuel carbon debt. *Science*, **319**, 1235–1238.
28. Finnan J (2007) Reed canary grass. In: Fact Sheet Tillage No. 7. Teagasc, Carlow, Ireland.
29. Finnan J, Burke B (2014) Nitrogen dynamics in a mature *Miscanthus × giganteus* crop fertilized with nitrogen over a five year period. *Irish Journal of Agricultural and Food Research*, **53**, 171–188.

30. Fischer G, Prieler S, van Velthuizen H, Berndes G, Faaij A, Londo M, de Wit M (2010) Biofuel production potentials in Europe: sustainable use of cultivated land and pastures, Part II: land use scenarios. *Biomass and Bioenergy*, **34**, 173–187.

31. Freibauer A, Rounsevell M, Smith P, Verhagen J (2004) Carbon sequestration in the agricultural soils of Europe. *Geoderma*, **122**, 1–23.

32. Garratt J (1992) The Atmospheric Boundary Layer. Cambridge University Press, Cambridge, UK.

33. Goulden M, Munger J, Fan S, Daube B, Wofsy S (1996) Measurements of carbon sequestration by long-term eddy covariance: methods and a critical evaluation of accuracy. *Global Change Biology*, **2**, 169–182.

34. Guo L, Gifford R (2002) Soil carbon stocks and land use change: a meta analysis. *Global Change Biology*, **8**,345–360.

35. Haslwanter A, Hammerle A, Wohlfahrt G (2009) Open-path vs. closed-path eddy covariance measurements of the net ecosystem carbon dioxide and water vapour exchange: a long-term perspective.*Agricultural and Forest Meteorology*, **149**, 291–302.

36. Houghton R, House J, Pongratz J *et al.* (2012) Carbon emissions from land use and land-cover change. *Biogeosciences*, **9**, 5125–5142.

37. Huyghe C, De Vliegher A, van Gils B, Peeters A (2014) Grasslands and Herbivore Production in Europe and Effects of Common Policies. Éditions Quae, Versailles, France.

38. Jackson R, Canadell J, Ehleringer J, Mooney H, Sala O, Schulze E (1996) A global analysis of root distributions for terrestrial biomes. *Oecologia*, **108**, 389–411.

39. Jaksic V, Kiely G, Albertson J, Oren R, Katul G, Leahy P, Byrne KA (2006) Net ecosystem exchange of grassland in contrasting wet and dry years. *Agricultural and Forest Meteorology*, **139**, 323–334.

40. Jones M, Donnelly A (2004) Carbon sequestration in temperate grassland ecosystems and the influence of management, climate and elevated CO_2. *New Phytologist*, **164**, 423–439.

41. Kaimal J, Finnigan J (1994) Atmospheric Boundary Layer Flows: Their Structure and Measurement. Oxford University Press, Oxford, UK.

42. Kaimal J, Izumi Y, Wyngaard J, Coté R (1972) Spectral characteristics of surface-layer turbulence. *Quarterly Journal of the Royal Meteorological Society*, **98**, 563–589.

43. Kandel T, Elsgaard L, Karki S, Laerke P (2013) Biomass yield and greenhouse gas emissions from a drained fen peatland cultivated with reed canary grass under different harvest and fertilizer regimes. *Bioenergy Research*, **6**, 883–895.

44. Kindler R, Siemens J, Kaiser K *et al.* (2011) Dissolved carbon leaching from soil is a crucial component of the net ecosystem carbon balance. *Global Change Biology*, **17**, 1167–1185.

45. Kljun N, Kormann R, Rotach M, Meixner F (2003) Comparison of the Langrangian footprint model LPDM-B with an analytical footprint model. *Boundary-Layer Meteorology*, **106**, 349–355.

46. Kljun N, Calanca P, Rotach M, Schmid H (2004) A simple parameterisation for flux footprint predictions. *Boundary-Layer Meteorology*, **112**, 503–523.

47. Kormann R, Meixner F (2001) An analytical footprint model for non-neutral stratification. *Boundary-Layer Meteorology*, **99**, 207–224.

48. Landström S, Lomakka L, Anderson S (1996) Harvest in spring improves yield and quality of reed canary grass as a bioenergy crop. *Biomass and Bioenergy*, **11**, 333–341.

49. Lavergne S, Molofsky J (2004) Reed canary grass (*Phalaris arundinacea*) as a biological model in the study of plant invasions. *Critical Reviews in Plant Sciences*, **23**, 415–429.

50. Lewandowski I, Clifton-Brown J, Scurlock JMO, Huisman W (2000) *Miscanthus*: European experience with a novel energy crop. *Biomass and Bioenergy*, **19**, 209–227.

51. Lewandowski I, Scurlock J, Lindvall E, Christou M (2003) The development and current status of perennial rhizomatous grasses as energy crops in the US and Europe. *Biomass and Bioenergy*, **25**, 335–361.

52. Lind S, Shurpali N, Peltola O *et al.* (2015) Carbon dioxide exchange of a perennial bioenergy crop cultivation on a mineral soil. *Biogeosciences Discussions*, **12**, 16673–16708.

53. Lloyd J, Taylor J (1994) On the temperature dependence of soil respiration. *Functional Ecology*, **8**, 315–323.

54. Long S (1983) C_4 photosynthesis at low temperatures. *Plant Cell and Environment*, **6**, 345–363.

55. Mander U, Jarveoja J, Maddison M, Soosaar K, Aavola R, Ostonen I, Salm J (2012) Reed canary grass cultivation mitigates greenhouse gas emissions from abandoned peat extraction areas. *Global Change Biology Bioenergy*, **4**, 462–474.

56. Mauder M, Foken T (2004) Documentation and instruction manual of the eddy covariance software package TK2. Universitat Bayreuth, Abt. Mikrometeologie, Arbeitsergebnisse Nr. 26, Germany.

57. McCalmont JP, Hastings A, McNamara NP, Richter GM, Robson P, Donnison IS, Clifton-Brown J (2016)Environmental costs and benefits of growing *Miscanthus* for bioenergy in the UK. *GCB Bioenergy*. doi:10.1111/gcbb.12294.

58. McLaughlin S, Walsh M (1998) Evaluating environmental consequences of producing herbaceous crops for bioenergy. *Biomass and Bioenergy*, **14**, 317–324.

59. Meyers T, Hollinger S (2004) An assessment of storage terms in the surface energy balance of maize and soybean. *Agricultural and Forest Meteorology*, **125**, 105–115.

60. Moncrieff J, Massheder J, Debruin H *et al.* (1997) A system to measure surface fluxes of momentum, sensible heat, water vapour and carbon dioxide. *Journal of Hydrology*, **189**, 589–611.

61. Monteith J (1977) Climate and the efficiency of crop production in Britain. *Philosophical Transactions of the Royal Society B: Biological Sciences*, **281**, 277–294.

62. Monteith J (1978) Reassessment of maximum growth rates for C_3 and C_4 crops. *Experimental Agriculture*,**14**, 1–5.

63. Moureaux C, Ceschia E, Arriga N, Béziat P, Eugster W, Kutsch W, Pattey E (2012) Eddy covariance measurements over crops. In: Eddy Covariance: A Practical Guide to Measurement and Data Analysis (edsAubinet M, Vesala T, Papale D), pp. 319–331. Springer Atmospheric Sciences, New York, NY.

64. Munro D, Oke T (1975) Aerodynamic boundary-layer adjustment over a crop in neutral stability. *Boundary-Layer Meteorology*, **9**, 53–61.

65. Naidu S, Long S (2004) Potential mechanisms of low-temperature tolerance of C_4 photosynthesis in*Miscanthus* × *giganteus*: an in vivo analysis. *Planta*, **220**, 145–155.

66. Oncley S, Foken T, Vogt R *et al.* (2007) The Energy Balance Experiment EBEX-2000. Part I: overview and energy balance. *Boundary-Layer Meteorology*, **123**, 1–28.

67. Osborne B, Saunders M, Walmsley D, Jones M, Smith P (2010) Key questions and uncertainties associated with the assessment of the cropland greenhouse gas balance. *Agriculture Ecosystems & Environment*, **139**,293–301.

68. Özdemir E, Härdtlein M, Eltrop L (2009) Land substitution effects of biofuel side products and implications on the land area requirement for EU 2020 biofuel targets. *Energy Policy*, **37**, 2986–2996.

69. Peichl M, Leahy P, Kiely G (2011) Six-year stable annual uptake of carbon dioxide in intensively managed humid temperate grassland. *Ecosystems*, **14**, 112–126.

70. Peichl M, Carton O, Kiely G (2012) Management and climate effects on carbon dioxide and energy exchanges in a maritime grassland. *Agriculture Ecosystems & Environment*, **158**, 132–146.

71. Poeplau C, Don A (2013) Sensitivity of soil organic carbon stocks and fractions to different land-use changes across Europe. *Geoderma*, **192**, 189–201.

72. Poeplau C, Don A (2014) Soil carbon changes under *Miscanthus* driven by C_4 accumulation and C_3decompostion - toward a default sequestration function. *Global Change Biology Bioenergy*, **6**, 327–338.

73. Poeplau C, Don A, Vesterdal L, Leifeld J, van Wesemael B, Schumacher J, Gensior A (2011) Temporal dynamics of soil organic carbon after land-use change in the temperate zone - carbon response functions as a model approach. *Global Change Biology*, **17**, 2415–2427.

74. Qin Z, Dunn JB, Kwon H, Mueller S, Wander MM (2016) Soil carbon sequestration and land use change associated with biofuel production: empirical evidence. *GCB Bioenergy* **8**, 66–80.

75. Reichstein M, Falge E, Baldocchi D *et al.* (2005) On the separation of net ecosystem exchange into assimilation and ecosystem respiration: review and improved algorithm. *Global Change Biology*, **11**,1424–1439.

76. Ruppert J, Mauder M, Thomas C, Luers J (2006) Innovative gap-filling strategy for annual sums of CO_2 net ecosystem exchange. *Agricultural and Forest Meteorology*, **138**, 5–18.

77. Saijonkari-Pahkala K (2001) Non-wood plants as raw material for pulp and paper. *Agricultural and Food Science*, **10**, 1–101.

78. Scott HD (2000) Soil Physics: Agricultural and Environmental Applications. Iowa State University Press,Ames, IA.

79. Shurpali N, Hyvonen N, Huttunen J, Biasi C, Nykanen H, Pekkarinen N, Martikainen P (2008) Bare soil and reed canary grass ecosystem respiration in peat extraction sites in Eastern Finland. *Tellus Series B-Chemical and Physical Meteorology*, **60**, 200–209.

80. Shurpali N, Hyvonen N, Huttunen J *et al.* (2009) Cultivation of a perennial grass for bioenergy on a boreal organic soil - carbon sink or source? *Global Change Biology Bioenergy*, **1**, 35–50.

81. Sims R, Mabee W, Saddler J, Taylor M (2010) An overview of second generation biofuel technologies. *Bioresource Technology*, **101**, 1570–1580.

82. Skinner R, Adler P (2010) Carbon dioxide and water fluxes from switchgrass managed for bioenergy production. *Agriculture Ecosystems & Environment*, **138**, 257–264.

83. Smith P, Lanigan G, Kutsch W *et al.* (2010) Measurements necessary for assessing the net ecosystem carbon budget of croplands. *Agriculture Ecosystems & Environment*, **139**, 302–315.

84. Soussana J, Allard V, Pilegaard K *et al.* (2007) Full accounting of the greenhouse gas (CO_2, N_2O, CH_4) budget of nine European grassland sites. *Agriculture Ecosystems & Environment*, **121**, 121–134.

85. Stoy P, Mauder M, Foken T *et al.* (2013) A data-driven analysis of energy balance closure across FLUXNET research sites: the role of landscape scale heterogeneity. *Agricultural and Forest Meteorology*, **171**, 137–152.

86. Twine T, Kustas W, Norman J *et al.* (2000) Correcting eddy-covariance flux underestimates over a grassland. *Agricultural and Forest Meteorology*, **103**, 279–300.

87. Vanderborght J, Graf A, Steenpass C *et al.* (2010) Within-field variability of bare soil evaporation derived from eddy covariance measurements. *Vadose Zone Journal*, **9**, 943–954.

88. Vellinga T, van den Pol-van Dasselaar A, Kuikman P (2004) The impact of grassland ploughing on CO_2 and N_2O emissions in the Netherlands. *Nutrient Cycling in Agroecosystems*, **70**, 33–45.

89. Vickers D, Mahrt L (1997) Quality control and flux sampling problems for tower and aircraft data. *Journal of Atmospheric and Oceanic Technology*, **14**, 512–526.

90. Wagle P, Kakani V, Huhnke R (2015) Net ecosystem carbon dioxide exchange of dedicated bioenergy feedstocks:

switchgrass and high biomass sorghum. *Agricultural and Forest Meteorology*, **207**, 107–116.

91. Wang D, Portis A, Moose S, Long S (2008) Cool C_4 photosynthesis: pyruvate P_i dikinase expression and activity corresponds to the exceptional cold tolerance of carbon assimilation in *Miscanthus × giganteus.Plant Physiology*, **148**, 557–567.

92. Webb E, Pearman G, Leuning R (1980) Correction of flux measurements for density effects due to heat and water-vapor transfer. *Quarterly Journal of the Royal Meteorological Society*, **106**, 85–100.

93. Willems A, Augustenborg C, Hepp S, Lanigan G, Hochstrasser T, Kammann C, Muller C (2011) Carbon dioxide emissions from spring ploughing of grassland in Ireland. *Agriculture Ecosystems & Environment*,**144**, 347–351.

94. Wilson K, Goldstein A, Falge E *et al.* (2002) Energy balance closure at FLUXNET sites. *Agricultural and Forest Meteorology*, **113**, 223–243.

95. Wrobel C, Coulman B, Smith D (2009) The potential use of reed canarygrass (*Phalaris arundinacea* L.) as a biofuel crop. *Acta Agriculturae Scandinavica Section B-Soil and Plant Science*, **59**, 1–18.

96. Xiong S, Kätterer T (2010) Carbon-allocation dynamics in reed canary grass as affected by soil type and fertilization rates in northern Sweden. *Acta Agriculturae Scandinavica Section B-Soil and Plant Science*, **60**,24–32.

97. Zenone T, Chen J, Deal M *et al.* (2011) CO_2 fluxes of transitional bioenergy crops: effect of land conversion during the first year of cultivation. *Global Change Biology Bioenergy*, **3**, 401–412.

98. Zenone T, Gelfand I, Chen J, Hamilton S, Robertson G (2013) From set-aside grassland to annual and perennial cellulosic biofuel crops: effects of land use change on carbon balance. *Agricultural and Forest Meteorology*, **182**, 1–12.

99. Zenone T, Fischer M, Arriga N *et al.* (2015) Biophysical drivers of the carbon dioxide, water vapor, and energy exchanges of a short-rotation poplar coppice. *Agricultural and Forest Meteorology*, **209**, 22–35.

100. Zeri M, Anderson-Teixeira K, Hickman G, Masters M, Delucia E, Bernacchi C (2011) Carbon exchange by establishing biofuel crops in Central Illinois. *Agriculture Ecosystems & Environment*, **144**, 319–329.

101. Zimmermann J, Dauber J, Jones M (2012) Soil carbon sequestration during the establishment phase of Miscanthus × giganteus: a regional-scale study on commercial farms using 13C natural abundance. *GCB Bioenergy*, **4**, 453–461.

102. Zona D, Janssens I, Aubinet M, Gioli B, Vicca S, Fichot R, Ceulemans R (2013) Fluxes of the greenhouse gases (CO_2, CH_4 and N_2O) above a short-rotation poplar plantation after conversion from agricultural land. *Agricultural and Forest Meteorology*, **169**, 100–110.

CHAPTER 7

IMPACTS OF TREATED WASTEWATER REUSE ON SOME SOIL PROPERTIES AND PRODUCTION OF GLADIOLUS COMMUNIS

Hebah I. Hasan, Anwar M. Battikhi, Mostafa M. Qrunfleh

Keywords: Treated wastewater; Potable water; Aggregate stability; Soil water retention curve; Pore size distribution curve; Gladiolus; Spikes; Corms; Cormels

INTRODUCTION

Jordan is classified as an arid to semi-arid region, with a land area of approximately 89 thousand km2 . Jordan has been experiencing an imbalance in the population-water resources equation. Whereas water resources in Jordan have fluctuated around a stationary average, the country's population has continued to rise due to the sudden influx of immigrants from neighboring countries due to conflicts and unrests. The current use of water in Jordan already exceeds renewable supply. The deficit is covered by the unsustainable practice of overdrawing from the highland aquifers, resulting in lowering water tables and declining quality (MWI, 2007). The reuse of treated wastewater for irrigation is a valuable strategy to maximize available water resources, but the quality of this water can present agricultural challenges.

A range of cultivated crops are grown using irrigated wastewater; mainly citrus, vegetables, field crops, bananas and cut flowers. The global flower industry is currently an important sector in the world's economy. Despite its small size, Jordan cut flower industry gained several advantages; mainly due to its geographical location and climate.

Gladiolus (family: iridaceae) is represented by 180 species. It is found throughout Africa and the Mediterranean area, with the greatest concentration in southern Africa (Wilfret, 1980). The gladiolus corm lives for one year only, but a new corm is formed each year on top of the old one. Little research was done on cut flower production using TWW irrigation. Moreover, it is important to evaluate the effects of the reuse of TWW on soil properties. Rusan et al. (2008) assessed the effect of TWW irrigation of cut flower on plant growth and soil and plant quality parameters. The plants were irrigated with fresh water (FW) or with TWW at three irrigation frequencies. The EC values for FW and TWW were 1.05 dS/m and 3.27 dS/m, respectively. The results indicated variation in soil salinity and other soil chemical properties with TWW irrigation. Yield responded positively to the TWW irrigation with no negative impact on the quality of the produce. Nirit et al. (2006) investigated the effect TWW irrigation of roses cultivated in two soil-less medium, perlite and choir, on rose's yields and quality. Water treatments used in the experiment were drinking water (DW) and TWW. The EC for water treatments are 1.15 dS/m and 2.5 dS/m respectively. The visible appearance of the plants, their growth, the quantity and size of the flowering stems were not affected by the irrigation treatments. Friedman et al.

Citation: Hebah I. Hasan, Anwar M. Battikhi and Mostafa M. Qrunfleh, Impacts of Treated Wastewater Reuse on Some Soil Properties and Production of Gladiolus Communis, DOI: 10.4172/2376-0354.1000111.

(2007) also examined the effects of DW and TWW irrigation (with EC equal to 0.6 dS/m for PW and 7.8 dS/m for TWW) on the chemical composition of soil and leaves, flower yield and postharvest performance of sunflower and Celosia. The results demonstrated a higher accumulation of Cl, Na, B, and NO_3 - in soils irrigated with TWW as compared to DW. The aim of this research was to evaluate the short-term influence of applying TWW on some soil chemical and physical properties and to study the yield and quality of Gladiolus communis, irrigated with TWW.

MATERIALS AND METHODS

Experiments Site and Experimental Design

The experiments were conducted in Al-Jubeiha Agricultural Research Station located at the University of Jordan. The site has an altitude of 991 m above sea level, and it is at 35.873 o E longitude and at 32.012o N latitude. The location is characterized by semi-arid climate with warm summers and temperature ranging from 25◦ C to 33◦ C and relatively short and cold winters with mean annual precipitation about 492.8 mm. The experiments were conducted inside the greenhouse (GH) and in the open field (OF). The experimental design in both sites was Randomized Complete Block Design (RCBD). In the GH experiment (winter season), the corms were planted in pots with dimension of 35cm height, 30cm upper diameter, and 20 cm the lower diameter. Ten pots were used for each treatment in each block. GH temperature and relative humidity were 30o C and 25% respectively. In the OF experiment (summer season), the field was divided into three blocks with dimension of 3.60m by 1.35m. In each block, there were five rows representing five water quality treatments. Each row is 1.35 m long and 20cm wide. Planting space between corms was 15 cm. So, each row in each block had 10 corms. The replicates were 1.0 m away from each other. Each treatment was 50cm away from the other treatment within the same block. Five water quality treatments were used as follows: 100% Household water (T1);100% Treated wastewater (T2); Mixing 25% HW+75%TWW (T3); Mixing 50% HW +50 % TWW (T4); Mixing 75% HW+25% TWW (T5).

Preparing the Soil before Planting

For the GH experiment, all pots were filled with soil from Al-Jubeiha Agricultural Research Station and distributed randomly in the greenhouse. For OF experiment, the land was prepared firstly by tillage. Then, boundaries of the field that would be used were defined. The soil was saturated with fresh water and left to equilibrate for 48 hrs to reach the moisture content at field capacity (F.C). After that, the rows were prepared with 20 cm width and 1.35 m length. The rows then were covered with black plastic mulch. Soil samples were randomly taken from the OF and GH for physical and chemical analysis. Samples were collected from the average depth of 35 cm. Three samples from each treatment/block were collected.

IRRIGATION METHOD

The soil water content was controlled by monitoring soil moisture with Hydra Probe soil sensor. The probe was inserted in each treatment row at a depth of 0-10 cm and the moisture level was measured directly based on the average of three readings for each treatment. Soil moisture was kept at F.C. So, daily measurements of water content were taken. The irrigation frequency was twice a week at GH experiment, and three times in a week at OF experiment. Since the planting date was in winter in GH, less evapotranspiration is expected. Also, temperature and relative humidity were controlled. But in the OF experiment, planting date was in summer, more evapotranspiration was expected. Higher temperature, less humidity, and wind effect, all increased water requirement by plants. The F.C was estimated before planting by saturating the soil, then, by letting it to drain for 24 hours, and then measuring the moisture content at 10cm average depth. Water content at F.C and other soil physical properties are shown in Table 1.The volume of water needed for irrigation was measured by subtracting the current water content from water content at F.C and multiplying it by the soil volume.

The volume of soil is equal to the volume of pot in GH experiment, but in OF experiment, the volume of soil was measured by taking the distance between planted corms as diameter, and 35 cm soil depth. Irrigation water was applied manually to the plants by using graduated cylinder. Amounts of water consumed by plants at each month for both experiments are presented in Table 2.

Table 1. Soil moisture content at F.C, texture, and bulk density for both experiments (GH and OF):

Soil measured parameters	GH	OF
$\theta_{F.C}$ (cm^3/cm^3)	0.48	0.44
Sand (%)	13.8	33.0
Silt (%)	48.0	29.3
Clay (%)	38.2	37.7
Bulk density (g/cm^3)	1.19	1.00

Table 2. Monthly average amounts of water consumed (liter per plant) in both experiments (GH and OF):

Experiment	Month	Water amount (liter per plant)
GH	October	1.80
	November	2.94
	December	3.62
	January	6.40
	February	2.00
	Total	16.76
OF	April	1.92
	May	3.80
	June	7.77
	July	6.00
	Total	19.49

Planting

The corms that were used in the experiments were brought from the available cultivars in the local market. White Purity cultivar was utilized. A corm, for each pot was used. The corms were planted at depths equal to twice of its height. Planting date for the GH experiment was on October 18, 2011. The plants took 4 months to produce flowers. Planting date for the OF experiment was on April 9, 2012. The plants took 3 months to produce flowers

Water Analyses

The TWW was brought from Baqa'a wastewater treatment plant which is wastewater treated through biological filters, and maturation ponds, thus allowing for effluent reuse. TWW was stored in a storage tank for both experiments. The mixing of TWW with HW was manually done, and then the mixed water was stored in containers. The mixture was shaken before each irrigation event. Average BOD of the TWW was 32 mg/l exceeding the allowable limit for flower planting. Water samples were analyzed for the following parameters since there are many papers that discuss the effects of these parameters on flower growth and quality: pH, EC, Ca, Mg, Cl, Na, K, N, P, Fe, Zn, Cd, Mn, Pb and Cu. The pH and EC were measured directly by pH/mV meter and professional Benchtop Conductivity meter respectively. Ca and Mg were measured by titration using EDTA method. Sodium and K were measured directly by Flame Photometer. Chloride was determined by titration with silver nitrate. N determination was done by Kjeldahl method. Total P was determined by ascorbic acid spectrophotometric method . Zn, Cd, Pb, Fe, Mn, and Cu were determined by Atomic Absorption Spectrometer .All water analyses were done according to Clesceri, et al (1989).

Soil Analyses Before and After Planting

Soil texture was measured by pipette method according to Day, (1965). Bulk density was measured by core method according to Blake, (1965). Aggregate stability was measured by determining Mean Weight Diameter (MWD) according to Angers and Mehuys, (1993). Soil moisture characteristic curve was measured by using ceramic plate extractor according to Richards, (1965). Pore size distribution curve was constructed by firstly measuring the slopes at soil moisture versus soil moisture tension, and then plotted the reciprocal of the slope as a function of pore radius. pH was measured directly by pH/mV meter according to Peech, (1965), and EC by Professional Benchtop Conductivity meter according to Bower and Wilcox (1965). Sodium and, potassium were measured directly by flame photometer according to Pratt, (1965). Chloride was determined by titration with silver nitrate according to Adriano and Doner, (1982). Total N determination was done by Kjeldahl method according to Bremner, (1965). Available P was determined by ascorbic acid spectrophotometric method according to Olsen and Dean (1965). Organic matter in the soil was determined by wet combustion method as described in Nelson and Sommers (1982). Microelements contents were determined by Atomic Absorption Spectrophotometer according to Liang and Karamanos, (1993).

Flower Yield and Quality

Flower production (spikes) and quality were tested. Total yield represented by the number of spikes per corm was recorded at harvest. Samples of spikes were taken in order to measure the effect of irrigating gladiolus by TWW on flower quality. Three random samples were collected from each treatment. In order to fix the place of cutting the spike, each spike was cut from below the fourth leaf. Spike length, spike diameters, floret length, floret diameters, floret number for each spike, new corms diameter and the number of newly formed cormels per corm were evaluated. Spike and floret length were measured directly by using the ruler. The spike length was measured from the bottom to the top of the spike. Measurement for floret length was taken on the floret that shows color. Spike, floret and new corm diameter were measured by using Vernier caliper. Spike diameter was measured from three places and the average of the three readings for each spike was taken. The measures were taken from the bottom, below the first floret, and the upper portion of the spike. The measurement of floret diameter was taken in the middle of the floret that was showing color.

Statistical Analyses

Both experiments were RCBD design. Three blocks (replicates) were used. The SPSS program was used for the statistical analysis of soil and plant parameters evaluated. The comparisons among treatments were based on Duncan Multiple Range Test (DMRT) for the 0.05 p level.

RESULTS AND DISCUSSION

Water Analysis

The results of water analysis throughout both seasons are presented in Table 3.Results showed that water treatments differ in EC, Na^+, K^+, Cl^-, N, and total P contents. This was expected since treated wastewater had higher concentration of salts, N, and P. Therefore, any change in plant quality parameters or soil properties will be attributed to the quality of irrigation water. Technical regulation for TWW reuse of in agriculture was used to compare with the quality of used TWW in the experiments and it was as follows: pH changes with time not significantly and ranged between 7 and 8 for all treatments. EC varied from 1 to 2 dS/m and it is within the acceptable range. Ca, Mg, and total N concentrations were lower than the highest acceptable values. Na, Cl, and total P concentration in T2 treatment were higher than acceptable value. K was higher in T2 treatment. All microelements were within the acceptable levels, except for Cd which was higher than the acceptable value. The standard values were obtained from The Institution for Standards and Metrology (ISM, 2006). Average BOD value was 32 mg/l and it is higher than the acceptable value which is 15 mg/l.

Table 3. Characterization of water used in different treatments of GH and OF experiments (average data).

Treatments	pH	EC	Ca	Mg	Na	K	Cl	N	P-Total	Zn	Fe	Cd	Mn	Cu	Pb
T1 (HW)	8.1	1.032	52	29	143	11	248	3.0	2	0.33	0.37	0.03	0.05	0.03	0.07
T2 (TWW)	7.7	1.952	55	34	292	42	447	9.7	235	0.97	0.17	0.05	0.09	0.06	0.14
T3 (0.25HW+0.75TWW)	7.7	1.724	54	32	258	37	375	6.3	176	0.67	0.19	0.04	0.07	0.05	0.12
T4 (0.50HW+0.50TWW)	7.6	1.505	53	34	222	30	345	4.2	108	0.62	0.25	0.04	0.06	0.04	0.12
T5 (0.75HW+0.25TWW)	7.8	1.269	55	30	183	22	300	5.2	47	0.46	0.29	0.04	0.06	0.04	0.10
*Standard values	6-9	-	230	100	230	-	400	70	30	5.00	5.00	0.01	0.20	0.20	0.20

EC, dS/m, element concentrations, ppm. *All standards values are presented in the institution for standards and metrology (ISM), Water Reclaimed domestic wastewater No.893/200

Table 4. Organic matter content , pH, EC, N, P, Ca, Mg, Na, K, Cl concentrations and SAR before planting (PB) and after planting (T1-T5 treatments), for (GH and OF).

Experiment	Treatments	OM (%)	N (%)	P (mg/kg soil)	pH	EC (dS/m)	Ca (mg/100g)	Mg (mg/100g)	Na (mg/100g)	K (mg/100g)	Cl (mg/100g)	SAR
GH	BP	1.86 a	0.93a	4.70 c	7.7 NS*	0.99 d	5.96 d	1.76 c	02.30 d	0.36 bc	20.41 d	1.07d
	T1(HW)	1.53b	0.48 b	7.41 abc	7.4NS	3.96 c	18.92 c	6.16 b	16.06 c	0.24 c	75.34 c	4.15c
	T2(TWW)	1.53b	0.54b	9.69 a	7.3NS	9.50 a	46.07 a	10.59 a	42.72 a	0.60 a	156.60 a	7.31a
	T3(0.25HW+0.75TWW)	1.51 b	0.53 b	8.64 ab	7.5NS	8.24 a	39.52 a	9.06 a	40.23 a	0.52 ab	145.61 ab	7.43a
	T4(0.50HW+0.50TWW)	1.50 b	0.49 b	6.70 abc	7.3NS	6.23 b	30.39 b	6.25 b	27.50 b	0.33 c	111.94 bc	5.85b
	T5(0.75HW+0.25TWW)	1.49 b	0.60 b	5.92 bc	7.5NS	4.25 c	20.53 c	5.04 b	17.90 c	0.27 c	81.22 c	4.56c
OF	BP	2.39NS	0.52 a	27.28 c	7.4NS	0.718 c	03.31 b	2.09NS	1.52 d	0.66 b	14.59 b	0.86 d
	T1(HW)	2.47NS	0.19 b	161.24 b	7.7NS	2.34 b	08.49 ab	3.50NS	6.83 b	1.81 a	32.59 a	4.04 b
	T2(TWW)	2.43NS	0.19 b	265.85 a	7.7NS	4.68 a	12.70 a	5.12NS	10.87 a	2.43 a	38.20 a	6.45 a
	T3(0.25HW+0.75TWW)	2.92NS	0.19 b	206.55 b	7.6NS	3.64 a	11.08 a	3.44NS	9.53 b	2.29 a	35.80 a	5.61 b
	T4(0.50HW+0.50TWW)	2.52NS	0.23 b	186.56 b	7.6NS	3.80 a	12.67 a	3.84NS	10.49 bc	2.26 a	35.27 a	5.23 bc
	T5(0.75HW+0.25TWW)	2.51NS	0.15 b	168.57 b	7.6NS	3.43 ab	12.67 a	4.31NS	9.14 c	2.12 a	31.14 a	4.35 c

Soil Physical and Chemical Properties

Soil used in GH experiment was silty clay loam, whereas for OF was clay loam. Noteworthy, irrigation water treatments significantly impacted some of the soil parameters evaluated. Change in aggregate stability before and after planting for GH and OF experiments is shown in Figures 1 and 2, respectively. No significant differences in aggregate stability were observed among treatments in the OF experiment, whereas it was reduced by 40% in all treatments in the GH experiment. The decomposition of soil structure observed in GH experiments due to the irrigation treatment could be attributed to the reduction in soil OM observed, since OM acts as a cementing agent in the soil; and leads to soil aggregate formation. These results are in line with the ones from Dal-Ferro et al. (2010) who found that reduction in soil organic matters reduce aggregate stability in soils. Increasing soil Na content resulted in an increase in SAR (7.3 fold in GH experiment; and about 6 fold in OF experiment). This leads also to the deterioration of the soil structure and aggregate stability. Since Na had relatively large size, single electrical charge, and hydration status, it tends to cause dispersion of soil particles. Results from this study agreed with findings from Nikos et al. (2002). Change in soil moisture characteristic curve before and after planting for GH and OF experiments is shown in Figures 3 and 4, respectively. Soil moisture characteristic curve is affected by many factors such as texture, clay content, and organic matter content in the soil. No significant differences in soil water retention curve after planting was found to occur between treatments. Pore size distribution curves for GH and OF experiments were plotted in Figures 5 and 6, respectively. Pore size distribution curves were not changed significantly between treatments. Pore size was mainly at the micropores range (Scott, 2000) for all treatment, with the maximum size being equal or less than 0.0973 μm

Soil chemical properties before and after planting are presented in Table 4, while microelements content is shown in Table 5. Soil salinity increased significantly for all treatments after planting with the highest value for TWW treatment. This was expected since TWW had high salts content .As a result, soil Na, Ca, Mg, and K increased for all treatments with the highest concentration for TWW irrigation treatment. Soils from T2 treatments became saline soils. K, Ca, Mg, and Cl- - content were significantly different among treatments in GH experiment, with the T2 values being the highest ones. Soil moisture was kept at F.C in order to study the maximum accumulation of salts in the soil. Also, the increasing percentage of salinity in GH experiment was greater than in the OF experiment. The volume of soil was limited and each plant was isolated in pot from other plants in the same treatment, so, overlapping did not occur for applied water. But in OF experiment, all plants were planted in the same row, so, overlapping occurred which led to little leaching. As results, accumulation of salts in GH experiment was greater than OF experiment. Results from this study agreed with Galavi et al. (2010), and Herpin et al. (2007). There was a reduction in soil organic matter and N% among all treatments in GH experiment after planting. This was expected since no fertilizers were added to the soil. Application of TWW and the conditions such as temperature and humidity in GH encouraged microbial activity for N mineralization causing significant decrease of soil OM (Herpin et al., 2007). But in OF experiment, there was no change in soil organic matter and reduction in soil N percentage. This result agreed with Jueschke et al., (2008), and did not agree with Galavi et al., (2010). Total P increased after planting in OF experiment significantly. This is expected also, since TWW has higher concentration of P than in the HW. Accumulation of P was greater in OF experiment and this may resulted from Ca and Mg accumulation since P was fixed by Ca and Mg carbonate in the soil. Accumulation of Ca and Mg in GH experiment was greater than OF experiment, P may have been fixed by Ca and Mg carbonate in GH experiment. Galavi et al. (2010), found the same results. Results in soil microelements analyses for GH experiment showed that Zn, Mn, Cd, Pb, and Cu content increase in all treatments, but no change in Fe concentration were found. Results in soil microelements analyses for OF experiment showed that Mn, Pb, and Cu content were significantly increased. These results were expected since TWW had certain concentrations of these elements as described in Table 3. Soil pH would limit the availability of microelements and will reduce accumulation in the soil. The same results were reported by Koupai et al. (2006). In OF experiment, results of soil physical and chemical properties are more variable than GH experiment, and this is attributed to spatial variability of the soil in the field which is not present in GH experiment. Soils from GH experiment were more homogeneous since the soil sieved, mixed well, and distributed in pots for the research.

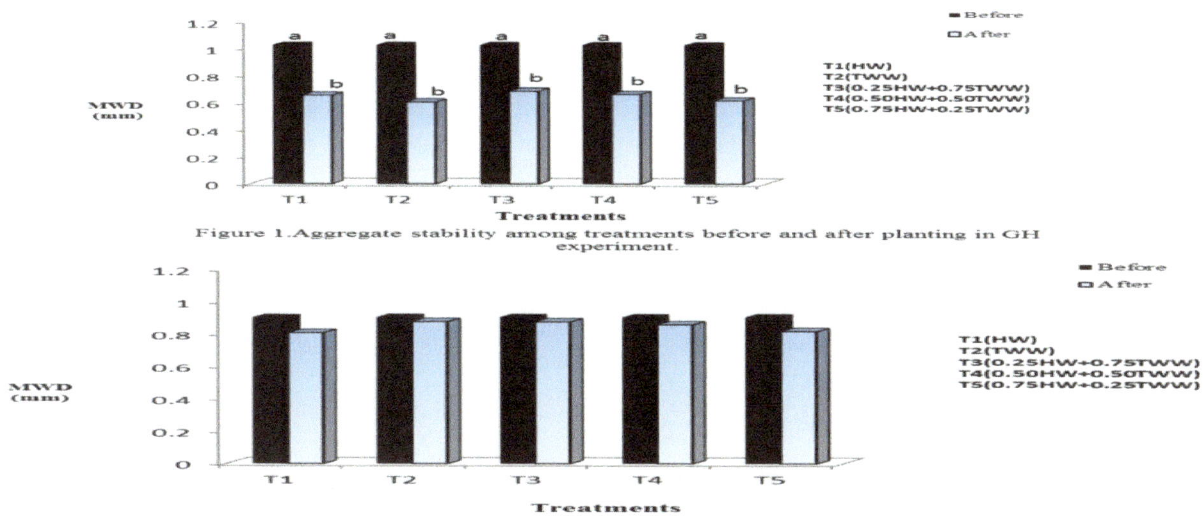

Figure 1: Aggregate stability among treatment before and after planning in GH experiment.

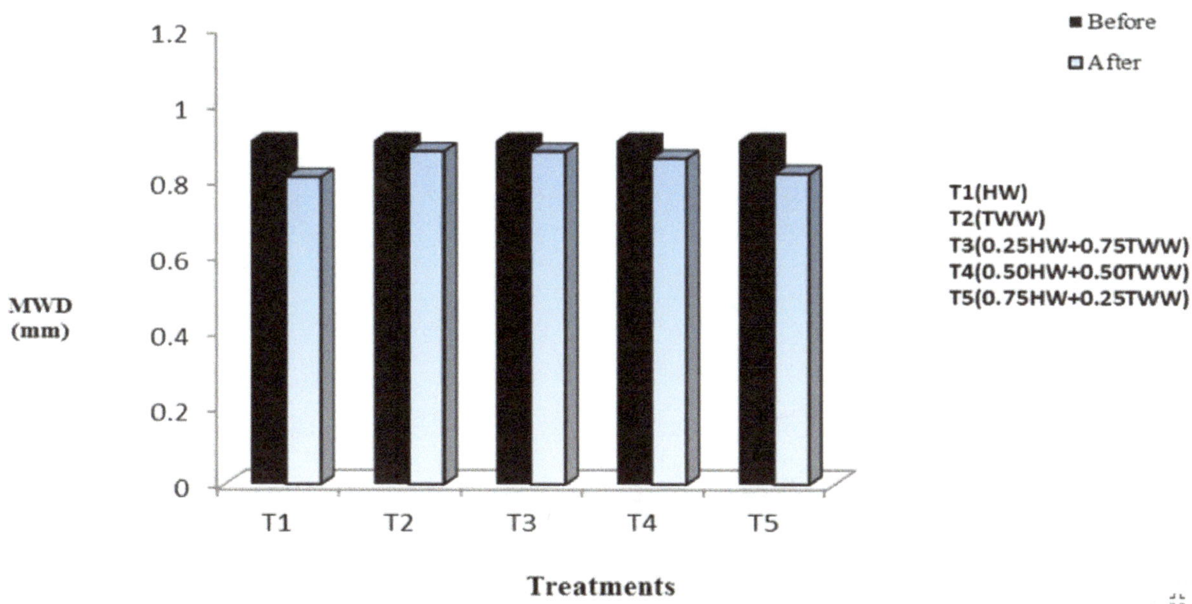

Figure 2: Aggregate stability among treatment before and after planning in OF experiment.

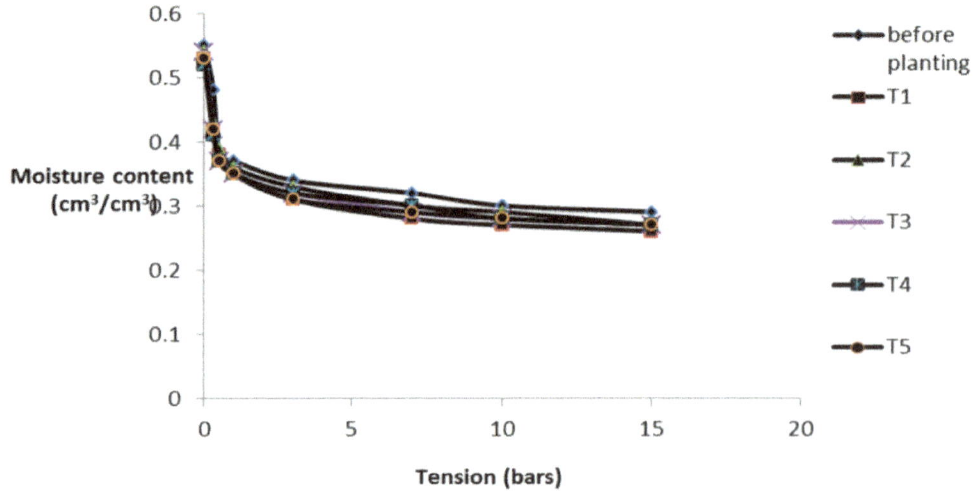

Figure 3: Soil moisture characteristic curve before and after planning for GH experiment.

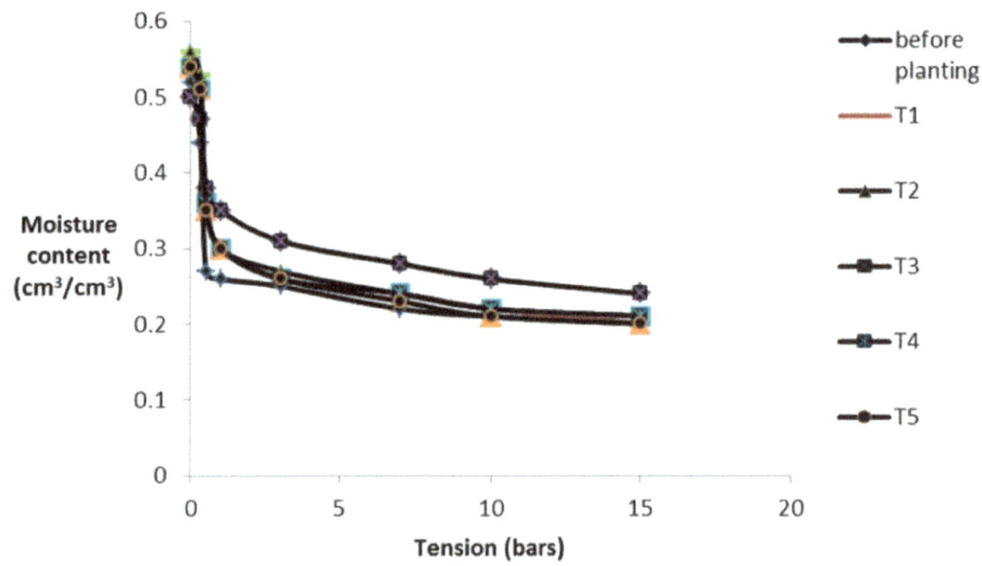

Figure 4: Soil moisture characteristic curve before and after planning for OF experiment.

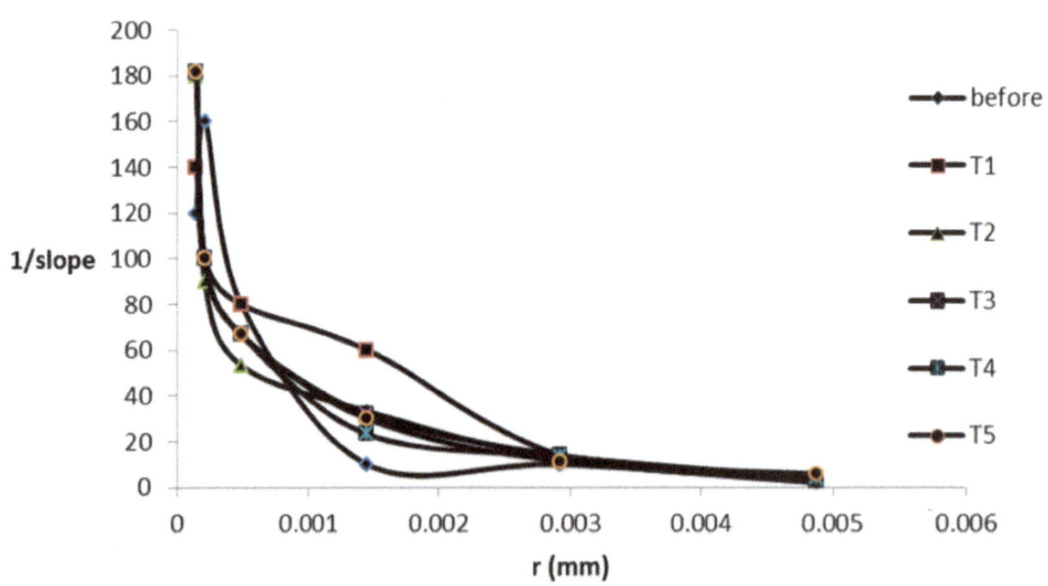

Figure 5: Pore sixe distribution curve before and after planning for GH experiment.

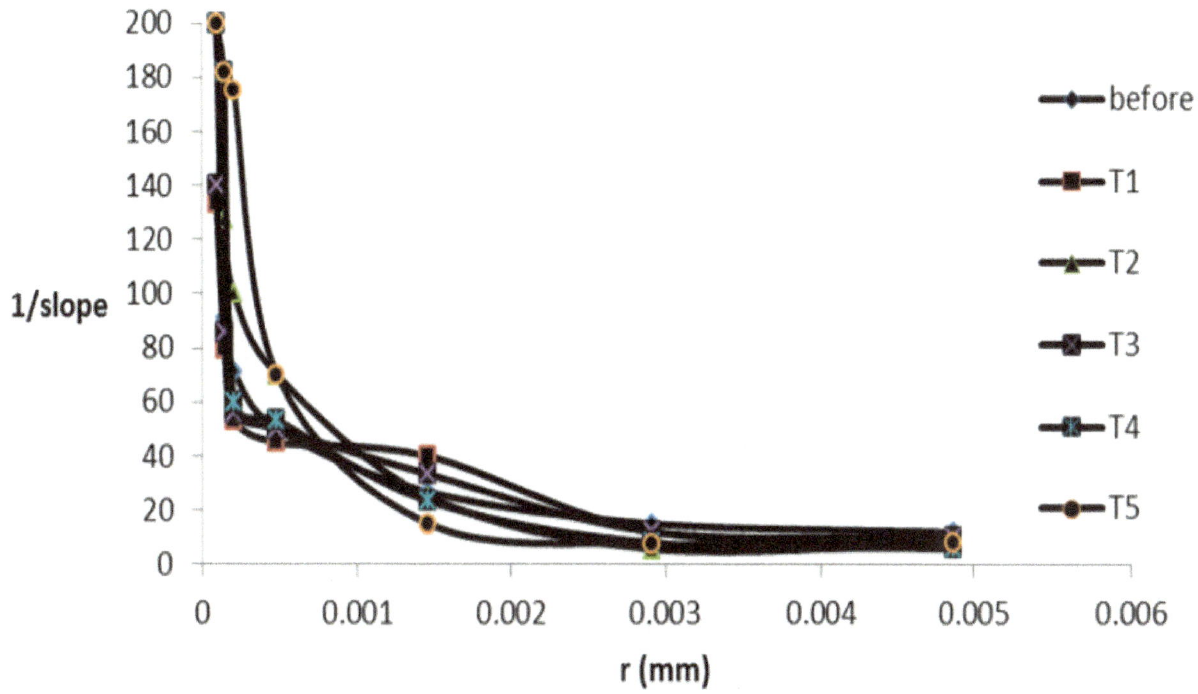

Figure 6: Pore sixe distribution curve before and after planning for OF experiment.

Plant Analyses

Spike quality parameters are presented in Table 6. Production did not change for all treatments. For spike length gave the lowest value for T2 (100%TWW). Salt content in water could have effects on the development of Gladiolus. The presence of high Na and Cl in water had probably affected plant development. It is known that Gladiolus is a salt sensitive horticultural plant. High salt concentration impaired plant growth because it reduces its ability to absorb water, and this had most probably affected spike length later. Also, the presence of K in water and its accumulation in soils could have an effect on spike length. K may increase the length of spikes to a certain level since it plays a major role in growth regulation, whereas at high concentration it may reduce it. Increasing P content would affect spike length for the plant to a certain limit since P plays an important role in root development. So, increasing P led to increase spike length. These results are in agreement with the published results of Lehri et al. (2011). Some microelements content could have an effect on spike length. Zn and Cu in GH experiment had increased significantly and would probably increase spike length. Increasing Zn will increase the utilization of carbon dioxide evolution and utilization of carbohydrate which led to increase spike length. The results agree with Halder et al. (2007, b). Spike diameters had been affected by water treatments. T2 gave the lowest value since it contained high salt concentrations. Corm size was affected by the presence of microelements content. Increasing Zn affects corm size positively as discussed before. Corm size increase significantly. This agrees with Singh et al. (2012) and Halder et al. (2007, b). Floret number per spike was affected significantly by water treatment for OF experiment. The number of florets was affected by many parameters such as N, P, K, and Zn. N deficiency may have reduced the number of floret. N is considered to be one of the major essential elements which regulate the cell and tissue function of the plants. Also, N controls the vegetation and reproduction stage in the plant. As a result of reducing N, the number of florets reduced. Zn application to the soil would increase the number of floret per spike since Zn has an effect in the increasing enzymatic reactions in the plants. This agrees with Halder et al. (2007, b). Numbers of cormels per corm also were affected significantly for both experiments. T2 gave the least number of cormels per corm. Increasing K and P to a certain limit would increase the average number of cormels per corm. But in our experiments, it decreases it. Balance between nutrients is essential for better performance. Results from Lehri et al. (2011) further agree with our findings. Some microelements content also have an effect to the number of cormels per corm; they are expected to increase the number of cormels to certain limit. Increasing Zn, Fe, and Cu would increase the number of cormels per corm. Also, the imbalance between nutrients reduces the number of cormels. Singh et al. (2012) and Halder et al. (2007, a) reported similar findings. Other quality parameters such as floret length, and floret diameter were not changed significantly

Table 5. Zn, Mn, Cd, Pb, Fe, and Cu concentrations in soil (ppm) before planting (BP) and after planting (T1-T5) for (GH and OF).

Experiment	Treatment	Zn	Mn	Cd	Pb	Fe	Cu
GH	BP	1.50 c	2.53 a	0.15 b	0.00 b	1.68NS*	1.46 c
	T1(HW)	3.80 a	1.46 bc	0.21ab	0.85 a	1.60NS	1.71 a
	T2(TWW)	2.66 b	1.88 abc	0.23 a	0.90 a	1.44NS	1.59 ab
	T3(0.25HW+0.75TWW)	3.14 ab	2.29 ab	0.24 a	0.96 a	1.39NS	1.56 bc
	T4(0.50HW+0.50TWW)	2.75 b	1.16 c	0.23 a	0.91 a	1.41NS	1.45 c
	T5(0.75HW+0.25TWW)	2.51 b	1.14 c	0.24 a	0.98 a	1.37NS	1.54bc
OF	BP	3.34NS	4.11 a	0.23NS	0.00 b	1.96NS	1.31 b
	T1(HW)	3.92NS	2.23 c	0.25NS	0.67 a	2.22NS	1.78 a
	T2(TWW)	3.40NS	3.24 bc	0.25NS	0.81 a	2.17NS	1.99 a
	T3(0.25HW+0.75TWW)	3.85NS	2.79 bc	0.24NS	0.74 a	2.29NS	1.97 a
	T4(0.50HW+0.50TWW)	3.72NS	2.74 bc	0.26NS	0.81 a	2.36NS	1.94 a
	T5(0.75HW+0.25TWW)	4.03NS	3.38 ab	0.27NS	0.90 a	2.61NS	1.94 a

Mean values followed by the same letter within the same column are not significantly different at 0.05 p level, as obtained from DMRT.*Not Significant.

Table 6. Spike quality parameters for GH and OF experiments.

Experiment	Treatment	Spike length (cm)	Spike diameter (mm)	Floret number/spike	Floret length (cm)	Floret diameter (cm)	Newly formed corms size (cm)	Number of cormels/corm
GH	T1(HW)	104.9 bc	8.23 a	13NS*	1.24NS	1.24NS	4.99 a	53 b
	T2(TWW)	102.7 c	7.60 c	13NS	1.26NS	1.26NS	4.76 b	53 b
	T3(0.25HW+0.75TWW)	108.7 ab	8.00 ab	13NS	1.28NS	1.28NS	4.77 b	72 a
	T4(0.50HW+0.50TWW)	106.1 bc	7.87 bc	13NS	1.28NS	1.28NS	4.96 a	65 a
	T5(0.75HW+0.25TWW)	111.9 a	8.10 ab	13NS	1.26NS	1.26NS	4.98 a	63 ab
OF	T1(HW)	86.63 a	8.00 a	13 a	6.47NS	1.31NS	2.71 a	7 a
	T2(TWW)	71.4 b	6.90 b	11 b	6.53NS	1.32NS	1.94 b	2 b
	T3(0.25HW+0.75TWW)	86.8 a	8.10 a	14 a	6.5NS	1.37NS	2.69 a	6 a
	T4(0.50HW+0.50TWW)	84.77 a	7.80 a	13 a	6.33NS	1.32NS	2.56 a	6 ab
	T5(0.75HW+0.25TWW)	85.27 a	7.90 a	13 a	6.57NS	1.38NS	2.55 a	5 ab

Mean values followed by the same letter within the same column are not significantly different at 0.05 p level, as obtained from DMRT. *Not Significant.

CONCLUSION

The present study evaluates the effect of using TWW as irrigation water on soil physical and chemical properties, and on spikes production and quality parameters of gladiolus. Applying TWW significantly affects soils physico-chemical properties, since no leaching requirements were added. Ca, Mg, Na, K, and Cl contents were increased when TWW was used. The maximum accumulations occurred for soils that were irrigated with 100% TWW. Soil physical properties such as aggregate stability were affected by TWW irrigation. TWW if mixed with HW in ratios 50%TWW+50%HW, and 25%TWW+75% HW gave results for plants quality better than100% TWW. Some quality parameters of Gladiolus, namely: spike length, spike diameter, floret number per spike, number of cormels per corm, and newly formed corms size, were affected by the quality of irrigation water. They gave the worst quality as mentioned before for plants when irrigated with 100% TWW.

RECOMMENDATIONS

Based on results obtained from both experiments, it is recommended to do the following:

REFERENCES

1. Adriano, D., Doner, H. 1982, Bromine, Chlorine, and Fluorine, In: Methods of Soil Analysis, edited by Page, A., Part2, Chemical and Microbiological Properties, 2nd edition, pp. 449-483.
2. Angers, D., Mehuys, G. 1993, Aggregate stability to water, In: Soil sampling and methods of analysis, edited by Carter M., Canadian Society of Soil Science, pp. 651-657.
3. Blake, G.R.. 1965, Bulk density, In: Methods of soil analysis, edited by Black C.A., part 1, Agronomy 9, American Society of Agronomy, Madison, Wisconsin, pp. 374-390.
4. Clesceri, L.S., Greenberg , A.E. Trussell , R.R. Franson, M.A. 1989, Standard Methods for the Examination of Water and Wastewater, Seventeenth Edition, Prepared and published jointly by: American Public Health Association, American Water Works Association, and Water Pollution Control Federation, American Public Health Association, 1015 fifteenth Street NW, Washington, DC 20005.

5. Dalferro, N., Berti, A. Matthews, P. Giardini, L. Morari, F.2010, Study of the Effect of Structure on Soil Aggregate Stability using 3D Network Model, In: World Congress of Soil Science, Soil Solutions for a changing world, 1-6 August 2010, Brisbane, Australia.

6. Friedman, H., Bernstein, N., Bruner, M., Rot, I., BenNoon, Z., Zuriel, A., Zuriel, R., Finkelstein, S., Umiel, N., Hagiladi, A. 2007, Application of Secondary Treated Effluents for Cultivation of Sunflower (Helianthus annuus L.) and Celosia (Celosia argentea L.) as Cut Flowers, Scientia Horticulturae, 115: 62-69.

7. Galavi, M., Jalali, A. Ramroodi, M.2010, Effect of Treated Municipal Wastewater on Soil Chemical Properties and Heavy Metal Uptake by Sorghum (Sorghum Bicolor L.), Journal of Agricultural Science, 2 (3): 235-241.

8. Halder, N.K., Ahmad, R., Sharifuzzaman, S.M., Arabagam, K.A., Siddiky, M.A. 2007, a. Effect of Boron and Zinc Fertilization on Corm and Cormel Production of Gladiolus in Grey Terrace Soils of Bangladesh, International Journal of Sustainable Crop Production, 2(5): 85-89.

9. Jueschke, E., Marschner, B., Tarchitzky, J., Chen, Y.2008. Effects of Treated Wastewater Irrigation on the Dissolved and Soil Organic Carbon in Israeli Soils, Water Science Technology, 57(5): 727-733.

10. Koupai, J.A., Fard, B.M., Afyuni, M., Bagheri, M.R. 2006. Effect of Treated Wastewater on Soil Chemical and Physical Properties in an Arid Region, Plant Soil Environment, 52(8): 335-344.

11. Lehri S.M., Kurd, A.A., Rind, M.A., Bangulzai, A. 2011. The Response of Gladiolus Tristis L. to N and P_2O_5 Fertilizers, Sarhad Journal of Agriculture , 27(2): 185-188.

12. Liang J., Karamanos, R. 1993, DTPA-Extractable Fe, Mn, Cu, and Zn, In : Soil sampling and methods of analysis, edited by Carter M., Canadian Society of Soil Science, pp. 87-90.

13. Nelson, D.W., Sommers, L.E. 1982, Total Carbon, Organic Carbon, and Organic Matter, In: Methods of Soil Analysis, edited by Page, A.L., Part 2, Chemical and Microbiological Properties, 2nd Edition, pp. 539-579.

14. Nikos, J.W., James, W.B., Krista, E.P. 2002, Basics of Salinity and Sodicity Effects on Soil Physical Properties, Department of Land Resources and Environmental Sciences, Montana State UniversityBozeman.

15. Olsen, S. R., Dean, L.A. 1965, Phosphorus, In: Methods of soil analysis, edited by Black, C.A., part 1, Agronomy 9, American Society of Agronomy. Madison Wisconsin, 1035-1048.

16. Peech, M., 1965, Hydrogen activity, In: Methods of soil analysis, edited by Black, C.A., part 1, Agronomy 9, American Society of Agronomy, Madison, Wisconsin, 914-925

17. Richards, S.J., 1965, Soil Suction measurements with tensiometers, In: Methods of soil analysis, edited by Black, C.A., part 1, Agronomy 9, American Society of Agronomy, Madison, Wisconsin, pp. 153-163

18. Rusan, M., Bashabsheh, I., Safi, M. 2008, Reuse of TWW for Cut Flowers Production and Impact on Soil and Plant Quality Parameters, In: International Conference on Construction and Building Technology ICCBT 6: pp. 63-78.

19. Scott, H.D., 2000, Soil Physics, Agricultural and Environmental Application, Iowa State University Press-Ames.

20. Singh, J.P., K. Kumar, P.N. Katiyar, 2012, Effect of Zinc, Iron, and Copper on Yield Parameters of Gladiolus, Hortiflora Research Spectrum ,1(1): 64-68.

21. The Institution for Standards and Metrology (ISM), 2006,Water-Reclaimed domestic wastewater, Technical Regulation, 3rd edition, JS 893:2006.

22. Wilfret, G., 1980, Gladiolus, In: Larson, R (Ed.), Introduction to Floriculture, Department of Horticultural Science, North Carolina State University, Raleigh, North Carolina. Academic Press.

CHAPTER 8 SOIL QUALITY FOR EVALUATING SMALL AGRICULTURAL LAND USE

Aweke M. Gelaw, B. R. Singh and R. Lal

Keywords: Soil quality; soil functions; land degradation; land use; Ethiopia

INTRODUCTION

Land degradation and declining soil fertility are critical problems affecting agricultural productivity and human welfare in Sub-Saharan Africa [1]. The main soil-environmental concerns in the region are nutrient depletion, loss of soil organic matter (SOM) and loss of soil functions (*i.e.*, productivity) [1,2]. In Ethiopia, total cultivated land has reached ~12 million hectares in mid-2013, but most of the soils are highly degraded [3]. Further, population growth and agricultural production are not growing *at par.* As a result, expansion to marginal lands and protected areas has become a common practice.

Tigray, the northernmost region in Ethiopia, is most known for its serious land degradation problems. Much of the woodland in Tigray started to disappear in the early 1960s under pressure from the rapidly growing population [4]. Hengsdijk *et al.* [5] wrote their observations as follows: "perhaps nowhere in the world land degradation and soil nutrient depletion are more evident than in the marginal highlands of Tigray". In the region, a short and variable rainy season in combination with degraded soils resulted in low soil productivity and frequent crop failures. As a result, the local population is structurally dependent on food aid [6]. If unattended to, land degradation and soil nutrient depletion would further reduce agricultural productivity and increase pressure on marginal environments, adversely affecting food security and livelihoods of smallholder farmers in the region [6].

Indeed, Tigray is not only known for its severe land degradation, but also for its vast environmental rehabilitation efforts in the last two decades [7]. Among the recent efforts towards enhancing agricultural development in the region, rainwater harvesting has been widely adopted [8] because supplementary irrigation is essential for crop production in arid regions as it increases soil water availability during dry spells [9]. Further, farmers in Tigray have a culture of selectively taking care of trees, which are remnants of the original woodlands. *Acacia albida* Del. (Syn. *Faiderbhia albida* (Del.) A Chev.) trees are among the most selected ones in the region. Nowadays, farmers grow these trees in and around their farmlands in order to improve soil fertility and increase crop yields [10].

Sustainability of agricultural systems is an important issue in Ethiopia. Many of the issues of agricultural sustainability are related to soil quality. Thus, its assessment and the direction of change with time is a primary indicator of whether agriculture

Citation: Aweke M. Gelaw, B. R. Singh and R. Lal, Soil Quality for Evaluating Small Agricultural Land Use, doi:10.3390/su7032322.

is sustainable [11,12]. Soil quality is a combination of soil physical, chemical and biological properties that are able to change readily in response to variations in soil conditions [13]. It may be affected by land use type and agricultural management practices because these may cause alterations in soil's physical, chemical and biological properties, which in turn results in change in land productivity [14,15]. Integrated soil quality indices based on a combination of soil properties provide a better indication of soil quality than individual parameters. Karlen and Stott [16] developed a soil quality index (SQI) based on four soil functions, namely the ability of the soil to: (1) accommodate water entry (WE); (2) facilitate water movement, and absorption (WMA); (3) resist surface degradation (RD); and (4) supply nutrients for plant growth (PNS). Each soil function was explained by a set of indicators. Several authors among them Glover *et al.* [17], Masto *et al.* [12] and Fernandes *et al.* [18] used a similar framework.

A soil quality index (SQI) helps to assess the soil quality of a given site or ecosystem and enables comparisons between conditions at plot, field or watershed level under different land uses and management practices. Several studies were conducted to assess fertility statuses of soils in SSA [1,2,3,4,5,6,7,8]; however, almost all were only based on evaluation of individual soil parameters. Therefore, this study was conducted at a typical semi-arid agricultural watershed in Eastern Tigray, Northern Ethiopia, with the following objectives:

(1) To evaluate effects of *F. albida* based agroforestry (AF), irrigation based *Psidium guajava* fruit production (IR) and a tree-less row-crop management (RF) (Figure 1) on selected physical, chemical and biological soil quality indicators and,

(2) To compute an overall integrated soil quality index (SQI) for each land use system and compare among the indices.

Figure 1. The three agricultural land use systems at a semi-arid watershed in Tigray, Northern Ethiopia, with dryland crop production (RF), *F. albida*-based agroforestry (AF) and irrigation-based *P. guajava* fruit production (IR).

The study was conducted to test the hypothesis that land use change from dry land rainfed cultivation (RF) to *F. albida*agroforestry (AF) and irrigation based *P. guajava* fruit production (IR) systems improves physical, chemical, and biological soil quality indicators and the overall integrated soil quality index.

MATERIALS AND METHODS

Descriptions of the Study Site

Mandae watershed is located in Eastern Tigray, Northern Ethiopia. Geographically, it is located between 15°26′00N to 15°32′00N latitude and 55°00′00E to 55°60′00E longitude, with an area of about 10 km^2, and an elevation of 1960 to 2000 m a.s.l. Average daily air temperature of the area ranges between 15 °C and 30 °C in winter and summer, respectively. Mean annual rainfall of the area is 558 mm, with a large inter-annual variation. Soils are classified as Arenosols, and associations of Arenosols with Regosols according to the World Reference Base for soil resources [19]. These soils are developed from alluvial deposits and Adigrat sandstones. Their textures are dominated by sand, loamy sand and sandy loam fractions [20]. Major land uses of the watershed include *Faidherbia albida* based agroforestry (27.7 ha), rainfed crop production (11.9 ha), open pasture (23.2 ha), and irrigation-based guava (*P. guajava)* fruit production (11.3 ha). Agricultural rotation in the area is usually maize (*Zea mays*)-teff (*Eragrostis tef*)-field beans (*Vicia faba*)-finger millet (*Eleusine coracana*) in the agroforestry and rainfed cultivation land use systems. Fallowing is not practiced in the area due to population pressure and scarcity of farmlands. Use of chemical fertilizers is minimal and land is prepared for cultivation by using a wooden plow with oxen. Crop residues and manures are used for animal feed and household fuel, respectively. No pesticides and other agricultural inputs are used in the area. Irrigation from shallow wells started in the area in late 1990s and currently most of the irrigated areas are covered by guava fruits. Smallholder mixed crop-livestock farming is a typical farming system of the region.

Soil Sampling and Analysis

Fifteen soil samples were collected in May 2010 from the surface (0–15 cm) layer of five sites randomly chosen at different locations from three agricultural land uses (AF, IR and RF). The summit position of the watershed was excluded to minimize confounding effects of slope and soil erosion. The samples were air-dried, mixed, ground, and passed through a 2-mm sieve for chemical analyses. Core samples were also collected from the same depth using 100 cm^3 volume stainless steel tubes (5-cm diameter and 5.1-cm height). Initial weights of the soil cores were measured in the laboratory immediately after collection. Simultaneously, soil moisture content was determined gravimetrically by oven drying the whole soil at 105 °C for 24 h to compute dry bulk density (ρ_b) [21]. No adjustment was made for rock volume because it was rather minimal. The major parts of the soil analyses were carried out at Mekelle University soil laboratory, Ethiopia. Soil organic carbon (SOC) and total nitrogen (TN) were analyzed at the Carbon Sequestration and Management Center (C-MASC) Laboratory (The Ohio State University, Columbus, OH, USA) using auto CN analyzer (Vario Max CN Macro Elemental Analyser, Elementar Analysensysteme GmbH, Hanau, Germany) by the dry combustion method [22]. Similarly, water stable aggregation (WSA) was measured at C-MASC soil physics laboratory by the wet sieving method [23]. Because soils did not show carbonates when tested with 10% HCl, it was assumed that the total C obtained in the analysis closely estimates soil organic carbon (SOC) concentration. Available P (Olsen) was analyzed using a standard Olsen method [24]. Cation exchangeable capacity (CEC) was estimated titrimetrically by ammonium distillation method [25]. Lastly, total porosity was calculated from particle density of 2.65 g/cm^3.

Microbial Biomass Carbon (MBC)

Another set of nine field-moist soil samples (40 g each) from the surface (0–15 cm) depth were collected in three replications from the three agricultural land uses (AF, IR and RF) in May 2012 for the determination of microbial biomass carbon (MBC). The samples were transported in an icebox to the Norwegian University of Life Sciences soil laboratory, Ås, Norway. The MBC analysis was carried out following the fumigation-extraction method [26,27]. At first, each sample was divided in to three subsamples, and one out of the three (10.0 g) was fumigated with ethanol-free chloroform for 24 h at 25 °C in an evacuated extractor. Afterwards, from the remaining two subsamples, one was used for moisture determination and the other treated as control for each plot. Fumigated and non-fumigated soils were extracted with 40-mL 0.5-mol·$L^{-1}$$K_2SO_4$ (1:4 soil:extractant) and shaken for 1-h on a reciprocal shaker. The extracts were filtered using Whatman No. 42 filter paper of 7-cm diameter and stored frozen at −15 °C prior to analysis. Finally, total organic carbon in the extracts was measured using Total Organic Carbon Analyzer (SHIMADZU) at NMBU laboratory, Ås, Norway. Microbial Biomass Carbon (MBC) was calculated as follows:

$$MBC= E_C / KE_C \quad (1)$$

where E_C = (organic C extracted from fumigated soils) − (organic C extracted from non-fumigated soils) and KE_C = 0.45 [28].

Soil Quality Assessment

Soil quality assessment tools need to be flexible in terms of selection of soil functions to be assessed and indicators to be measured to ensure that assessments are appropriate for specific management goals [29]. Effects of land use on soil quality were assessed following the framework suggested by Karlen and Stott [16]. We followed this framework because of its flexibility, ease of use and its potential for interactive use. It is the same approach that became the Soil Management Assessment Framework (SMAF) [30]. It uses selected soil functions, which are weighted and integrated according to the following expression:

$$SQI = WE(wt)+WMA(wt)+RD(wt)+PNS(wt) \quad (2)$$

where, wt is a numerical weighting for each soil function.

These numerical weights were assigned to each soil function according to their importance in fulfilling the overall goals of maintaining soil quality under specific conditions of this study. According to Karlen and Stott [16], the sum of weights for all soil functions must equal 1.0. Karlen and Stott [16] assigned equal weight to each soil function. However, different weight values of 0.2, 0.2, 0.2 and 0.4 were assigned for this study for WE, WMA, RD, and PNS, respectively (Table 1). For this study, PNS was assigned with more value than other functions, because use of chemical fertilizers was minimal in the area and hence nutrient supply was considered the most important production constraint. Further, sustaining crop production is the major goal of soil management strategies in most developing countries including Ethiopia. The PNS function was further divided into three second-level functions viz. nutrient storage, nutrient cycling and nutrient availability (Table 1).

Table 1. Soil quality indexing framework (adapted from Glover *et al.* [17]).

Function	Weight	Indicator Level 1	Weight	Indicator Level 2	Weight	Source for Indicators/Weights
Accommodate Water Entry	0.20	WSA	0.40			[17,31]
		BD	0.20			[17]
		POR	0.20			[12]
		SOC	0.20			[12]
Facilitate Water Movement and Availability	0.20	POR	0.60			[12,17,31]
		SOC	0.40			[17,31]
Resist Surface Degradation	0.20	WSA	0.60			[17,31]
		Microbial Processes	0.40	MBC	0.60	[12,17,31]
				SOC	0.20	[12,17,31]
				TN	0.20	[12;31]
Supply Plant Nutrient	0.40	Nutrient Storage	0.40	CEC	0.40	[12]
				SOC	0.40	[12]
				TN	0.20	[12]
		Nutrient Cycling	0.20	SOC	0.40	[12,31]
				MBC	0.20	[12,31]
				TN	0.40	[31]
		Nutrient Availability	0.40	SOC	0.20	[12]
				pH	0.20	[31]
				TN	0.20	[12]
				AVP	0.20	[12]
				AVK	0.20	[12]

An ideal soil would fulfill all the functions considered important, and would have an integrated SQI of 1.0 under the proposed framework. However, as a soil fails to meet the ideal criteria, its SQI would decrease, with zero being the lowest rating. Associated with each soil function are soil quality indicators that influence, to varying degrees, the specific soil function. Threshold values for each soil quality indicator were set based on the range of values measured in natural ecosystems (the adjacent grass pasture in our case) and on critical values in the literature (Table 2). Glover *et al.* [17] also used adjacent grass pasture areas to determine critical values for a study conducted in Washington State, USA. After finalizing the thresholds, the soil property values recorded under the three agricultural land use systems were transformed into unit-less scores (between 0 and 1), using the following equation [12]:

$$\text{Non-linear score(Y)} = \frac{1}{(1+e^{-b(x-A)})} \tag{3}$$

where, x is the soil property value, A the baseline or value of the soil property where the score equals 0.5 and b is the slope of the tangent to the curve at the baseline.

Table 2. Relative importance of the different soil properties used for the soil quality indexing.

Soil Quality Indicator	Weight	Soil Function
Soil organic carbon	0.264	Accommodate water entry Facilitate Water movement and availability Resist Surface structure degradation Supply plant nutrients
Aggregate Stability	0.200	Accommodate water entry Facilitate Water movement and availability Resist surface structure degradation
Bulk density	0.040	Accommodate water entry
Porosity	0.160	Accommodate water entry Facilitate water movement and availability
Microbial biomass carbon	0.064	Resist surface structure degradation Supply plant nutrients
Cation exchange capacity	0.064	Supply plant nutrients
Total Nitrogen	0.112	Supply plant nutrients Resist surface structure degradation
Available phosphorus	0.032	Supply plant nutrients
Available Potassium	0.032	Supply plant nutrients
pH	0.032	Supply plant nutrients
Total	1.00	

The score for each indicator was calculated after establishing the baseline, the lower, and the upper threshold values (Table 3). Threshold values are soil property values where the score equals one (upper threshold) when the measured soil property is at the most favorable level; or equals zero (lower threshold) when the soil property is at an unacceptable level. Baseline values are generally regarded as minimum target values [12]. There are two baselines for "Optimum" curves, lower base line and upper base line, which corresponds to 0.5 score of the growth and death curves, respectively [12].

Table 3. Scoring function values and references used for evaluating the soil quality indices (adapted from Masto *et al.* [12]).

Indicator	Scoring Curve	Depth (cm)	LT	UT	LB	UB	OPT	Slope	Source of Threshold/ Baseline Values
Physical properties									
BD (Mgm^{-3})	Less is better	0–15 cm	1.0	2.0	1.5	-	-	−2.0832	[31]; Adjacent grass pasture
WSA (>0.5 mm)	More is better	0–10 cm	0.0	40.0	20.0	-	-	0.0339	Adjacent grass pasture
TP (V%)	Optimum	0–15 cm	20.0	80.0	40.0	60.0	50.0	0.0644	[12,31]; Adjacent grass pasture
Chemical Properties									
CEC (cmol (+) kg^{-1})	More is better	0–15 cm	0.0	18.0	9.0	-	-	0.0757	[12]; Adjacent grass pasture
pH (1:2.5)	Optimum	0–15 cm	3.0	9.0	5.0	8.0	7.0	0.5332; −0.496	[18]
TN ($kgha^{-1}$)	More is better	0–15 cm	0.0	2000.0	1000.0	-	-	0.0007	[12]; Adjacent grass pasture
AVP ($kgha^{-1}$)	More is better	0–15 cm	0.0	50.0	25.0	-	-	0.0226	[12]
AVK ($kg{\cdot}ha^{-1}$)	More is better	0–15 cm	0.0	400.0	200.0	-	-	0.0036	[12]
Biological Properties									
SOC (gkg^{-1})	More is better	0–15 cm	0.0	10.0	5.0	-	-	0.1341	[12]; Adjacent grass pasture
MBC ($mgkg^{-1}$)	More is better	0–15 cm	0.0	300.0	150.0	-	-	0.0042	[12]; Adjacent grass pasture

Using this non-linear scoring curve equation, three types of standardized scoring functions typically used for soil quality assessments were generated: (1): More is better"; (2) "Less is better"; and (3) "Optimum" as per earlier studies [12,16,17,18,31,32]. The equation defines a "More is better" scoring curve for positive slopes, a "Less is better" curve for negative slopes, and an "Optimum" curve is defined by the combination of both positive and negative slopes. These scoring curves are presented in detail by many authors [17,18,31,32,33,34].

Statistical Analyses

Effects of different land use systems on soil quality indicators, functions and integrated quality indices were subjected to one-way ANOVA. Excel spreadsheet was used for transforming soil quality indicator values into unit-less scores. Differences between means of parameters were considered significant at the 0.05 level using the Tukey's studentized (HSD) test. The data were analyzed using R version 3.02 software package [35].

RESULTS AND DISCUSSION

Soil Physical Quality Indicators

Bulk density ranged from 1.48 $Mg{\cdot}m^{-3}$ in AF to 1.57 $Mg{\cdot}m^{-3}$ in both IR and RF land use systems (Table 4). However, there was no significant difference in BD among land uses. Although soils under AF land use contained SOC concentration twice more than that under RF, the detrimental effects of tillage may have offset the beneficial effects of SOC on BD [17,32]. Soils under AF land use also had the highest percentage of water stable aggregates (WSA) of 17.3%, but it was not significantly higher than that under IR and RF land uses. Addition of more organic matter from leaf and root litters from the *F. albida*trees in AF than the other land uses likely explains the improved WSA in AF [36]. Similarly, a study by Gelaw *et al.* [37] at the same site found that soils under natural grazing lands adjacent to cultivated lands were well structured, and contained higher SOC concentrations. Total porosity (TP) ranged from 35.5% in RF to 43.5% and 44.9% in AF and IR land uses, respectively. However, the difference among land uses was not statistically significant. Similarly, the detrimental effects of tillage may have offset the beneficial effects of SOC on TP [17,32,37].

Table 4. Effects of land use systems on selected soil physical, chemical and biological quality indicators at Mandae watershed in eastern Tigray, north Ethiopia.

Soil Quality Indicator	Land Use			
	RF	AF	IR	
Physical				
BD ($Mg \cdot m^{-3}$)	1.57 (0.03)	1.48 (0.05)	1.57 (0.02)	NS
WSA (>0.5 mm)	11.3 (1.8)	17.3 (2.5)	13.6 (3.6)	NS
TP (V%)	35.4 (3.6)	43.5 (2.0)	44.9 (2.7)	NS
Chemical				
CEC (cmol (p+) kg^{-1})	5.4 (1.0) ^{b}	11.5 (0.8) ^{a}	4.8 (1.8) ^{b}	**
pH	6.6 (0.3) ^{b}	6.4 (0.2) ^{b}	8.0 (0.03) ^{a}	***
TN ($kg \cdot ha^{-1}$)	809.7 (134.6) ^{b}	1568.6 (85.4) ^{a}	1042.7 (244.6) ^{a,b}	*
AVP ($kg \cdot ha^{-1}$)	24.4 (10.7)	39.1 (4.3)	39.8 (4.7)	NS
AVK ($kg \cdot ha^{-1}$)	216.5 (56.9) ^{b}	1019.1 (161.0) ^{a}	297.7 (71.8) ^{b}	***
Biological				
SOC ($g \cdot kg^{-1}$)	3.2 (0.7) ^{b}	6.4 (0.3) ^{a}	5.9 (1.1) ^{a,b}	*
MBC ($mg \cdot kg^{-1}$)	75.5 (24.1)	95.9 (10.3)	100.1 (31.3)	NS

RF, Dryland crop production; AF, *Faidherbia albida* based agroforestry; IR, irrigation based fruit production; ± Mean values followed by standard errors in the parentheses; values with different letters are significantly different.* $p < 0.05$; ** $p < 0.01$; NS = not significant (Tukey's test, $p = 0.05$).

Soil Chemical Quality Indicators

CEC of the soils studied ranged from the highest under AF (11.5 cmol p+ kg^{-1}) to the lowest under IR (4.8 cmol p+ kg^{-1}). It was significantly higher ($p < 0.01$) under AF than that under IR and RF land uses (Table 4). Generally, CEC was low with an exception of some improvements under AF land use. Rabia *et al.* [20] also reported similar results for the same area. Accordingly, up to 90% of soil samples from this area had extremely-low (<5)-to-low (5–15 cmol p+ kg^{-1}) CEC values [20]. EC values of the soils were also much lower than the FAO salinity hazard levels for most crops [20] (Table 4).

In general, Arenosols have neutral pH values [38]. However, soils under IR land use showed a significantly higher ($p < 0.001$) pH value than that under other land uses, and it was slightly alkaline. The source of this slight alkalinity development in the soil under IR land use could be from the supplemental irrigation. Similar results were also reported by Rabia *et al.*[20].

Soils under AF contained the highest total nitrogen (TN) stock (1568.6 $kg \cdot ha^{-1}$), and it was significantly higher ($p < 0.05$) than that in IR and RF land uses (Table 4). Hadgu *et al.* [10] reported similar results in their study in central Tigray, Northern Ethiopia, which compared TN contents of soils under canopies of *F. albida* and eucalyptus trees with those from tree-less fields. Similarly, available potassium (AVK) was significantly higher ($p < 0.001$) under AF than that under other land uses (Table 4). In contrast, available phosphorus (AVP) contents did not differ among land uses. The higher AVK under AF than that under other land uses could be related to the recycling of nutrients in the aboveground biomass, root biomass or through the recycling of depositions by cattle, which gather for shade under the tree-canopies during sunny days [39]. Sanchez [40] also reported a significant increase both in soil K content and sorghum (*Sorghum bicolar*) yield on soils under the canopy of *F. albida* trees from that on soils 15-m away in two parklands in Burkina Faso. Results presented here are also in accord with reports by Nair [41] that microsite enrichment qualities of trees such as *F. albida* in West Africa and *P. cineraria* in India have long been recognized in many traditional farming systems.

Soil Biological Quality Indicators

Both SOC and MBC are among principal soil parameters, which affect biological processes and soil quality. The highest SOC concentration was measured in AF (6.4 $g \cdot kg^{-1}$) followed by that in IR (5.9 $g \cdot kg^{-1}$), and the lowest was in RF (3.2 $g \cdot kg^{-1}$) (Table 4). Thus, SOC was significantly higher ($p < 0.05$) in AF than that in RF land use. However, it did not statistically differ between AF and IR, and between IR and RF land uses (Table 4). On the other hand, MBC was slightly higher in soils under IR (100.1 $mg \cdot kg^{-1}$) than that under AF and RF, but the differences were not statistically significant (Table 4). Higher MBC values under IR than that under AF and RF may be explained by less disturbance of soils under IR than those under the other intensively tilled land uses. The intensity of tillage in IR was less than that under AF and RF land uses. Besides, irrigation farms under guava fruits were not convenient for oxen plowing. Weed control and irrigation in IR land use were also practiced by hand. Soil organic carbon in intensively cultivated soils has less physical protection than that in less cultivated soils because tillage disrupts macroaggregates and exposes previously protected SOM microbial processes [14,37]. Similarly, Franchini *et al.* [42] reported an increase in MBC under no-till (NT) than that under conventional tillage systems (CT) receiving more plant residues in Southern Brazil. The lower MBC regardless of more plant residue addition under CT was due to higher CO_2-emissions, which implies little conversion of carbon from plant residues into MBC [42]. Indeed, parameters associated with soil microbiological activities are sensitive, considered rapid indicators of effects of soil management, and are useful as indicators of soil quality [42].

Soil Quality Indicators Integration and Assessment

For this study, four soil functions contributed to the overall soil quality index (SQI) (Table 1). They were weighted according to their relative importance in fulfilling the goals of maintaining soil quality in the area. Thus, the major driving soil parameters for the integrated SQI were SOC (26.4%), WSA (20.0%), TP (16.0%), TN (11.2%), MBC (6.4%) and CEC (6.4%). Those six soil quality indicators together contributed for more than 80% of the variability in the overall SQI (Table 2). Further, BD contributed 4.0% followed by AVP, AVK and pH with a contribution of each 3.2% to the overall SQI. Regarding the soil's function for plant nutrient supply, SOC, TN, and CEC contributed 32%, 24%, and 16% of the PNS function, respectively. Available P, AVK and pH each contributed 8% of the soil's function for plant nutrient supply. The soil's MBC contribution to this function was minimal (4%). Overall, SOC alone contributed for more than 25% and 30% of SQI and PNS values, respectively.

Integration of the soil property values into SQI using the framework resulted in a significantly higher ($p < 0.05$) score in AF than in RF land use system for its ability to accommodate water entry (Table 5). The relatively higher WSA, TP and SOC values of the soil under AF land use than those in the soil under RF were largely responsible for the improvement in its ability to accommodate water entry in AF (Table 4). Glover *et al.* [17] also reported higher scores for soil's ability to accommodate water entry because of higher WSA and lower BD under integrated and organic management systems than those under a conventional system in Washington State, USA. Regarding the soil's ability to facilitate water movement and availability, AF also scored significantly higher ($p < 0.05$) value than RF because of the relatively higher TP and SOC values in AF (Table 5). These results indicated that AF land use improved the soil's ability to hold and release water mainly due to its higher SOC content (Table 4). However, land use had no significant effect on soil's resistance to surface degradation (Table 5). This may be a clear indication of the detrimental effects of tillage on soil structure [17,32,37]. In contrast, AF scored significantly higher ($p < 0.05$) value for the soil's ability to supply plant nutrients than RF largely due to higher levels of AVK, CEC, SOC, TN and AVP in the rooting zones of AF land use (Table 5). The score for the soil under IR land use was not significantly different from that under RF (Table 5). Further, the score for nutrient storage capacity of soils under AF land use was significantly higher ($p < 0.05$) than that under RF, but it was not significantly different from that under IR (Figure 2). However, nutrient cycling was not significantly affected by land use regardless of some improvements in AF. Trees in agroforestry systems can improve nutrient cycling and increase soil chemical fertility through bringing up nutrients from deeper layers and minimizing leaching hazards [41]. In contrast, nutrient availability was affected by land use. Thus, AF scored significantly higher ($p < 0.01$) value for its capacity in nutrient availability than that in other land uses (Figure 2).

Table 5. Soil quality ratings for the different land uses at the watershed.

Soil Function	Land Use			
	RF	AF	IR	
Accommodate Water Entry (0.20)	0.09 (0.00) ^{b}	0.11 (0.002) ^{a}	0.10 (0.004) ^{a,b}	*
Facilitate Water Entry and Availability (0.20)	0.10 (0.004) ^{b}	0.12 (0.004) ^{a}	0.11 (0.004) ^{a,b}	*
Resist Surface Degradation (0.20)	0.09 (0.003)	0.11 (0.002)	0.09 (0.005)	NS
Source of Plant Nutrients (0.40)	0.19 (0.01) ^{b}	0.24 (0.004) ^{a}	0.21 (0.015) ^{a,b}	*
Integrated Soil Quality Index (1.00)	0.47 (0.01) ^{b}	0.58 (0.01) ^{a}	0.51 (0.02) ^{a,b}	**

RF, Dryland crop production; AF, *Faidherbia albida* based agroforestry; IR, irrigation based fruit production; ± Mean values followed by standard errors in the parentheses; values with different letters are significantly different. * $p < 0.05$; ** $p < 0.01$; NS = not significant (Tukey's test, $p = 0.05$).

Finally, the integrated SQI calculated for the land uses using the framework by Karlen and Stott [16] were in the following order: 0.58 (AF) > 0.51 (IR) > 0.47 (RF) (Table 5). Soil quality index differed significantly ($p < 0.01$) between AF and RF land use systems (Table 5). Similarly, Karlen *et al.* [31] reported an improvement in soil quality rating from 0.45 to 0.86 in over ten-year period by retention or addition of crop residues on a no-till (NT) continuous corn in Wisconsin, USA. In another study, Karlen *et al.* [32] reported a significant improvement in SQI ratings from 0.48 and 0.49 under plow and chisel, respectively, to 0.68 under NT using selected physical, chemical and biological soil quality indicators on Rozetta and Palsgrove silt loam soils in Wisconsin, USA. Stott *et al.* [43] in a recent study on Vertisols in Texas using the SMAF model also reported an improvement in overall SQI ranging from 75% to 94% of an optimum when compared with similar soils after 57 years of different agricultural management systems.

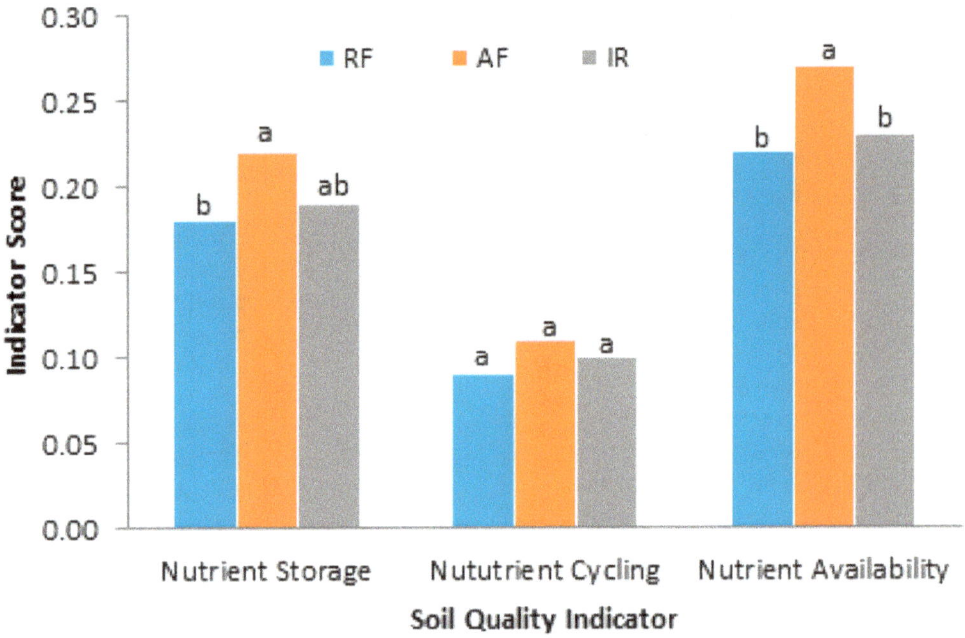

Figure 2. Effects of three agricultural land use systems (RF, AF and IR) on nutrient supplying capacities of soils at the watershed.

Regardless of a significant improvement in AF than that in RF land use, SQI ratings in all the three land use systems were very small compared with an ideal soil (Table 5). This result was in agreement with findings from other authors [5,44] who reported that low organic matter and nutrient stocks are typical characteristics of soils in Tigray, mainly due to nutrient mining because of crop harvests and complete removal of crop residues for feed and fuel. One fundamental principle of sustainability is to return to the soil the nutrients removed through harvests and other loss pathways [45], and one of the main tenets of agroforestry is that trees enhance soil fertility [45,46]. This is supported by observations of higher crop yields near *F. albida* tree canopies in Ethiopia [10,47,48,49] and elsewhere [50,51], which showed the potentials of agroforestry systems in improving soil quality and productivity of smallholder farms in Ethiopia and the wider region.

CONCLUSIONS

Relatively higher WSA, TN and SOC concentrations measured in soils under AF land use resulted in improved water entry, movement and availability than those under IR and RF. Soil's ability to supplying plant nutrients was also improved under AF than under RF land use largely due to higher levels of AVK, CEC, SOC, TN and AVP in the rooting zones of AF land use. However, there was no significant improvement in the soil's resistance to surface degradation in all land uses, which may be because of the detrimental effects of tillage. Further, when selected physical, chemical, and biological soil quality indicators were integrated into an overall SQI, AF land use received a higher soil quality rating (0.58) than that of RF (0.47). Thus, the result of this study highlighted the potentials of *F. albida* based AF systems for improving soil quality and productivity of smallholder farms in the area. Further, it demonstrated the effectiveness of the soil quality indexing framework in the study area and beyond to assess soil quality and thus recognized that changes in soil and crop management are needed for a more efficient and sustainable use of soil resources.

ACKNOWLEDGMENTS

The International Foundation for Science (IFS) (Grant No. C/4687) provided financial support for the fieldwork. Some laboratory facilities were provided by Mekelle University, Ethiopia. Practical laboratory work for WSA, SOC and TN determination was performed at the laboratory facility of the Carbon Management and Sequestration Center (C-MASC), at The Ohio State University (OSU), Columbus, OH, USA. Financial assistance from the Norwegian State Education Loan Fund for the first author is gratefully acknowledged. The department of Environmental Science (IMV), Norwegian University of Life Sciences (NMBU) is also acknowledged for paying the publishing fee.

AUTHOR CONTRIBUTIONS

Aweke M. Gelaw: Conception of the idea, designing the experiment, collecting and analyzing the data, and writing the article; Bal Ram Singh: Conception of the idea, Supervision and reviewing the article; Rattan Lal: Conception of the idea, supervision and reviewing the article.

CONFLICTS OF INTEREST

There is no conflict of interest among the authors.

REFERENCES

1. Sanchez, P.A.; Shepherd, K.D.; Soule, M.J.; Place, F.M.; Buresh, R.J.; Izac, A.M.N.; Mokwunye, A.U.; Kwesiga, F.R.; Ndiritu, C.G.; Woomer, P.L.; *et al.* Soil Fertility Replenishment in Africa: An Investment in Natural Resource Capital. In *Replenishing Soil Fertility in Africa*; Buresh, R.J., Sanchez, P.A., Calhoun, F., Eds.; SSSA and ICRAF: Madison, WI, USA, 1997; pp. 1–46.
2. Smaling, E. An Agroecological Framework for Integrating Nutrient Management, with Special Reference to Kenya. Ph.D. Thesis, Agricultural University of Wageningen, Wageningen, The Netherlands, 1993; p. 250.

3. Agricultural Transformation Agency (ATA). Status of soil resources in Ethiopia and priorities for sustainable management. In Proceedings of the Global Soil Partnership (GSP) for Eastern and Southern Africa Launching Workshop, Nairobi, Kenya, 25–27 March 2013.
4. Eweg, H.P.A.; van Lammeren, R.; Deurloo, H.; Woldu, Z. Analysing degradation and rehabilitation for sustainable land management in the highlands of Ethiopia. *Land Degrad. Dev.* **1998**, *9*, 529–542.
5. Hengsdijk, H.; Meijerink, G.W.; Mosugu, M.E. Modeling the effect of three soil and water conservation practices in Tigray, Ethiopia. *Agric. Ecosyst. Environ.* **2005**, *105*, 29–40.
6. Belete, T. Efforts for Sustainable Land Management in Tigray: The role of Extension. In *Policies for Sustainable Land Management in the Highlands of Tigray, Northern Ethiopia*; Gebremedhin, B., Pender, J., Ehui, S., Haile, M., Eds.; International Food Policy Research Institute (IFPRI): Mekele, Ethiopia, 2002.
7. Girmay, G. Land Use Change Effects in Northern Ethiopia: Runoff, Soil and Nutrient Losses, Soil Quality, and Sediment as Nutrient Sources. PhD Thesis, Norwegian University of Life Sciences, Ås, Norway, 2009; p. 140.
8. Moges, G.; Hengsdijk, H.; Jansen, H.C. Review and quantitative assessment of *ex situ* household rainwater harvesting systems in Ethiopia. *Agric. Water Manag.* **2011**, *98*, 1215–1227.
9. Feng, Z.; Wang, X.; Feng, Z. Soil N and Salinity leaching after the autumn irrigation and its impact on ground water in Hetao irrigation district, China. *Agric. Water Manag.* **2005**, *71*, 131–143.
10. Hadgu, K.M.; Kooistra, L.; Rossing, W.A.H.; van Bruggen, A.H.C. Assessing the effect of Faidherbia albida based land use systems on barley yield at field and regional scale in the highlands of Tigray, Northern Ethiopia. *Food Secur.* **2008**,*1*, 337–350.
11. Karlen, D.L.; Mausbach, J.W.; Doran, J.W.; Cline, R.G.; Harris, R.F.; Schuman, G.E. Soil quality: A concept, definition and framework for evaluation. *Soil Sci. Soc. Am. J.* **1997**, *61*, 4–10.
12. Masto, R.E.; Chhonkar, P.K.; Singh, D.; Patra, A.K. Soil quality response to long-term nutrient and crop management on a semi-arid Inceptisol. *Agric. Ecosyst. Environ.* **2007**, *118*, 130–142.
13. Brejda, J.J.; Moorman, T.B.; Karlen, D.L.; Dao, T.H. Identification of regional soil quality factors and indicators: I. Central and southern high plains. *Soil Sci. Soc. Am. J.* **2000**, *64*, 2115–2124.
14. Islam, K.R.; Weil, R.R. Land use effects on soil quality in a tropical forest ecosystem of Bangladesh. *Agric. Ecosyst. Environ.* **2000**, *79*, 9–16.
15. Sanchez-Maranon, M.; Soriano, M.; Delgado, G.; Delgado, R. Soil quality in Mediterranean mountain environments: Effects of land use change. *Soil Sci. Soc. Am. J.* **2002**, *66*, 948–958.
16. Karlen, D.L.; Stott, D.E. A framework for evaluating physical and chemical indicators of soil quality. In *Defining Soil Quality for a Sustainable Environment*; Doran, J.W., Coleman, D.C., Bezdicek, D.F., Stewart, B.A., Eds.; ASA and SSSA: Madison, WI, USA, 1994; pp. 53–72.
17. Glover, J.D.; Reganold, J.P.; Andrews, P.K. Systematic method for rating soil quality of conventional, organic, and integrated apple orchards in Washington State. *Agric. Ecosyst. Environ.* **2000**, *80*, 29–45.
18. Fernandes, J.C.; Gamero, C.A.; Rodrigues, J.G.L.; Mirás-Avalos, J.M. Determination of the quality index of a Paleudult under sunflower culture and different management systems. *Soil Tillage Res.* **2011**, *112*, 167–174.
19. WRB; International Union of Soil Science Working Group. *World Soil Resources Reports No. 103*; FAO: Rome, Italy, 2006.
20. Rabia, A.H.; Afifi, R.R.; Gelaw, A.M.; Bianchi, S.; Figueredo, H.; Huong, T.L.; Lopez, A.A.; Mandala, S.D.; Matta, E.; Ronchi, M.; *et al.* Soil mapping and classification: A case study in the Tigray Region, Ethiopia. *JAEID* **2013**, *107*, 73–99.
21. Blake, G.R.; Hartge, K.H. Bulk density. In *Methods of Soil Analysis. Part 1*, 2nd ed.; Klute, A., Ed.; American Society of Agronomy—Soil Science Society of America: Madison, WI, USA, 1986; pp. 363–375.
22. Nelson, D.W.; Sommers, L.E. Total carbon, Organic carbon and Organic matter. In *Methods of Soil Analysis. Part 3*; Book Series 5; Sparks, D.L., Page, A.L., Helmke, P.A., Loeppert, R.H., Soltanpour, P.N., Tabatabai, M.A., Johnston, C.T., Sumner, M.E., Eds.; SSSA: Madison, WI, USA, 1996; pp. 961–1010.
23. Yoder, R.E. A direct method of aggregate analysis of soils and a study of the physical nature of erosion losses. *J. Am. Soc. Agron.* **1936**, *28*, 337–351.
24. Olsen, S.R.; Cole, C.V.; Watanabe, F.S.; Dean, L.A. *Estimation of Available Phosphorous in Soils by Extraction with Sodium Bicarbonate*; USDA: Washington, DC, USA, 1954.
25. Chapman, H.D. Cation exchange capacity. In *Methods of Soil Analysis*; Black, C.A., Evans, D.D., Ensminger, L.E., White, J.L., Clark, F.E., Eds.; American Society of Agronomy: Madison, WI, USA, 1965; Volume 9, pp. 891–901.
26. Brookes, P.C.; Landman, A.; Pruden, G. Chloroform fumigation and release of soil N: A rapid direct extraction method to measure microbial biomass N in soil. *Soil Biol. Biochem.* **1985**, *17*, 837–842.
27. Vance, E.D.; Brookes, P.C.; Jenkinson, D.S. An extraction method for measuring soil microbial biomass C. *Soil Biol. Biochem.* **1987**, *19*, 703–707.
28. Wu, J.; Joergensen, R.G.; Pommerening, B.; Chaussod, R.; Brookes, P.C. Measurement of soil microbial biomass Carbon by fumigation-extraction an automated procedure. *Soil Biol. Biochem.* **1990**, *22*, 1167–1169.
29. Wienhold, B.J.; Karlen, D.L.; Andrews, S.S.; Stott, D.E. Protocol for Soil Management Assessment Framework

(SMAF) soil indicator scoring curve development. Renew. *Agric. Food Syst.* **2009**, *24*, 260–266.

30. Andrews, S.S.; Karlen, D.L.; Cambardella, C.A. The soil management assessment framework: A quantitative soil quality evaluation method. *Soil Sci. Soc. Am. J.* **2004**, *68*, 1945–1962.
31. Karlen, D.L.; Wollenhaupt, N.C.; Erbach, D.C.; Berry, E.C.; Swan, J.B.; Eash, N.S.; Jordahl, J.L. Residue effects on soil quality following 10-years of no-till corn. *Soil Tillage Res.* **1994**, *31*, 149–167.
32. Karlen, D.L.; Wollenhaupt, N.C.; Erbach, D.C.; Berry, E.C.; Swan, J.B.; Eash, N.S.; Jordahl, J.L. Long-term tillage effects on soil quality. *Soil Tillage Res.* **1994**, *32*, 313–327.
33. Wymore, A.W. Model-Based Systems Engineering. In *An Introduction to the Mathematical Theory of Discrete Systems and to the Tricotyledon Theory of System Design*; CRC: Boca Raton, FL, USA, 1993.
34. Hussain, I.; Olson, K.R.; Wander, M.M.; Karlen, D.L. Adaptation of soil quality indices and application to three tillage systems in southern Illinois. *Soil Tillage Res.* **1999**, *50*, 237–249.
35. R Core Team. R: A Language and Environment for Statistical Computing. R Foundation for Statistical Computing: Vienna, Austria, 2012. Available online: http://www.R-project.org. (Accessed on 9 March 2014).
36. Tisdale, J.M.; Oades, J.M. Organic matter and water stable aggregates in soils. *J. Soil Sci.* **1982**, *33*, 141–161.
37. Gelaw, A.M.; Singh, B.R.; Lal, R. Organic carbon and nitrogen associated with soil aggregates and particle sizes under different land uses in Tigray, Northern Ethiopia. *Land Degrad. Dev.* **2013**.
38. Hartemink, A.E.; Huting, J. Land Cover, Extent, and Properties of Arenosols in Southern Africa. *Arid Land Res. Manag.***2008**, *22*, 134–147.
39. Arevalo, L.A.; Alegre, J.C.; Bandy, D.E.; Szott, L.T. The effect of cattle grazing on soil physical and chemical properties in a silvopastoral system in the Peruvian Amazon. *Agrofor. Syst.* **1998**, *40*, 109–124.
40. Sanchez, P.A. Science in agroforestry. *Agrofor. Syst.* **1995**, *30*, 5–55.
41. Nair, P.K.R. *An Introduction to Agroforestry*; Kluwer Academic Publishers: London, UK, 1993; p. 489.
42. Franchini, J.C.; Crispino, C.C.; Souza, R.A.; Torres, E.; Hungria, M. Microbiological parameters as indicators of soil quality under various soil management and crop rotation systems in southern Brazil. *Soil Tillage Res.* **2007**, *92*, 18–29.
43. Stott, D.E.; Cambardella, C.A.; Karlen, D.L.; Harmel, R.D. A Soil Quality and Metabolic Activity Assessment after Fifty-Seven Years of Agricultural Management. *Soil Sci. Soc. Am. J.* **2013**, *77*, 903–913.
44. Girmay, G.; Singh, B.R.; Mitiku, H.; Borresen, T.; Lal, R. Carbon Stocks in Ethiopian Soils in Relation to Land Use and Soil Management. *Land Degrad. Dev.* **2008**, *19*, 351–367.
45. Sanchez, P.A. Tropical soil fertility research, towards the second paradigm. In *Transactions 15th World Congress of Soil Science*; International Society of Soil Science and Mexican Society of Soil Science: Acapulco, Mexico, 1994; Volume 1, pp. 65–88.
46. Palm, C.A. Contribution of agroforestry trees to nutrient requirements of intercropped plants. *Agrofor. Syst.* **1995**, *30*, 105–124.
47. Poschen, P. An evaluation of the Acacia albida-based agroforestry practices in the Hararghe highlands of Eastern Ethiopia. *Agrofor. Syst.* **1986**, *4*, 129–143.
48. Kamara, C.S.; Haque, I. Faidherbia albida and its effects on Ethiopian highland Vertisols. *Agrofor. Syst.* **1992**, *18*, 17–29.
49. Asfaw, Z.; Ågren, G.I. Farmers' local knowledge and topsoil properties of agroforestry practices in Sidama, Southern Ethiopia. *Agrofor. Syst.* **2007**, *71*, 35–48.
50. Kwesiga, F.; Coe, R. The effect of short-rotation Sesbania sesban planted fallows on maize yields. *For. Ecol. Manag.***1994**, *64*, 199–208.
51. Sanchez, P.A.; Palm, C.A. Nutrient cycling and agroforestry in Africa. *Unasylva* **1996**, *47*, 24–28.

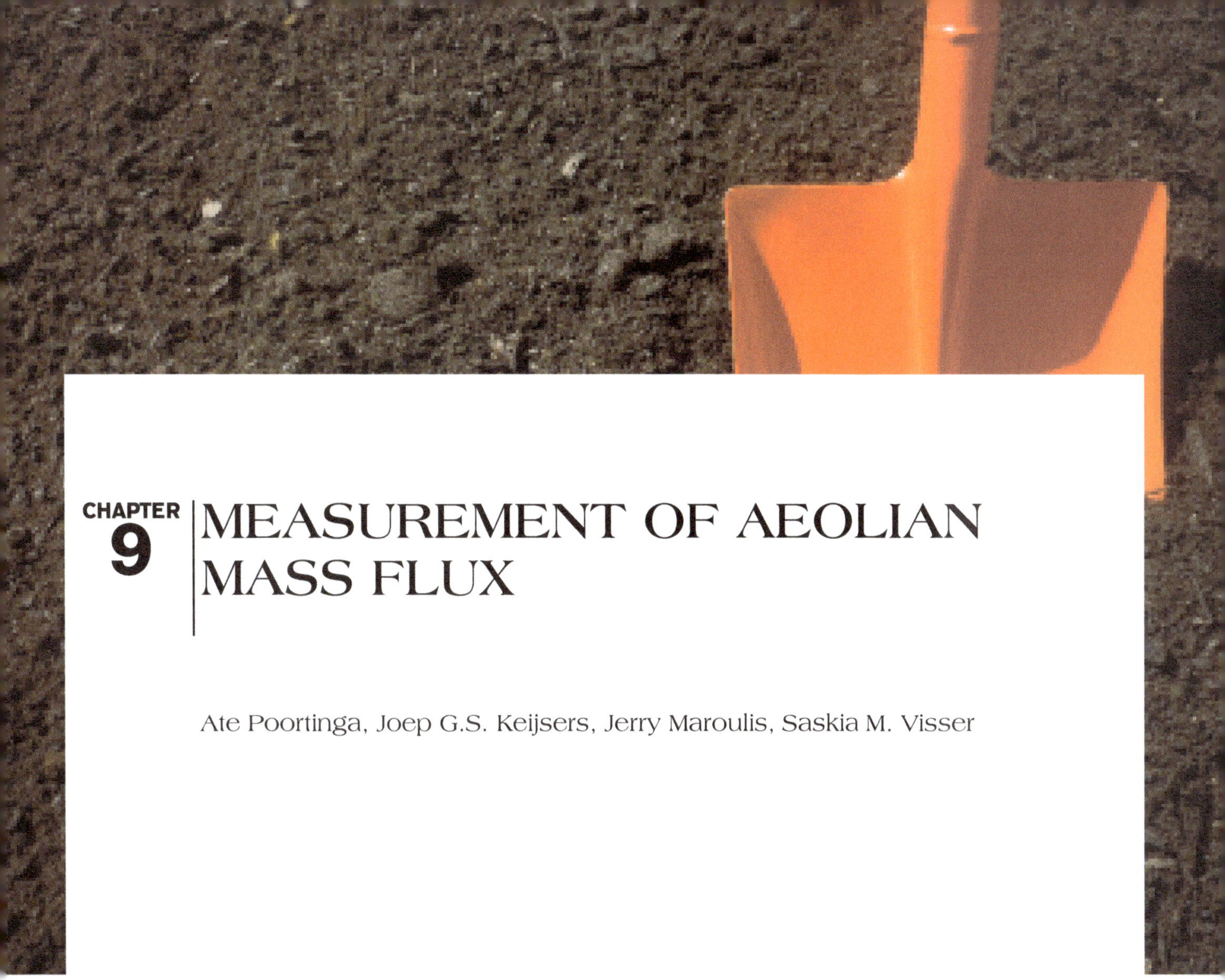

CHAPTER 9 MEASUREMENT OF AEOLIAN MASS FLUX

Ate Poortinga, Joep G.S. Keijsers, Jerry Maroulis, Saskia M. Visser

INTRODUCTION

Aeolian sediment transport is an important geomorphological process that shapes a number of landscapes including coastal (e.g., Arens, 1996; Wal, 2000; Jackson & Nordstrom, 2011), drift sand (e.g., Riksen et al., 2006;Riksen & Goossens, 2007), deserts (e.g., Bagnold, 1941; Wiggs, 2001), and also agricultural areas (e.g., Visser & Sterk, 2007; Chepil & Woodruff, 1963;Visser & Sterk, 2007). Along sandy coasts, aeolian processes drive the morphological development of coastal dunes that protects the hinterland against flooding. Maintaining the natural aeolian dynamics allows vegetation to flourish in different successive stages, creating an appealing area for tourism and recreation (Poortinga et al., 2011). In agricultural areas, however, aeolian processes are often erosive, as fertile top soil is highly susceptible to wind erosion (Nanney, Fryrear & Zobeck, 1993). Therefore, an in-depth understanding of the physical processes of wind-driven sediment transport is critically important.

It is widely recognized that aeolian sediment transport is highly variable in space and time (Baas & Sherman, 2005; Ellis et al., 2012). Despite our detailed understanding of the physics of windblown sand (e.g., Bagnold, 1941; Kok et al., 2012; Pähtz et al., 2013), accurately quantifying aeolian sediment patterns remains a challenge. Measurements of aeolian sediment budgets might improve our understanding, but often have limited spatial and temporal resolution. Approaches used to measure aeolian sediment transport include passive sediment traps (Rasmussen & Mikkelsen, 1998;Dong, Sun & Zhao, 2004; Sterk & Raats, 1996; Basaran et al., 2011;Mendez, Funk & Buschiazzo, 2011), active samplers such as acoustic samplers (Spaan & van den Abeele, 1991; Yurk, Hansen & Hazle, 2013;Ellis, Morrison & Priest, 2009; Schönfeldt, 2012), laser particle counters (Hugenholtz & Barchyn, 2011b; Sherman et al., 2011; Hugenholtz & Barchyn, 2011a; Li, Sherman & Ellis, 2011), piezoelectric samplers (Baas, 2004; Stout, 1998), pressure sensitive samplers (Ridge et al., 2011) and terrestrial laser scanners (Nield & Wiggs, 2011). The physics of wind blown sand are often studied in the controlled environment of a wind tunnel (e.g.,Youssef et al., 2008; Van Pelt, Peters & Visser, 2009; Goossens, Offer & London, 2000; Butterfield, 1999) but also directly in the field (e.g., Namikas, 2003; Ellis et al., 2012). However, results from wind tunnel studies cannot be directly translated into field situations, due to differences in turbulence spectrum, wind profile above the bed and variability in environmental factors such as surface moisture, wind direction and velocity, bed elevation, vegetation, sediment composition, lag deposits, surface crusts and fetch. Despite recent progress in rapid data acquisition,

Citation: Poortinga A, Keijsers JGS, Maroulis J, Visser SM. (2014), Measurement of aeolian mass flux, PeerJ 2:e454 https://doi.org/10.7717/peerj.454.

where aeolian sediment flux data are collected at high temporal resolution, passive sediment catchers are still frequently used to study aeolian sediment flux.

Passive sediment traps consist of various compartments located at different elevations. Sediment captured within these compartments, provides valuable information about the vertical sediment flux distribution (Ni, Li & Mendoza, 2003; Dong et al., 2003; Butterfield, 1999), which is frequently used to estimate total sediment transport (Sterk, Herrmann & Bationo, 1996; Sterk et al., 2012; Sterk & Spaan, 1997; Visser, Sterk & Ribolzi, 2004). Aeolian mass fluxes are quantified by applying non-linear curve fitting through the sediment measurements within the different compartments. However, passive sediment traps have some inherent uncertainties (average of 10%) depending on the specific distribution of sediment within the compartments and their elevation above the surface; whereas sediment mass, inlet diameter, vertical position of the catchers, trapping efficiency, horizontal spacing between catcher arrays and wind direction were also identified as potential sources of error (Tidjani et al., 2011). Moreover, variations in elevation from the lowest compartment to the ground (referred to as base elevation hereafter) may also change during the experimental measurement. The vertical distribution of the aeolian mass flux is also important here. When the largest fraction of sediment is transported close to the surface, uncertainties related to the lowest compartment become more important for estimation of total flux; even though vertical flux distribution might also vary through time.

The aim of this study was to characterize aeolian mass flux from wind tunnel and field data by comparing passive trap and high-frequency saltiphone data. Uncertainties caused by the distribution of the different trapping compartments and the influence of the base elevation were analysed for both wind tunnel and field situations. This paper will firstly examine uncertainties resulting from the distribution of different sediment trapping compartments and the influence of base elevation on fast-temporal data acquisition from saltiphones applied in a wind tunnel study. Whereas passive sediment traps collect sediment transport data for each experiment, fast-temporal data is provided continuously throughout the experiment. The latter providing a more detailed analysis of sediment flux. Secondly, the paper will explore the implications of our findings to field studies by testing data gathered from two published studies and a newly acquired dataset. The newer dataset, which utilized three customized sediment catchers, was also used to investigate the variability in vertical sediment flux and total sediment transport.

MATERIALS AND METHODS

Data Collection

Specific details about the locations and data collection methods used in the two published field studies can be found in Farrell et al. (2012) (Table 1) andVisser, Sterk & Snepvangers (2004). Results and data from the wind tunnel study can be found in Poortinga et al. (2013a) and Poortinga et al. (2013b), respectively. The data collection procedure used for the new dataset is presented below, while the data of the present study and Visser, Sterk & Snepvangers (2004) can be obtained from http://dx.doi.org/10.1016/j.envsoft.2003.12.010.

Table 1: Type, number and temporal resolution of instruments used during the field experiment on Ameland. The spatial distribution is shown in Fig. 1B.

Instrument	Number	Use	Temporal resolution
Anemometers	3	Wind velocity profile	1 min
Windvane	1	Wind direction	1 min
Tipping bucket	1	Amount of rainfall	1 min
Saltiphones	2	Transport intensity	1 min
MWAC's	18	Sediment flux	Event

Study Area

The research took place from November to December 2010 on a beach at the north-western end of Ameland, one of the West Frisian barrier islands located in the northern extremity of The Netherlands (Fig. 1A). The site is characterized by strong wind and wave dynamics in constructing bedforms and embryonic dune development. Human influence on this part of the beach is minimal compared to the middle section of the island. The study area is located east of a sand bar, which attached to the island in the mid 1980s, causing a progressive, attenuating sand wave to the East (Cheung, Gerritsen & Cleveringa, 2007), resulting in relatively wide beaches (>150 m).Figure 1C was taken from the top of the foredune and shows the experimental site at low tide.

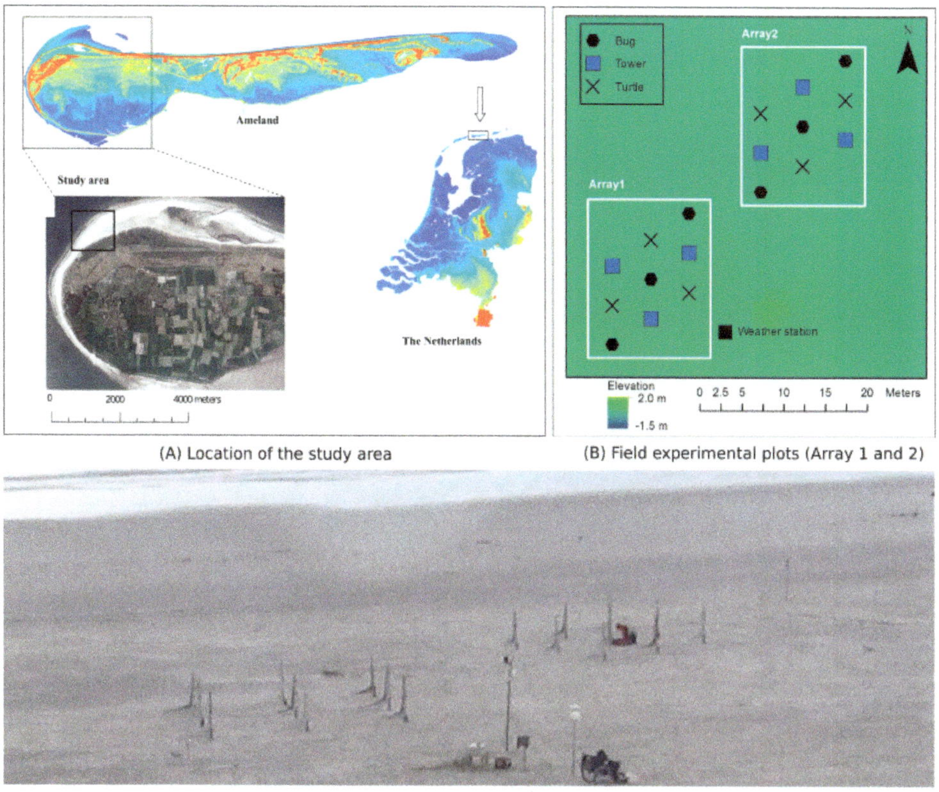

(A) Location of the study area

(B) Field experimental plots (Array 1 and 2)

(c) Photograph showing experimental setup

Figure 1: (A) Location of study area in northern Netherlands, on the island of Ameland and in the aerial photo of the western portion of Ameland. (B) Field experimental setup with specific equipment configuration. (C) Experimental plot at a location along the beach.

Measurement of Sand Size

Data was obtained on sediment characteristics, sediment transport and a number of significant meteorological parameters. Surface sands are largely composed of unconsolidated quartz grains with some feldspar and a small fraction of heavy minerals (Wal, 2000). To determine sediment size, samples were taken from the beach surface at a number of representative locations across the beach and mixed into one large sample. This sample was dried and sieved in fractions of 50, 100, 250, 500, 1000 and 2000 µm. The median diameter was found to be 180 µm.

Measurement of Sediment Flux

Sediment flux was measured using the Modified Wilson and Cook sediment catchers (MWAC). These catchers are designed and used for capturing sediment ranging from dust to sand. This instrument has been extensively tested in numerous studies (e.g., Van Pelt, Peters & Visser, 2009; Sterk & Raats, 1996; Goossens, Offer & London, 2000; Poortinga et al., 2013a;Youssef et al., 2008), where efficiencies between 42% and 120% were reported. The original design (Wilson & Cooke, 1980) contained six plastic bottles with glass inlets and outlets, placed horizontally at six heights between 0.15 and 1.52 m. These bottles were mounted on a rotating pole with a wind vane. Later studies (e.g., Sterk & Raats, 1996) used the same principle, but placed the bottles vertically instead of horizontally (Fig. 2A). Under beach conditions, aeolian sediment transport is governed by saltation, which seldom reaches heights above 15–20 cm. A traditional MWAC sediment catcher would therefore only capture sediment in the lower two or three bottles. This generates significant uncertainty in the analysis, as sediment flux is calculated based solely upon the fitting of an exponential curve through only two or three data points. Therefore, three different designs based upon the traditional MWAC were introduced (Fig. 2), but with all bottles mounted below 25 cm. The first design, nicknamed the "Bug" (Fig. 2B), consists of two stacks of three bottles opposite each other, with their inlets at the same height. The bottles are fixed to a wooden plate with an iron thread to ensure the vertical distance between the two inlets is 5 cm. This design allows for the collection of more measurement points in case of any small horizontal variations in sediment flux, thereby reducing any uncertainty in flux calculations. The second design "Turtle" (Fig. 2C) consists of two bottles on each side, located at various heights. In this design, the bottles are fixed in the original clips, resulting in a larger vertical spacing of 8 cm. In both designs, the horizontal distance between the inlets is 22 cm. The "Tower" design (Fig. 2D) represents the more traditional setup with three or four bottles stacked above each other. The vertical spacing between the bottles is 5 cm, as the bottles are fixed to a wooden plate with iron thread instead of the conventional clips.

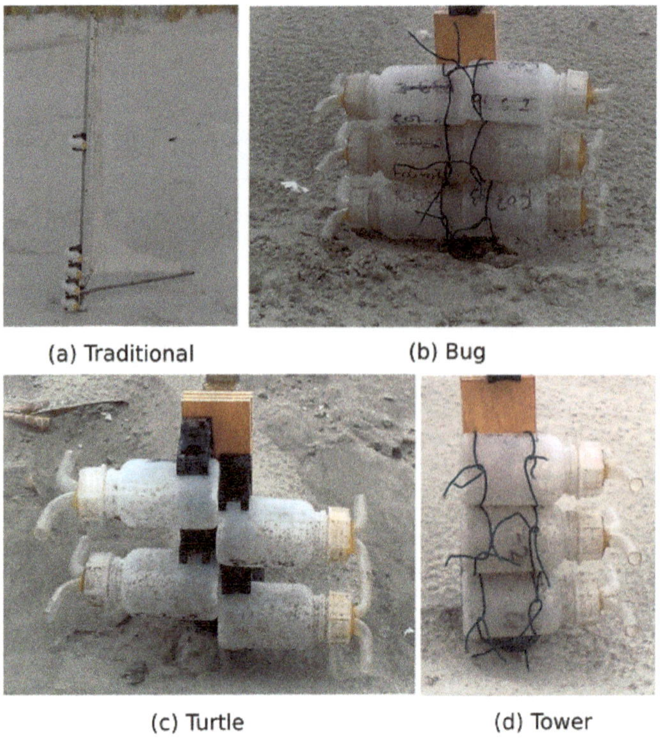

Figure 2: The traditional MWAC design with the 3 new modified designs. The traditional design (A) consists of 5 bottles distributed over 1 m. The "Bug" design (B) consists of a total of 6 bottles (3 on each side), the "Turtle" design (C) consists of 4 bottles (2 on each side) in the original clips, while the "Tower" design (D) consists of 3 bottles mounted above each other. Each bottle measures 10 cm high with a diameter of 4.5 cm.

To evaluate the differences between the three new designs, they were placed in a 3 × 3 grid and separated by a distance of 3 m. After the first event, an additional array of MWACs was installed 8 m from the first array in order to obtain more measurements (Fig. 1B). To ensure careful monitoring of the experiment, this second array was only installed when environmental conditions were favourable. Each array contained three catchers of each type. They were placed in a relatively flat and homogeneous part of the beach to ensure that the measured sediment flux was uniformly distributed. The elevation of the bottles relative to ground level was measured to an accuracy of 1 mm in order to account for changes in base elevation due to ripples; this was done after installation and before removal of the bottles.

Weather Data

A meteorological station with four anemometers was arranged as a vertical array on a tower, which included a wind vane, tipping bucket and two saltiphones (Spaan & van den Abeele, 1991), installed on the beach in the middle of the study area (Fig. 1B), recording every minute to a CR10 Campbell datalogger (Table 1) throughout the period of investigation. The on-site meteorological station contained 3 anemometers, measuring wind speed (ms^{-1}) at elevations of 0.54, 1.15 and 1.76 m. Pulses from the anemometer were averaged over the recording period and registered as average wind velocities per minute. Wind direction was measured using the wind vane at a height of 2.5 m, while the tipping bucket recorded rainfall to an accuracy of 0.2 mm.

In order to capture the temporal variability in transport intensity, two saltiphones were placed close to the surface at different locations in the experimental area (Fig. 1B). The saltiphones were also connected to a CR10 datalogger with a digital pulse output signal. For every second, the cumulative number of hits for that second were recorded.

Data Analysis

Vertical Distribution of Aeolian Mass Flux

When using passive sediment traps, sediment is trapped in different compartments that are located at different elevations above the surface. Sediment from each compartment was weighed and then plotted against elevation from which a non-linear regression was calculated to estimate total sediment transport. Despite various thoughts on whether to use an exponential, power of five parameter regression curve, the recent literature (Ellis et al., 2009) suggests that an exponential decay function (Eq. (1)) is most appropriate to describe aeolian sediment transport.

$$q_z = q_0 e^{-\beta z} \tag{1}$$

Curve fitting using Eq. (1) enables us to determine the coefficients q_0 and β, also referred to as the portion of creep (q_0) and decay (β), where z (m) represents the elevation and q_z (kg m^{-2}) the amount of sediment at elevation$z$. Regression coefficients q_0 and β can subsequently be used to calculate the total amount of sediment transport Q (kg m^{-1}). This is done by the integral of Eq. (1) over the height of the saltation layer (taken as 1 m). Sediment fluxes were expressed in kg m^{-1} rather than kg $m^{-1}s^{-1}$, as transport was highly intermittent during some events.

The cumulative transport function (CTF) of an aeolian mass flux (q_c) can be described by Eq. (2), using coefficient

β from Eq. (1). Figure 3 illustrates the relative sediment flux (black line) and the CTF (green line). When studying the characteristics of aeolian sediment flux, the CTF is preferred to the relative sediment flux, as this function is independent of the number of measurement points. Moreover, only coefficient β is used in the calculation, and therefore, the shape of the CTF is determined by the specific mass distribution between the different compartments and not by their elevation above the ground.

$$q_c = 1 - e^{-\beta z} \tag{2}$$

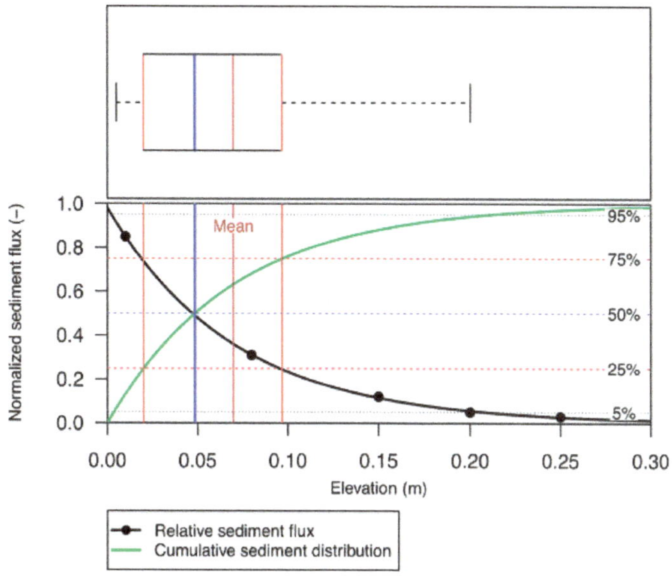

Figure 3: The vertical distribution of relative aeolian sediment flux (points), the non-linear regression (Eq. (1)) fitted through the data-points and the cumulative sediment distribution calculated from regression coefficient $\boldsymbol{\beta}$. The median (blue, Eq. (4)), mean (brown, Eq. (3)), upper and lower quantile (red, Eqs. (5) and (6)) are also shown as a boxplot.

Coefficient β (Eq. (1)) can also be used to determine the mean (Eq. (3)), median (Eq. (4)), and lower (Eq. (5)) and upper quartile (Eq. (6)). Figure 3shows the distribution function as a box-plot (top) and also for the relative sediment flux and CTF (bottom). The difference between the mean (brown line) and median (blue line) is the mean is calculated by the integral and the median by the point where the integral is 0.5. The median splits the CTF into two equal parts, whereas the mean describes the point where the CTF would balance. As the median is less sensitive to outliers compared to the mean, we make use of the median.

$$\bar{q_z} = \frac{1}{\beta} \tag{3}$$

$$q_{z50} = \frac{ln(2)}{\beta} \tag{4}$$

$$q_{z25} = \frac{ln(\frac{4}{3})}{\beta} \tag{5}$$

$$q_{z75} = \frac{ln(4)}{\beta}. \tag{6}$$

Dong et al. (2003) performed a series of wind tunnel experiments to investigate the flux profile of wind-blown sand. They determined the cumulative mass distribution from the measured data. Moreover, they used the equation $q_z = q_0 e^{-b/z}$, where the regression coefficient β (here given as$b$) is divided by elevation (Eq. (1)). Regression parameter β (as in Eq. (1)) can be calculated by $\beta = 1/b$. The q_{z50} for the different sediment size fractions and wind velocities, using β, is shown in Fig. 4.

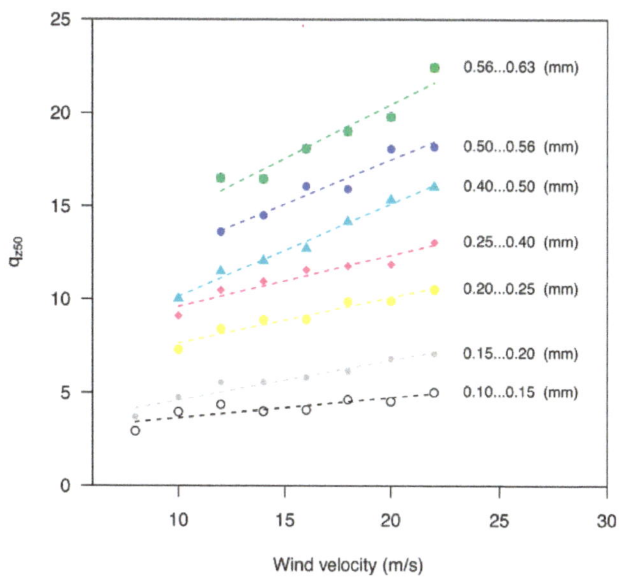

Figure 4: The q_{z50} **for different sediment fractions and wind velocities.** Data were recalculated from Dong et al. (2003).

Figure 4 shows the variation in q_{z50} for the various sediment size fractions over a range of wind speeds, especially where coarser sediments are transported at higher elevations, with q_{z50} increasing with wind velocity.

Uncertainties in Estimation of Aeolian Mass Flux

Ellis et al. (2009) identified three common methodological inconsistencies and thus sources of uncertainty in measuring aeolian sediment transport using passive traps. These include: (1) inconsistent representation of sediment trap elevations; (2) erroneous or sub-optimal regression analysis; and (3) inadequate or ambiguous bed elevation measurements.

In addition, the number of trapping compartments and location of the lowest sediment trap are also important considerations. Results from Dong & Qian (2007) (Table 1) were used to illustrate how base elevation and number of traps affects sediment flux estimation. They made use of a WITSEG sampler (Dong, Sun & Zhao, 2004), which is a vertically integrated wedge-shaped trap with 60 different compartments, where the lowest orifice can be aligned with the surface. The high data density of the WITSEG is advantageous when interested in a detailed description of the vertical mass distribution.

Dong & Qian (2007) determined the relative sediment flux (using Eq. (7)), where the relative sediment flux (qr_z) at height (z) is calculated by dividing the measured sediment flux (qz) by the total amount of sediment (Q) collected within all compartments. The dimensionless relative height (Zr) was calculated by dividing the actual height (z) by the maximum height (Z; 0.6 m in their study). After fitting a non-linear regression (Eq. (1)) through the relative sediment flux data, they found a linear correlation between the regression coefficients q_0 (portion of creep) and β (decay function).

$$qr_z = \frac{qz}{Q}, Zr = \frac{z}{Z}. \tag{7}$$

Figure 5 displays the dimensionless regression coefficients q_0 and β. Using elevation data of the different compartments, we calculated the relative regression coefficient q_0 for a sequence of β's, while changing the elevation from the base (lines with different colors), but using the same distribution of compartments. Measurements using the WITSEG were taken between 0 and 1 cm, which is in agreement with the experiments. Here, it is important to note the difference in shape between the different base elevation lines. When measurements are taken close to the surface, the correlation betweenq_0 and β is almost linear, for the domain under consideration. However, when moving further away from the surface, the relationship becomes log-linear (Fig. 5), which has major implications in terms of generating uncertainty in the estimation of q_0. Where measurements are taken further away from the surface, a small error in the calculation of β has even greater impacts upon estimating q_0 compared to measurements taken closer to the surface. An under- or overestimation in the q_0 regression parameter can have a significant effect on determining the total mass flux.

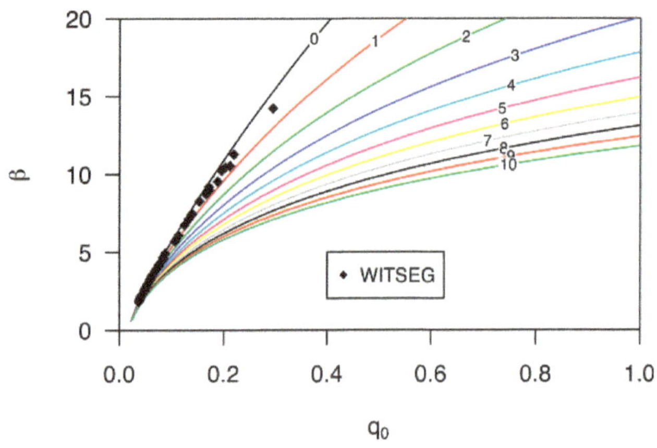

Figure 5: The regression coefficients q_0 **and beta calculated from the relative sediment flux** (Eq. (7)) for the WITSEG (data from Dong & Qian, 2007; Table 1). The coloured lines represent the relation between the q_0 and beta for different base elevations (shown on the plotted line in cm).

The vertical cumulative mass distribution of the aeolian mass flux was investigated for each of the previous studies as well as for the newly collected dataset. The spatial variability for q_{z50} and Q was also investigated for this newly acquired dataset. Due to the relative limited number of datapoints and the desire to maintain the original values in the interpolated maps, we used a simple inverse-distance weighting algorithm, with a minimum of three and a maximum of eight neighbours for spatial interpolation.

RESULTS AND DISCUSSION

Wind Tunnel Data

The wind tunnel study of Poortinga et al. (2013a) was used to investigate the CTF of an aeolian saltation cloud (data available in Poortinga et al., 2013b). Sediment loss was measured in three different ways using passive sediment catchers, saltiphones and a balance. The normalized sediment flux and CTF, calculated from four saltiphones at various heights above the surface (the highest at 25 cm), are presented in Fig. 6. The BEST, MWAC new and MWAC old represent the three different types of catchers used in the experimental runs. The fit between the non-linear regression line and calculated sediment flux had an average $r^2 = 0.99$, with a minimum $r^2 = 0.96$. The data was divided into high wind velocities (Figs. 6A, 6C, 6E) and low wind velocities (Figs. 6B, 6D, 6F), where the s50, s60 and s80 represent different sediment sizes with a *d*50 of 285, 230 and 170 µm, respectively. At higher wind velocities, more sediment is transported closer to the surface for the s50 and s60 sediment with a qz50$^-$ of 6.71 and 7.94 cm (s50) and 5.17 and 5.61 (s60) for the high and low wind velocities, respectively. In contrast, the s80 (fine) sediment, qz50$^-$ of 3.68 and 2.84 cm was measured for high and low wind speeds, respectively. The s80 (finer) sediment is transported more readily closer to the surface than coarser sediment fractions (s50 and s60), which is in agreement with the literature (Farrell et al., 2012; Dong et al., 2003; Dong & Qian, 2007).

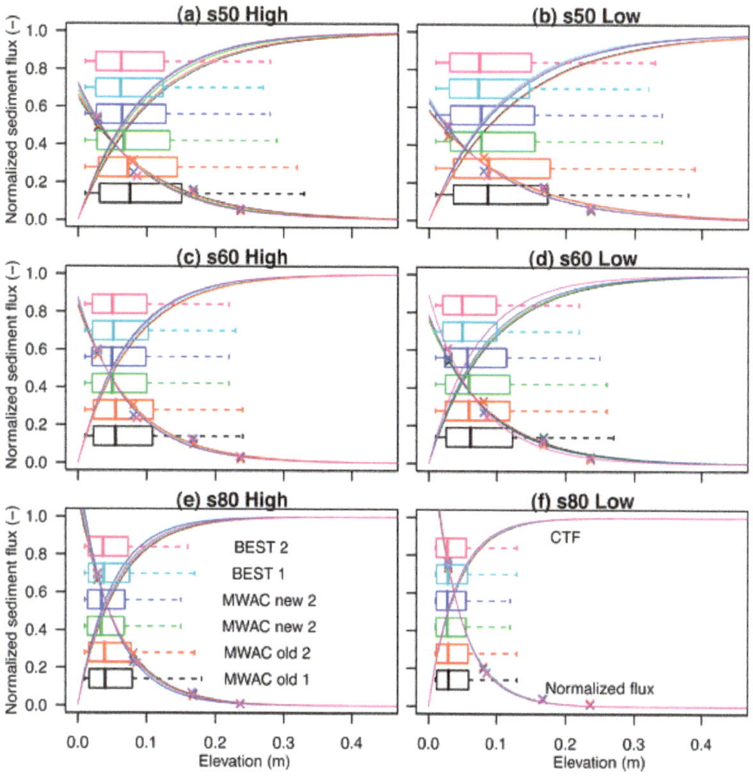

Figure 6: The relative sediment flux (Eq. (7)) and CTF (Eq. (2)) for three different types of sediment (s50, s60 and s80, with d_{50} **285, 230 and 170 μm, respectively), three different sediment catchers (MWAC old, MWAC new and BEST) and exposed to high (A, C, E) and low (B, D, F) wind velocities.** The box plots indicate the median, upper and lower quantile.

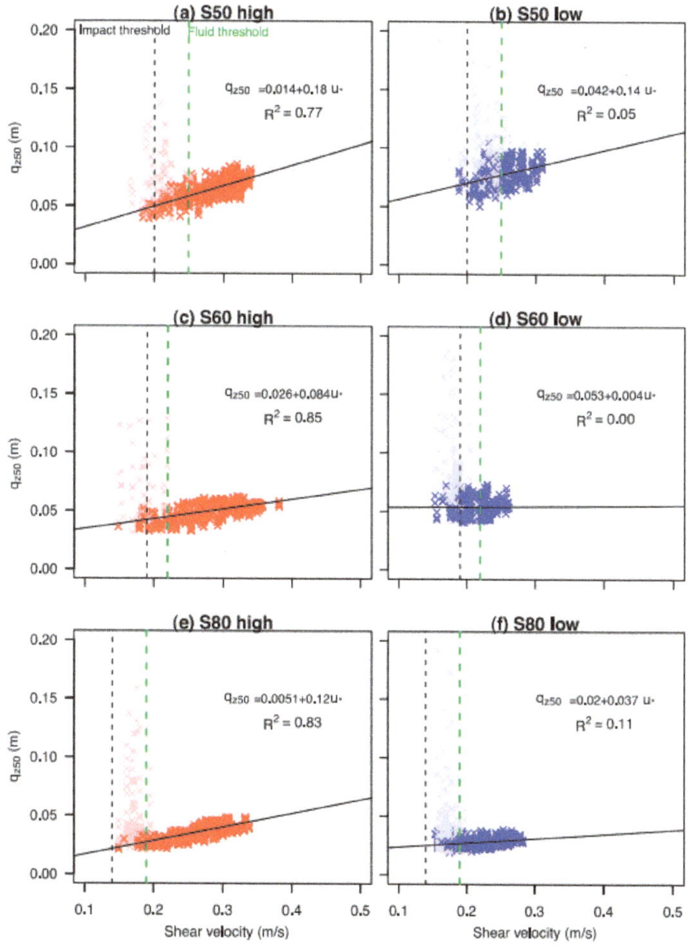

Figure 7: The shear velocity versus the q_{z50} **for three different types of sediment sizes (s50, s60 and s80) under high (A, C, E) and low wind velocities (B, D, F).**

Saltiphones were also used to rapidly acquire aeolian sediment flux data to enable a detailed investigation of the vertical sediment dynamics. Non-linear regression (Eq. (1)) was applied to the data points, and for all fluxes and data with $R^2 > 0.98$, the β was used to calculate q_{z50} (Eq. (4)). In Fig. 7, $q_{z50}$is plotted against shear velocity for experiments under high (Figs. 7A, 7C,7E) and low wind velocities (Figs. 7B, 7D, 7F). As shown in Fig. 6, finer sediment has a lower q_{z50} compared to coarser sediment. Despite considerable scatter, the median q_{z50} values increase with increasing shear velocities. In the region between the impact (vertical black dotted line: Fig. 7) and fluid thresholds (green dotted line), the scatter is considerable. This scatter (indicated with an alpha color), represents measurements with low sediment flux which is more pronounced in low wind velocities. A linear regression curve was calculated for the high and low wind shear velocities (straight line). While there is some correlation between shear velocity andq_{z50} under high shear velocities (see R^2 of the linear regression in Fig. 7), theR^2 under low wind velocities is rather low. All plots show a positive correlation between shear velocity and median q_{z50}.

In the wind tunnel experiment, use was made of three different types of sand catchers: the MWAC old, MWAC new and BEST (cf. Poortinga et al., 2013a). The findings in Fig. 7 were used to validate the results of these sediment catchers: q_{z50} of the measured sediment flux was calculated for each experiment; and, q_{z50} based on the mean shear velocity during the experiment, was calculated using data from Fig. 7. Figures 8A–8C showsq_{z50} based on the values of the sediment catchers (before) and the values calculated from the saltiphones data (after). Differences between the two values increase from coarse to finer sediment. For measurements using BEST, the differences are generally larger than the other catchers (Fig. 8).

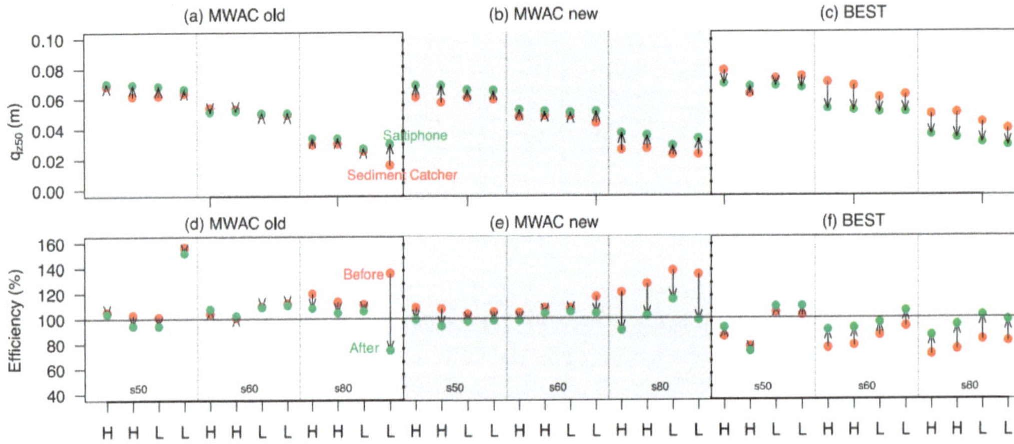

Figure 8: The q_{z50} **from** saltiphones (green) compared with q_{z50} **from the sediment catchers (red) (A–C) and the recalculated efficiency (D–F) using the** q_{z50} **as a reference.** Data is shown for three different sediment catchers (MWAC old, MWAC new and BEST), three different types of sediment (s50, s60 and s80) under high (H) and low (L) wind velocities. During the experiment, the sediment catcher and saltiphones were located next to each other. The arrows (A–C) indicate the shift in q_{z50} used to calculate the new base elevation. The arrows (D–F) indicate the change in efficiency.

To test whether q_{z50} calculated from the saltiphones (Figs. 8A–8C) provides a better approximation of total sediment flux, q_{z50} was used as a reference point to reposition the base elevation; with the difference between the sediment catcher guiding the repositioning of the traps. The sediment flux was recalculated using this new base elevation. New sediment fluxes were then compared with sediment loss measured by a balance. Figures 8D–8Fshows the efficiency of the initial sediment flux estimation (red) and the newly calculated sediment flux (green), where 100% is an exact match with the balance. Some 29 of the 36 measurements were shown to indicate an improvement (Fig. 8). In general, improvements are considerable, and a decrease in efficiency is minimal. For finer sediment, improvements were even higher when compared to coarser sediment.

The relative sediment flux of the saltiphones and sediment catchers were used to determine q_0 and β (Fig. 9). When the lowest saltiphones were located at 3 cm, we found a strong linear relationship between q_0 and β. Finer sediment had a larger range of regression coefficients, with higher values for β given that a higher proportion of sediment is transported closer to the surface. The intercept of the linear regression increases with coarser sediment, whereas the slope of the regression decreases. The difference between the intercept and slope of the s50 and s60 sediment is small. For passive sediment catchers, there is good agreement between the calculated base elevation and the experimental results (Fig. 9). The BEST catcher was located 1.5 cm from the surface, whereas the lowest trap of the MWAC catchers was located between 4 and 5 cm. Mean measurement error was 1.3 mm with a maximum of 2.4 mm.

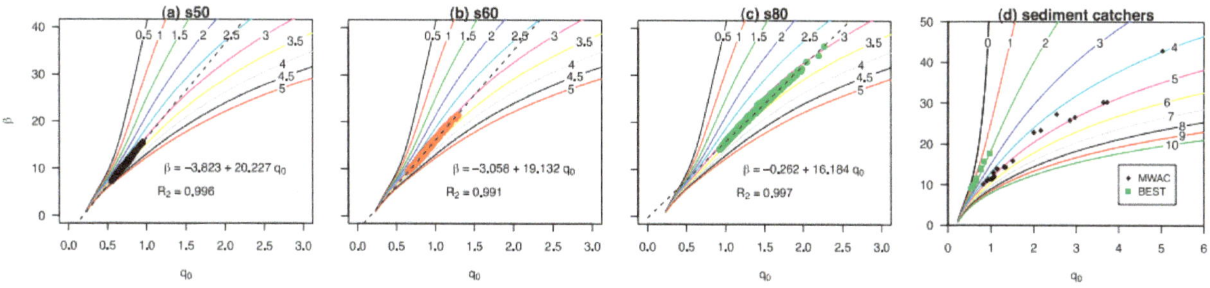

Figure 9: The q_0 **and** β, calculated from the relative sediment flux (Eq. (7)) for different sediment sizes (s50, s60, s80) using saltiphones data (graph A–C), and for all measurements using passive sediment traps (graph D). The lines represent different base elevations.

The disagreement in vertical flux distribution between the saltiphones and sediment catchers, and also between sediment loss measured by the balance and the calculated flux from the sediment catchers, is mainly caused by the specific configuration of the sediment catcher. For instance, when applying an exponential regression function, the elevation, orientation and measurement accuracy of the lowest bottle largely determines the result as finer sediment is more susceptible to errors compared to coarser sediment.Figure 10 presents the experimental outputs when using the MWAC and BEST catchers with s80 (fine) sediment, highlighting the measured relative sediment fluxes (black dots), including the exponential regression (Eq. (1)) and a linear regression. The BEST catcher contains one data-point below the q_{z50} while the MWAC has none. The influence of the lowest data-point is significant, as it determines the intersection with the y-axis and thus the total sediment flux. As for fine sediment, errors will be more pronounced as a larger portion of the mass is transported close to the surface, there are small inconsistencies in the orientation of the catcher, and thus measurement issues occur with the elevation or difference in efficiency under different mass flux density. Applying a linear function to the points close to the surface, and a power function for the higher located points (cf. Poortinga et al., 2013a), will therefore give more coherent results, as the effect of the lowest point on the total mass flux is reduced. Moreover, Ni, Li & Mendoza (2003) showed that saltating grains follow an exponential decay function, whereas creeping and reptating grains deviate from it. The mathematical description might therefore also be a source of uncertainty.

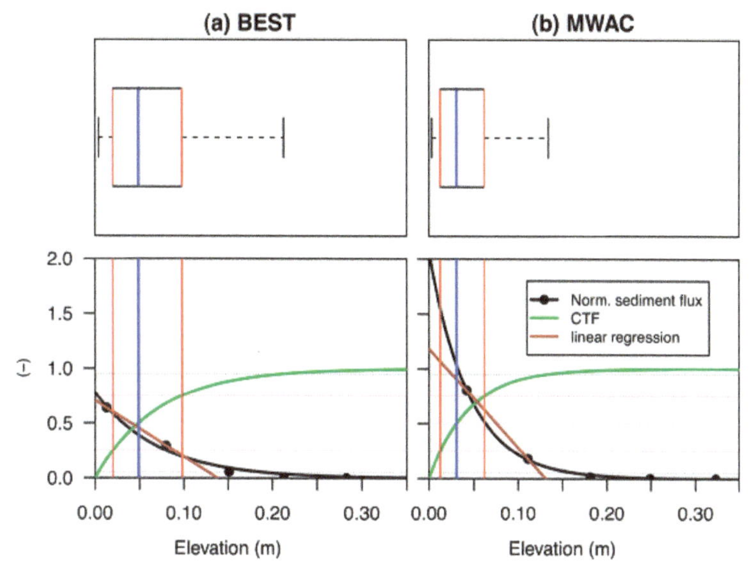

Figure 10: The vertical distribution of the relative aeolian sediment flux for the BEST and MWAC sediment catchers. The dots identify individual measurements for the s80 sediment size, while the non-linear regression curve is shown in black and the CTF in green. Furthermore, a linear function was plotted through the two points located closest to the surface (brown). The q_{z50} (blue:Eq. (4)), $\bar{q}$ (brown: Eq. (3)), upper and lower quantile (red: Eqs. (5) and (6)) are also shown as a boxplot.

Field Data

Where wind tunnel studies are limited in the replication of complex turbulent wind structures as seen in the field, field studies do not have the advantage of a controlled environment where specific parameters can be fixed. Surface moisture and bedform development, for instance, are known as important limiting factors in sediment transport, and can negatively affect measurements. Data from Farrell et al. (2012) were used in a re-analysis because their short-lived experiments contained several data points close to the surface. For the sub-environment Cow Splat Flat Fine (CSFF), q_{z50} were arranged according to date (Fig. 11A) and q_0 and β were calculated for the relative flux (Fig. 11B). For this sub-environment, the calculated elevation from the surface strongly agreed with the measured values. The variation inq_{z50} was best explained when arranging them according to measurement date; where no relation was found with shear velocity (ranging from 0.45 to 0.54 ms^{-1}) or grain size. A logical explanation would be the effects of surface characteristics such as surface moisture and incipient bedform development. However, this is far from conclusive, as it was also found that q_{z50} increased with decreasing R^2 (ranging from 0.968 to 0.999). The same study also took three measurements at the beach sub-environment over two consecutive days. These measurements received specific attention, as they were taken at a wet and immobile foreshore without visible bedform deformation. We found q_{z50} values of 3.7, 4.5 and 3.1 cm with R^2 of 0.966, 0.890 and 0.997, respectively. As this dataset only contains three data-points with varying R^2, it is difficult to draw conclusions based on q_{z50} or measured base elevation.

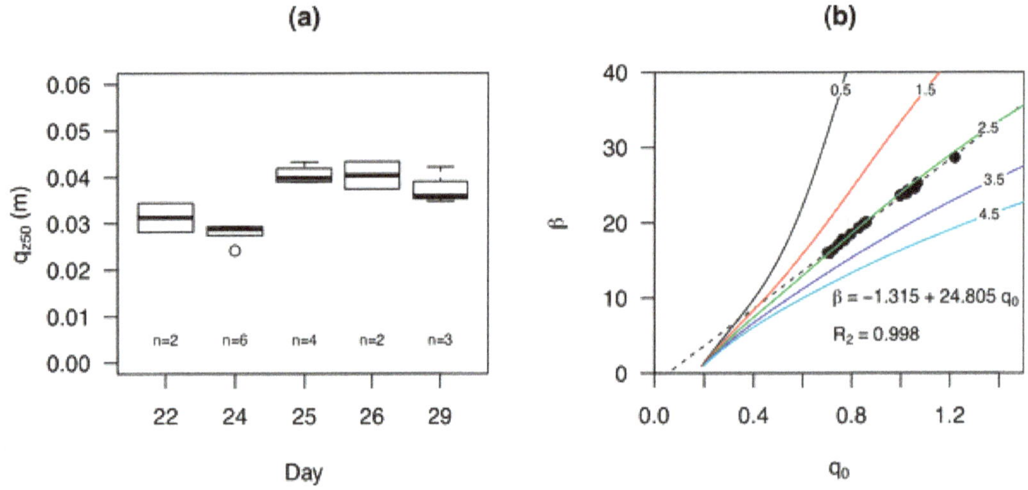

Figure 11: The q_{z50} **for experiments performed (at the sub-environment Cow Splat Flat Fine (CSFF)) on different days (A) and the** q_0 **and** β **for all events combined (B).** Coloured lines represent different base elevation. Data was obtained from Farrell et al. (2012).

Visser, Sterk & Snepvangers (2004) conducted experiments on three different geomorphic units: degraded, valley and dune. Besides sand, the soils in this area also contained considerable quantities of silt: 19.4, 15.9 and 13.0% and clay 21.6, 5.1 and 3% for the degraded, valley and dune site, respectively. The study obtained results for 11 different events in the year 2001, with 17 MWAC catchers installed at each site. In order to remove uncertainty from the data while maintaining an acceptable number of data points, measurements with $R^2 < 0.95$ were removed from the dataset. This differs from previous studies, where an $R^2 < 0.98$ was used as measurements were taken over longer periods. Furthermore, when comparing the different units (Figs. 12A, 12C, 12E), it was found that q_{z50} is highest for the degraded site, followed by the valley and dune site. The degraded and valley site have higher fractions of silt and clay, which are transported over higher elevations. However, surface crusts might also cause saltating particles to reach higher elevations. The variation in q_{z50} within an event is generally low for the degraded and valley site, but slightly higher for the degraded site. The variation between events is also small, except for the events on 10 and 13 July (Fig. 12). Here the values for q_{z50} are high and have a large variation. During these events, large amounts of dust were transported through the study area. No clear relation was found between q_{z50} and wind velocity.

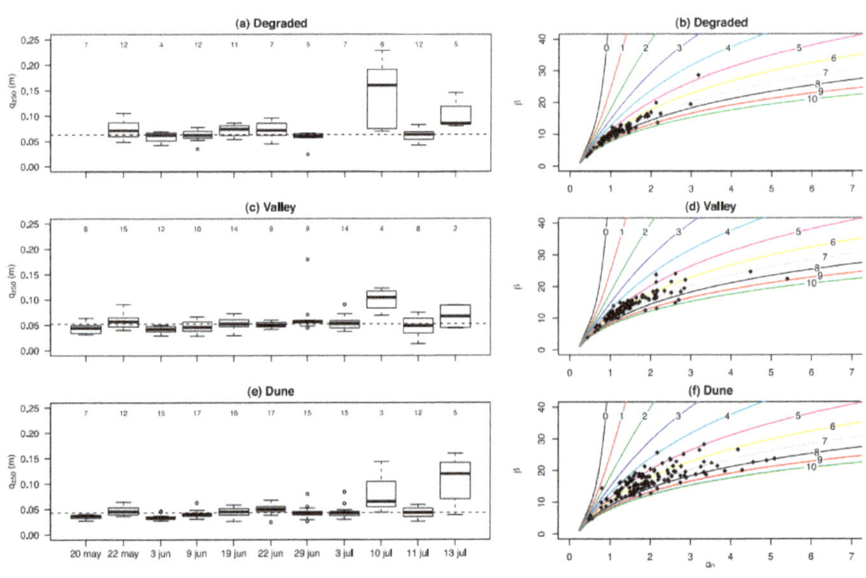

Figure 12: The q_{z50} **for 12 events in three different geomorphic units (A, C, E) and the relation between** q_0 **and** β **(B, D, F), where the lines represent different base elevations.** Data was obtained from Visser, Sterk & Snepvangers (2004). Numbers indicate the number of measurements included.

Figures 12B, 12D, 12F shows q_0 and β, calculated from the relative sediment flux. As surface elevation varied for the different measurements, the points are plotted on different curves. Due to a lower decay rate (coefficient β) at the degraded site, points are still closely related. However, at the dune site, sediment is transported closer to the surface, resulting in higher decay rates. As the lines spread with higher decay rates, there is higher spread in points. Compared to the degraded site, there is larger uncertainty in q_0 for the dune site, as small errors in β will result in larger errors in q_0 (Fig. 12).

Figure 13 displays the uncertainty in q_0 for one event (May 22) at the dune site. Here, the measured elevation is shown in red; where elevation is based on the relative q_o and β, shown in green. For this event, all calculated elevations are lower compared to the measured. This indicates that there is a likely error in the measured base elevation, leading to an overestimation ofq_0. As expected, errors in q_0 are largest for the dune site, followed by the valley and degraded site. However, the larger error in q_0 does not directly correspond to a larger error in base elevation. Elevation here was estimated using $z = ln(q_z/q_0)/\beta$ (Eq. (1)), with the higher decay rates at the dune site having a more pronounced effect on Eq. (1) than the larger range of q_0. For flux estimation, on the other hand, small changes in elevation have a much larger impact, as a greater portion of transport takes place close to the surface.

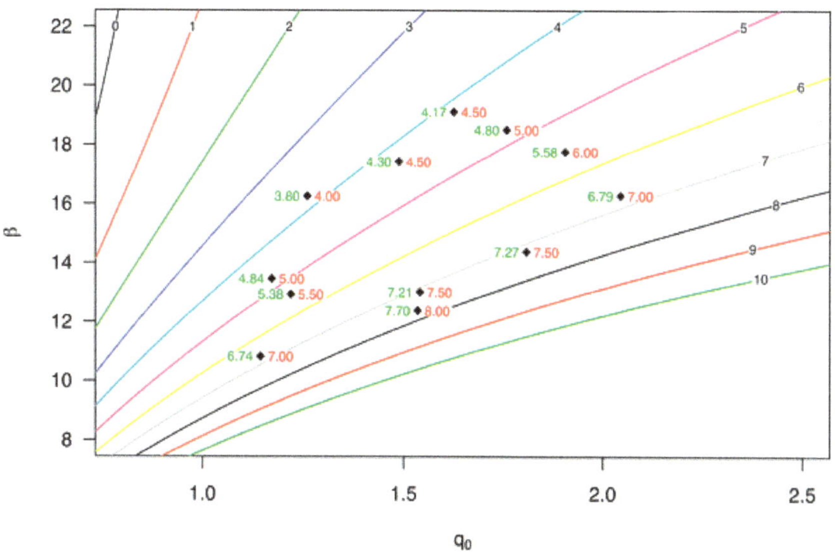

Figure 13: The q_0 **and** β **calculated from the relative sediment flux for the Dune site (May 22).** The lines represent different base elevations. The red numbers are the measured elevations whereas green values are the calculated elevations based on relative q_0 and β.

The Field Experiment

New data were collected for six different events (Fig. 14). The duration of the experiments varied from a few hours to two days, whereas saltation was measured for several hours. The averaged shear velocity during these saltation periods varied between 0.30 ms^{-1} for event 1 and 0.41 ms^{-1} for event 3. Wind directions predominantly came from the E-NE while only event 3 had variable wind conditions (Fig. 14). During events 2, 3 and 4, rainfall was recorded. Saltiphone data were also included as an indication of the degree of saltation activity.

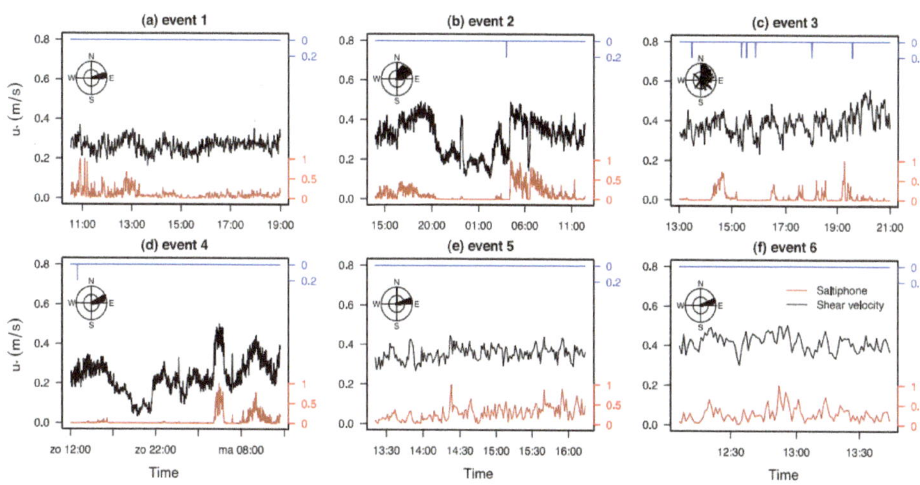

Figure 14: Shear velocity, wind direction, rainfall and normalized saltation activity for six events. The normalized saltation activity was determined using the total count of the saltiphones, divided by the maximum count during the event.

Of the three catchers (Turtle, Bug and Tower) used in the experiments, two of them contained compartments on both sides of the catchers. The impact of this horizontal variation on sediment flux was investigated by evaluatingR^2 as a non-linear regression (Eq. (1)) was applied to all measurements with at least four data-points. For the bug sampler, a non-linear regression was applied to both sides of the catcher, where the middle bottle of the opposite side was included. The results (Fig. 15A) show that the Tower has the best correlation, followed by the Turtle. The Bug has the poorest performance but contains more measurements. Disturbance of the airflow might have caused the decrease in performance. In general, most measurements have a very high correlation, indicating only minor impacts of horizontal variability. However, to exclude the effect of horizontal variability and other sources of uncertainty, only measurements with an $R^2 > 0.98$ were included for further analysis.

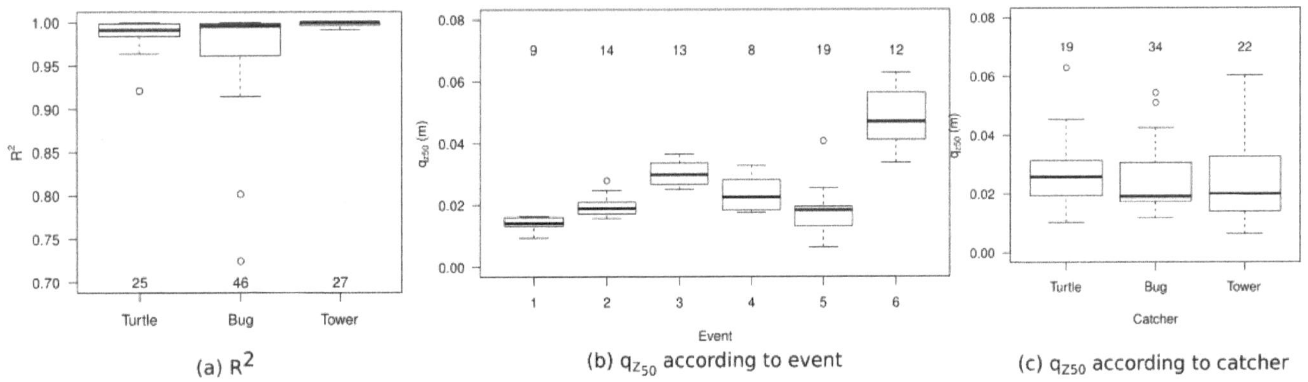

Figure 15: (A) The R^2 **for the three sediment catchers, (B) the** q_{z50} **for every event and (C) the** q_{z50} **according to catcher.** Numbers indicate the number of measurements included.

The q_{z50} values are shown in Fig. 15B. Values for event 6 are higher compared to other events, which is most likely caused by the frozen surface. During events experiencing rainfall, sediment was generally transported over higher elevations. However, the configuration of the traps on the catcher were also found to have an impact. Figure 15C shows that the the Turtle design gives generally higher values for q_{z50} compared to the other two designs. This is caused by point density being close to the surface. For the Bug and Tower designs, regression coefficient β is based on one point close to the surface, whereas the Turtle has two data points. The range in q_{z50} is larger for the Tower compared to the Bug, as the Bug has two data points at approximately the same elevation, with the measurement being refuted when these points do not match. Measured base elevation was in good agreement with the calculated base elevation, with an average difference of 0.7 mm and a maximum of 5 mm; with the Turtle displaying the largest variation, followed by the Bug and the Tower.

An Inverse Distance Weighting (IDW) algorithm was used to investigate the spatial variability of q_{z50} and Q (Fig. 16). We selected events 2, 3, 5 and 6, as during these experiments, two arrays of catchers were used. For all events, the lower-located array has lower values for q_{z50} compared to the higher array. Based on our observations, we can confirm that the surface of the upper array was generally wetter than the lower array. This is in line with other findings by Nield & Wiggs (2011). Farrell et al. (2012) also found that sediment is transported over higher elevations on wet surfaces. The spatial variability in saltation height (and thus surface characteristics) shows no alignment with the total transported sediment. Furthermore, the large variability in sediment flux between the different events, suggests there is also large variability in total sediment transport within individual events. Peak values are eight times higher than the lowest values within measurement plots. In general, there is good agreement in measured sediment flux between points located close to each other. However, within meters of these measurements we can see major differences in total sediment flux. Besides the limiting effect of surface moisture on aeolian sediment transport (Namikas & Sherman, 1995; Cornelis & Gabriels, 2003; Neuman, 2003), the variability in sediment flux can be attributed to the presence of aeolian streamers (Baas & Sherman, 2005; Baas, 2008) and/or fetch length (Davidson-Arnott, MacQuarrie & Aagaard, 2005; Bauer et al., 2009;Delgado-Fernandez, 2010).

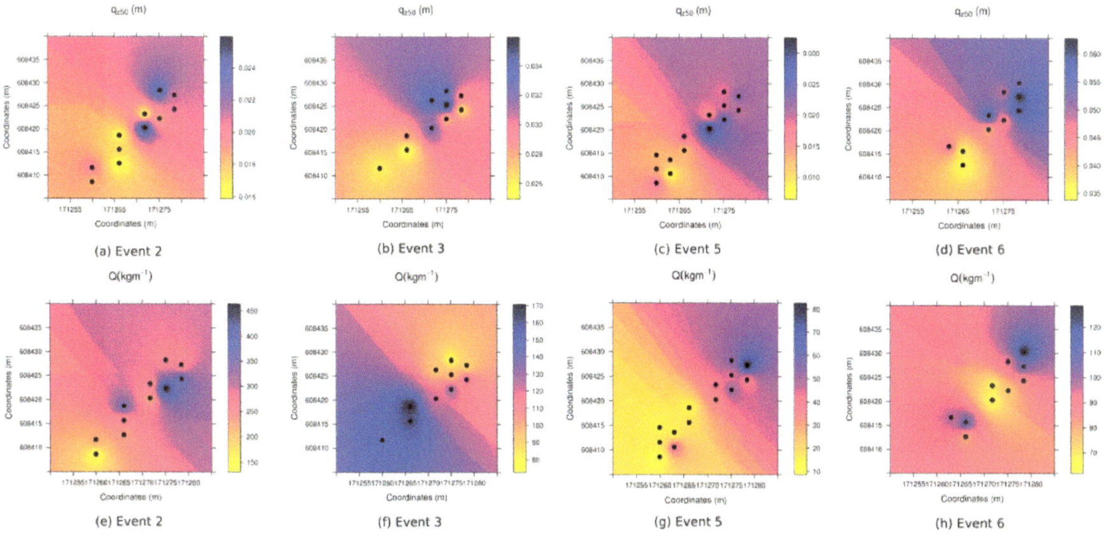

Figure 16: The spatial distribution of q_{z50} **(A–D) and** Q **(E–H).**

Differences in vertical sediment flux as found in the wind tunnel studies have limited validity for field studies, as surface conditions were found to have an important impact on saltation. Wet, frozen or crusted surfaces increase saltation height, as particles retain a higher proportion of their impact energy (Farrell et al., 2012). This effect was regarded as localized due to the spatial variability of the surface. Moreover, saltation trajectories have a scattered pattern between impact and fluid threshold. This may impact results from the field, as during some events, transport was highly intermittent due to fluctuations in wind speed (Davidson-Arnott & Bauer, 2009; Stout & Zobeck, 1997). However, additional rapidly-acquired field data are necessary to study this phenomena in more detail.

CONCLUSION AND RECOMMENDATIONS

Using fast-temporal data on aeolian sediment transport in a wind tunnel, we found that q_{z50} displays a scattered pattern between the impact and fluid threshold, but shows a linear increase with shear velocities above the fluid threshold. Furthermore, it was shown that errors that originate from the distribution of compartments and the location of the lowest sediment trap can be identified using relative sediment flux. In field situations, shear velocity was not found to be the most important controlling factor in vertical sediment flux characterization. Instead, surface moisture was an important control, although particle characteristics of the source area should also be considered. Errors have a more pronounced effect on sediment flux estimation for fine compared to coarse sediment, as fine sediment fractions have a larger portion transported closer to the surface. In order to reduce uncertainty, it is recommended to locate multiple traps closer to the surface.

ACKNOWLEDGEMENTS

We would like to acknowledge the following people who have made the completion of this manuscript possible. Pierre Jongerius and Corjan Nolet, thank you for all your hard work in assisting us with the field data collection. Three anonymous reviewers are also thanked for their invaluable comments.

REFERENCES

1. Arens SM. 1996. Rates of aeolian transport on a beach in a temperate humid climate.*Geomorphology* 17(1–3):3-18 Response of Aeolian Processes to Global Change
2. Baas AC. 2004. Evaluation of saltation flux impact responders (safires) for measuring instantaneous aeolian sand transport intensity. *Geomorphology* 59(1–4):99-118
3. Baas AC. 2008. Challenges in aeolian geomorphology: investigating aeolian streamers. *Geomorphology* 93(1):3-16
4. Baas ACW, Sherman D. 2005. Formation and behavior of aeolian streamers. *Journal of Geophysical Research* 110:F03011
5. Bagnold RA. 1941. *The physics of blown sand and desert dunes*. London: Methuen.
6. Basaran M, Erpul G, Uzun O, Gabriels D. 2011. Comparative efficiency testing for a newly designed cyclone type sediment trap for wind erosion measurements.*Geomorphology* 130(3–4):343-351
7. Bauer B, Davidson-Arnott R, Hesp P, Namikas S, Ollerhead J, Walker I. 2009.Aeolian sediment transport on a beach: surface moisture, wind fetch, and mean transport. *Geomorphology* 105(1):106-116
8. Butterfield GR. 1999. Near-bed mass flux profiles in aeolian sand transport: high-resolution measurements in a wind tunnel. *Earth Surface Processes and Landforms*24(5):393-412
9. Chepil W, Woodruff N. 1963. The physics of wind erosion and its control. *Advances in Agronomy* 15:211-302
10. Cheung KF, Gerritsen F, Cleveringa J. 2007. Morphodynamics and sand bypassing at Ameland Inlet, The Netherlands. *Journal of Coastal Research* 23(1):106-118
11. Cornelis W, Gabriels D. 2003. The effect of surface moisture on the entrainment of dune sand by wind: an evaluation of selected models. *Sedimentology* 50(4):771-790

12. Davidson-Arnott R, Bauer B. 2009. Aeolian sediment transport on a beach: thresholds, intermittency, and high frequency variability. *Geomorphology* 105(1–2):117-126

13. Davidson-Arnott RG, MacQuarrie K, Aagaard T. 2005. The effect of wind gusts, moisture content and fetch length on sand transport on a beach. *Geomorphology*68(1):115-129

14. Delgado-Fernandez I. 2010. A review of the application of the fetch effect to modelling sand supply to coastal foredunes. *Aeolian Research* 2(2):61-70

15. Dong Z, Liu X, Wang H, Zhao A, Wang X. 2003. The flux profile of a blowing sand cloud: a wind tunnel investigation. *Geomorphology* 49(3):219-230

16. Dong Z, Qian G. 2007. Characterizing the height profile of the flux of wind-eroded sediment. *Environmental Geology* 51:835-845

17. Dong Z, Sun H, Zhao A. 2004. Witseg sampler: a segmented sand sampler for wind tunnel test. *Geomorphology* 59(1–4):119-129

18. Ellis J, Li B, Farrell E, Sherman D. 2009. Protocols for characterizing aeolian mass-flux profiles. *Aeolian Research* 1(1–2):19-26

19. Ellis JT, Morrison RF, Priest BH. 2009. Detecting impacts of sand grains with a microphone system in field conditions. *Geomorphology* 105(1):87-94

20. Ellis JT, Sherman DJ, Farrell EJ, Li B. 2012. Temporal and spatial variability of aeolian sand transport: implications for field measurements. *Aeolian Research*3(4):379-387

21. Farrell E, Sherman D, Ellis J, Li B. 2012. Vertical distribution of grain size for wind blown sand. *Aeolian Research* 7:51-61

22. Goossens D, Offer Z, London G. 2000. Wind tunnel and field calibration of five aeolian sand traps. *Geomorphology* 35(3):233-252

23. Hugenholtz CH, Barchyn TE. 2011a. Discussion of: Sherman, DJ; Li, B; Farrell, EJ; Ellis, JT; Cox, WD; Maia, LP and Sousa, Phgo, 2011. Measuring aeolian saltation: a comparison of sensors. In: Roberts TM; Rosati JD and Wang P, eds., *Proceedings, symposium to honor Dr. Nicholas Kraus*, Journal of Coastal Research, Special Issue No. 59, pp. 280–290. *Journal of Coastal Research* 27(6):1202-1205

24. Hugenholtz CH, Barchyn TE. 2011b. Laboratory and field performance of a laser particle counter for measuring aeolian sand transport. *Journal of Geophysical Research* 116:F01010

25. Jackson NL, Nordstrom KF. 2011. Aeolian sediment transport and landforms in managed coastal systems: a review. *Aeolian Research* 3(2):181-196

26. Kok JF, Parteli EJ, Michaels TI, Karam DB. 2012. The physics of wind-blown sand and dust. *Reports on Progress in Physics* 75(10):106901

27. Li B, Sherman DJ, Ellis JT. 2011. Reply to: Hugenholtz, CH and Barchyn, TE, discussion of: Sherman, DJ; Li, B; Farrell, EJ; Ellis, JT; Cox, WD; Naia, LP and Sousa Phgo, 2011. measuring aeolian saltation: a comparison of sensors. In: Roberts TM, Rosati JD, Wang P. eds. *Proceedings, symposium to honor Dr. Nicholas Kraus*, Journal of Coastal Research, Special Issue No. 59, pp. 280–290. *Journal of Coastal Research* 27(6):1206-1208

28. Mendez M, Funk R, Buschiazzo D. 2011. Field wind erosion measurements with big spring number eight (BSNE) and modified wilson and cook (MWAC) samplers.*Geomorphology* 129(1–2):43-48

29. Namikas SL. 2003. Field measurement and numerical modelling of aeolian mass flux distributions on a sandy beach. *Sedimentology* 50(2):303-326

30. Namikas SL, Sherman DJ. 1995. A review of the effects of surface moisture content on aeolian sand transport. In: *Desert aeolian processes*. Netherlands: Springer. 269-293

31. Nanney R, Fryrear D, Zobeck T. 1993. Wind erosion prediction and control. *Water Science & Technology* 28(3–5):519-527

32. Neuman CM. 2003. Effects of temperature and humidity upon the entrainment of sedimentary particles by wind. *Boundary-Layer Meteorology* 108(1):61-89

33. Nield JM, Wiggs GF. 2011. The application of terrestrial laser scanning to aeolian saltation cloud measurement and its response to changing surface moisture. *Earth Surface Processes and Landforms* 36(2):273-278

34. Ni JR, Li ZS, Mendoza C. 2003. Vertical profiles of aeolian sand mass flux.*Geomorphology* 49(3):205-218

35. Pähtz T, Kok JF, Parteli EJ, Herrmann HJ. 2013. Flux saturation length of sediment transport. *Physical Review Letters* 111(21):218002

36. Poortinga A, van Minnen J, Keijsers J, Riksen M, Goossens D, Seeger M. 2013a.Measuring fast-temporal sediment fluxes with an analogue acoustic sensor: a wind tunnel study. *PLoS ONE* 8(9):e74007

37. Poortinga A, Van Minnen J, Riksen M, Seeger M. 2013b. Measuring fast-temporal sediment fluxes with an analogue acoustic sensor: a wind tunnel study

38. Poortinga A, Visser SM, Riksen MJ, Stroosnijder L. 2011. Beneficial effects of wind erosion: concepts, measurements and modeling. *Aeolian Research* 3(2):81-86

39. Rasmussen KR, Mikkelsen HE. 1998. On the efficiency of vertical array aeolian field traps. *Sedimentology* 45(4):789-800

40. Ridge JT, Rodriguez AB, Fegley SR, Browne R, Hood D. 2011. A new 'pressure sensitive' method of measuring aeolian sediment transport using a gauged sediment trap (gast) *Geomorphology* 134(3):426-430

41. Riksen M J, Goossens D. 2007. The role of wind and splash erosion in inland drift-sand areas in the netherlands. *Geomorphology* 88(1–2):179-192

42. Riksen M, Ketner-Oostra R, Turnhout C, Nijssen M, Goossens D, Jungerius P,Spaan W. 2006. Will we lose the last active inland drift sands of Western Europe? the origin and development of the inland drift-sand ecotype in the netherlands. *Landscape Ecology* 21(3):431-447

43. Schönfeldt H-J. 2012. High resolution sensors in space and time for determination saltation and creep intensity.

Earth Surface Processes and Landforms 37(10):1065-1073

44. Sherman DJ, Li B, Farrell EJ, Ellis JT, Cox WD, Maia LP, Sousa PH. 2011.Measuring aeolian saltation: a comparison of sensors. *Journal of Coastal Research: Special Issue* 59:280-290
45. Spaan W, van den Abeele G. 1991. Wind borne particle measurements with acoustic sensors. *Soil Technology* 4(1):51-63
46. Sterk G, Herrmann L, Bationo A. 1996. Wind-blown nutrient transport and soil productivity changes in southwest niger. *Land Degradation and Development*7(4):325-335
47. Sterk G, Parigiani J, Cittadini E, Peters P, Scholberg J, Peri P. 2012. Aeolian sediment mass fluxes on a sandy soil in Central Patagonia. *Catena* 95:112-123
48. Sterk G, Raats P. 1996. Comparison of models describing the vertical distribution of wind-eroded sediment. *Soil Science Society of America Journal* 60(6):1914-1919
49. Sterk G, Spaan W. 1997. Wind erosion control with crop residues in the sahel. *Soil Science Society of America Journal* 61(3):911-917
50. Stout JE. 1998. Effect of averaging time on the apparent threshold for aeolian transport. *Journal of Arid Environments* 39(3):395-401
51. Stout J, Zobeck T. 1997. Intermittent saltation. *Sedimentology* 44(5):959-970
52. Tidjani AD, Bielders C, Rosillon D, Ambouta K-M. 2011. Uncertainties in plot-scale mass balance measurements using aeolian sediment traps. *Soil Science Society of America Journal* 75(2):708-718
53. Van Pelt R, Peters P, Visser S. 2009. Laboratory wind tunnel testing of three commonly used saltation impact sensors. *Aeolian Research* 1(1–2):55-62
54. Visser S, Sterk G. 2007. Nutrient dynamics — wind and water erosion at the village scale in the sahel. *Land Degradation and Development* 18(5):578-588
55. Visser S, Sterk G, Ribolzi O. 2004. Techniques for simultaneous quantification of wind and water erosion in semi-arid regions. *Journal of Arid Environments* 59(4):699-717
56. Visser SM, Sterk G, Snepvangers JJ. 2004. Spatial variation in wind-blown sediment transport in geomorphic units in northern Burkina Faso using geostatistical mapping.*Geoderma* 120(1):95-107
57. Wal VD. 2000. Grain-size-selective aeolian sand transport on a nourished beach.*Journal of Coastal Research* 16(3):896-908
58. Wiggs G. 2001. Desert dune processes and dynamics. *Progress in Physical Geography* 25(1):53-79
59. Wilson SJ, Cooke RU. 1980. Wind erosion. In: Kirkby MJ, Morgan RPC, eds. *Soil erosion*. Chichester: Wiley. 217-225
60. Youssef F, Erpul G, Bogman P, Cornelis W, Gabriels D. 2008. Determination of efficiency of vaseline slide and wilson and cooke sediment traps by wind tunnel experiments. *Environmental Geology* 55(4):741-750
61. Yurk BP, Hansen EC, Hazle D. 2013. A deadtime model for the calibration of impact sensors with an application to a modified miniphone sensor. *Aeolian Research* 11:43-54

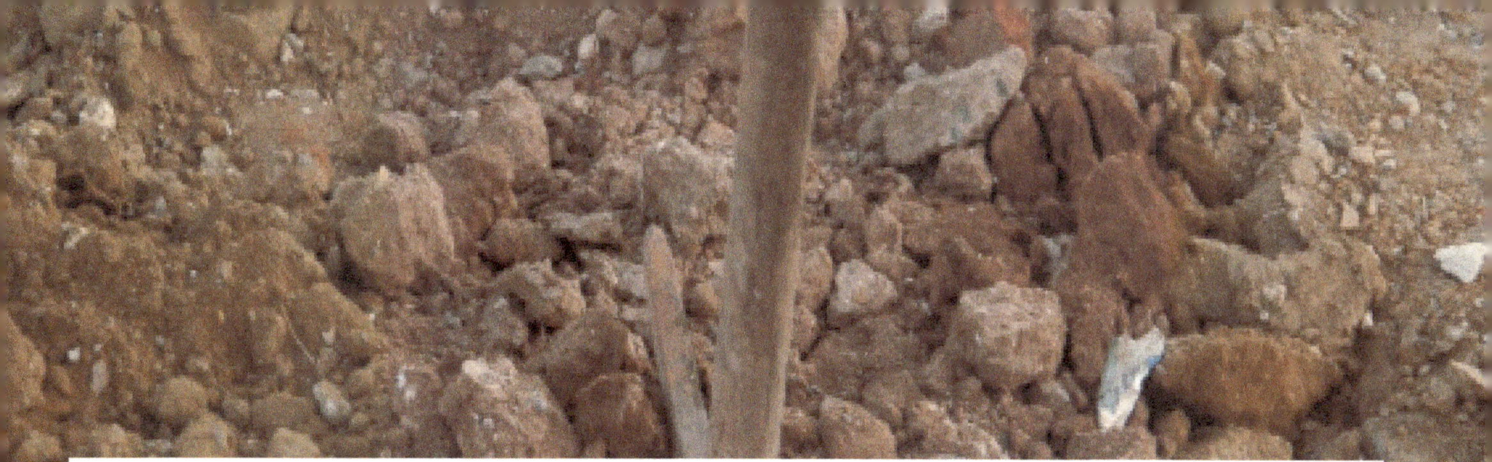

CHAPTER 10

MONITORING OF SOIL WATER CONTENT AND MULTIPLE SOIL-AIR-VEGETATION PARAMETERS

Giovanni Bitella, Roberta Rossi, Rocco Bochicchio, Michele Perniola and Mariana Amato

Keywords: Arduino; multi-sensor platform; open-hardware; microcontroller; soil water content; contactless temperature sensor; infrared thermometry; wireless data

INTRODUCTION

The need to increase water sustainability in agriculture has led to investments in precision irrigation techniques as well as a growing interest in phenotyping root traits for drought tolerance. Precision irrigation scheduling is based on actual site-specific real-time crop water requirements or measurements which can be logged and used to trigger irrigation when a threshold soil water content is reached [1,2]. Optimal sensors number and distances is crucial to obtain representative estimates of soil water availability at field-scale. Most soil hydraulic properties exhibit spatial correlation structures with ranges between 10 and 20 m [3] therefore a high number of sensors would be required to reflect this fine scale of variability. Phenotyping of root traits is an emergent, fast-growing field [4] and in particular phenotyping for the selection of drought-resilient genotypes [5] is receiving considerable attention as inferred by citation data increasing by 40% per year since the interest started to appear [6] and by the large investments in phenotyping [7]. Hi-tech phenotyping facilities are only available to a limited number of researchers, and do not satisfactorily address field-related issues. In fact field phenotyping is considered the real bottle-neck of effective variety selection [8].

Monitoring changes in soil water content at different depths through the rhizosphere gives key information on plant water-uptake rates and strategies but also on parameters such as rooting depth (RD) which can be estimated indirectly by soil-water depletion curves in the profile during a water-supply-limited period [9]. Soil can be considered "a dynamic fresh-water reservoir" [10] with a spatially variable water holding capacity. Therefore strategies for increasing efficiency in water use require a high spatial coverage and time-dynamics of soil water content information.

Soil water content can be either measured directly (*i.e.*, by gravimetric method) or indirectly by non-destructive (highly repeatable) measures of another parameter (*i.e.*, soil permittivity, temperature, resistivity) which is affected by soil water content change, this surrogate data variability is then converted to water content using either physically-based laws or empirical calibration curves. Soil water content sensors, including neutron probes and large scale methods such as ground penetrating radar, have been reviewed by Bittelli [11], the most used equipment for water monitoring at local scale are the dielectric sensors which measure soil dielectric

Citation: Bitella, G.; Rossi, R.; Bochicchio, R.; Perniola, M.; Amato, Monitoring of Soil Water Content and Multiple Soil-Air-Vegetation Parameters. Sensors 2014, 14, 19639-19659. doi:10.3390/s141019639.

permittivity using capacitance or time domain technologies (TDR): These sensors buried at different depths and at multiple locations provide useful soil water content time series. While TDR has long been the best established technology [12], capacitance probes are increasingly adopted for their simplicity of use, the low cost [13] and their sensitivity to changes in soil permittivity comparable to that of the TDR [14–16]. The main drawback of dielectric sensors is their sensitivity to soil electrical conductivity and temperature, which often requires specific calibration/correction procedures [17–19]. The simultaneous availability of other data, such as soil temperature would be therefore necessary in some cases (e.g., for correction of temperature-dependent capacitive probe readings) and desirable in many contexts such as root research. Soil temperature not only affects root growth development and turnover but, together with water availability, also root architectural traits such as root gravitropic angle [20], which is an important trait linked to water acquisition efficiency [21–23]. Moreover low root zone temperature can worsen the effects of other abiotic and biotic stresses such as salinity [24] and pathogen infections rate [25]. Temperature data in the root zone are therefore relevant to improve root data interpretation relative to other sources of variation or when temperature becomes itself an experimental factor. Soil temperature data are especially needed in contexts where decupling between air and soil temperature is expected, such as under greenhouse horticulture [24]. Coupling soil/atmosphere data with plant-based measurements such as canopy temperature (CT) can be used for non-destructive high throughput assessment of plant water stress and drought avoidance strategies [26,27]. Crop water status can be indirectly assessed by measuring canopy temperature and its reduction relative to air temperature. The cooling effect of transpiration makes canopy temperature inversely proportional to stomatal conductance, leaf water potential and turgor [28], these authors found CT negatively correlated to rice grain yield in drought stressed plants. CT was also used to develop crop-temperature-based water stress indices [29,30]. CT can either be measured through thermocouples or remotely through infrared thermometry (IRT), which measures the long-wave infrared radiation emitted by an object that is a function its temperature. IRT measurements have advantages over other methodologies for the assessment of plant water status, such as leaf water potential measurements by leaf chamber, which is relatively slow and therefore unsuitable to screen large breeding nurseries [26]. It is also faster than contact-based probes and is suited for on-the-go measurements [31]. For the mentioned reasons IRT has been widely used in agronomic research providing a reliable, simple, and fast tool for selecting cultivars with greater dehydration avoidance mechanisms [26,27,32] as well as for irrigation scheduling [33]. There are several high quality industrial devices available on the market however their high cost and complexity has limited their adoption in agricultural production and has progressively led to the development of alternative low-cost instruments [34,35]. Mahan and Yeater [34] compared the performances of some high quality industrial and low-cost IRT sensors in agricultural settings and they proved that the measurements from the two categories were similar over a 13–35 °C range, which is the range experienced by crops in most temperate regions, hence low-cost devices might be a sound alternative in many agricultural applications. In any agricultural field experiment, independently from the specific target, air temperature and humidity are the most frequently measured environmental parameters since they determine evapotranspiration demand, and control crop phenology and growth. Such measurements complement agronomy trials giving information on environmental effects superimposing experimental factors. The necessity of spatially distributed weather measurements, especially desirable in complex terrains, gave impulse to the development of new generation low-cost sensors increasingly used to obtain high spatial coverage [36]. Multi-sensor devices are increasingly needed, along with the relevant technologies to store, retrieve and remotely transmit a large and widespread amount of data. To this end, when a large number of measurements sites (*nodes*) are involved the use of Wireless Sensor Networks (WSN) is becoming a "dominant option" [37]. Wireless transmission technologies reviewed by Ruiz-Garcia and co-authors [38] include Zigbee and Bluetooth protocols and modem-based connections such as GPRS and GSM, often coupled with internet applications [39].

The cost of data loggers, multiplexers and sensors, however, limits the number of sensors and the measurements frequency. Recent advances in the field of "open hardware" components open new alleys for the development of measurement stations that can be connected with wireless transmission devices in order to fill a gap in low-cost tools for high spatio-temporal coverage of soil water data and multi-sensor acquisition. "Open hardware" philosophy in analogy with the concept of "open source software" aims at providing free and transparent access to hardware design, projects and code so that users can easily share, customize and upgrade their systems [40,41].

The objective of this work was the design of a low cost "open hardware" platform for the measurement, recording and wireless transmission of data on water content at different depths, air, soil and canopy temperatures.

The system is based on an open source microcontroller board of the series Arduino [42], handling both analog and digital sensors. Arduino platforms are the most famous and fast spreading electronic prototype development, and they have already been used in research for a variety of applications [43–47]. The circuit board, all the electronic components and software to program the microcontroller are freely available. Ease of use/programming is matched with the affordable prices of the electronic components, thus belonging to the group of so-called democratizing technology [48]. The platform is equipped with low-cost, new generation, dielectric sensors now available on the market, and this reduces the cost of data acquisition even further. Measurements at several sites and depths therefore become affordable, and this allows the space-time density of data required for applications aimed at water saving.

EXPERIMENTAL SECTION

Platform

The prototype platform we developed features several components (Figure 1) organized in 3 units depicted in schemes I and II:

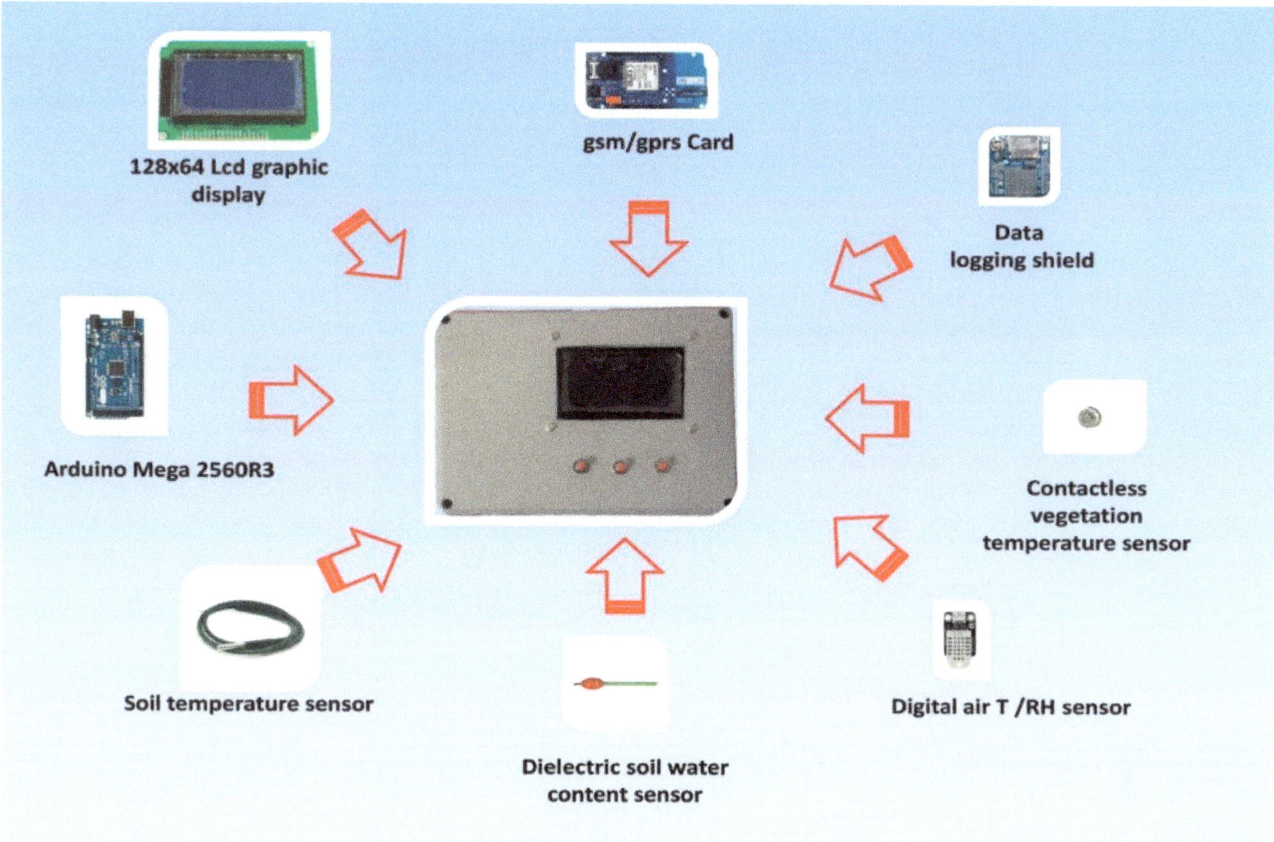

Figure 1. Components of platform.

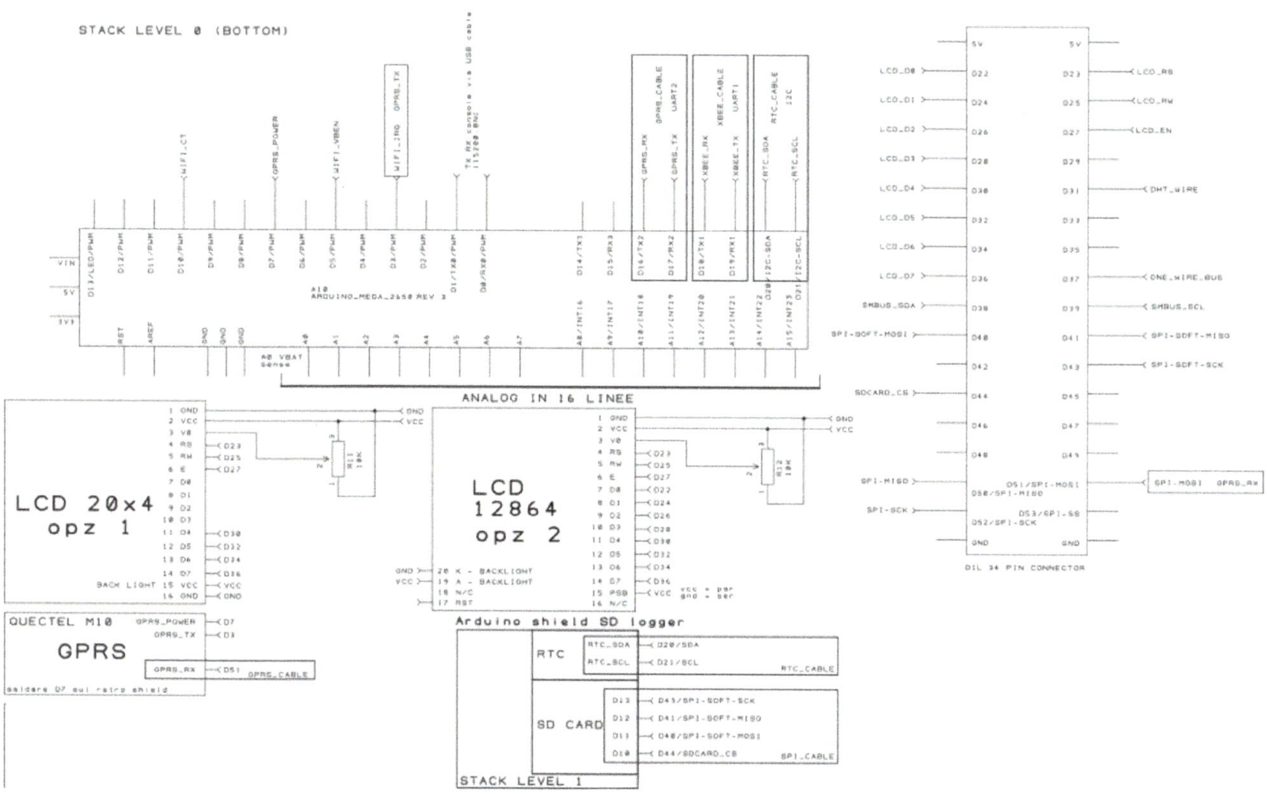

Scheme 1. Schematic electric diagram of the platform.

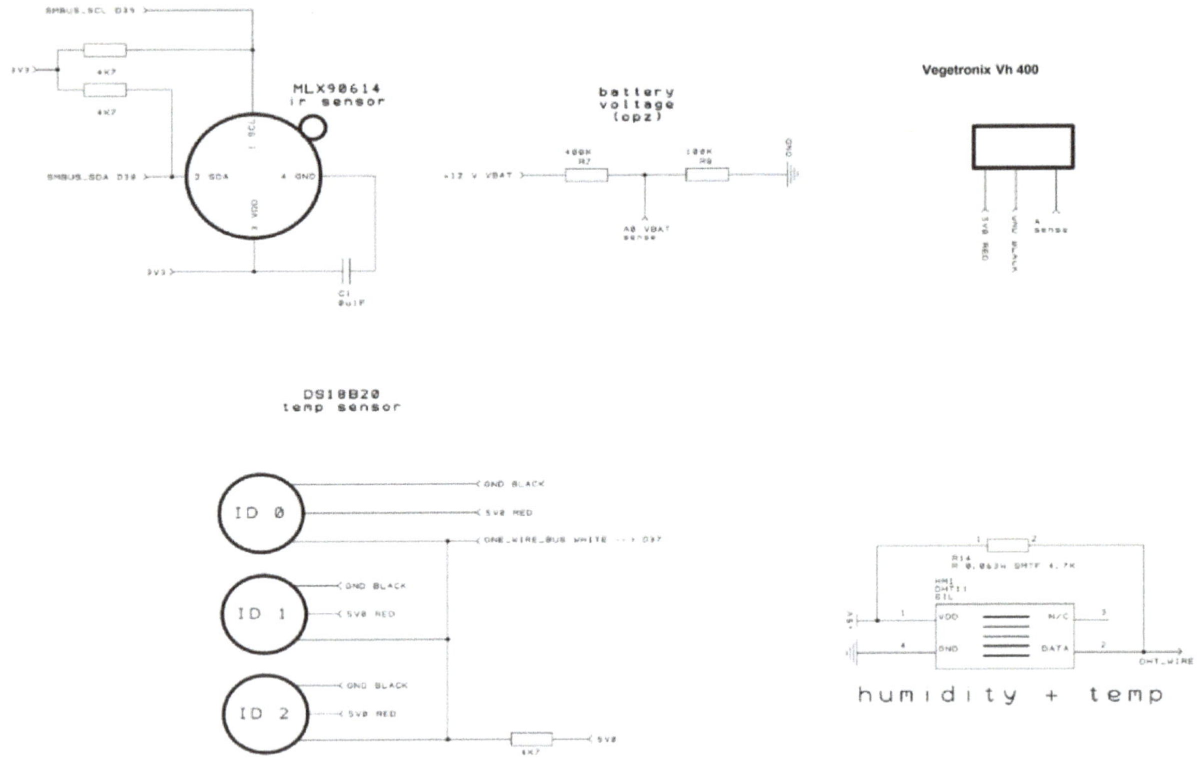

Scheme 2. Schematic electric diagram of sensors.

- Microcontroller board;
- Expansion boards:
  - – Data logging shield
  - – GPRS card for wireless transmission
  - – Optional LCD display for real-time data displaying;
- Sensors:

Platform components are shown in Figure 1, the system architecture is depicted in both Figure 2 (schematic representation of a measurement unit) and the schematic electric diagrams (Schemes 1 and 2). Schemes 1 and 2 are the electrical schematic diagrams of the circuit board. The electric diagram is an abstract representation that shows how all the platform components (sensors, data-loggers and micro-controller) are connected between themselves through input and output pins and where all the auxiliary electronic components (such as resistors and capacitors) are placed. These schemes can be used as a reference to reproduce the system.

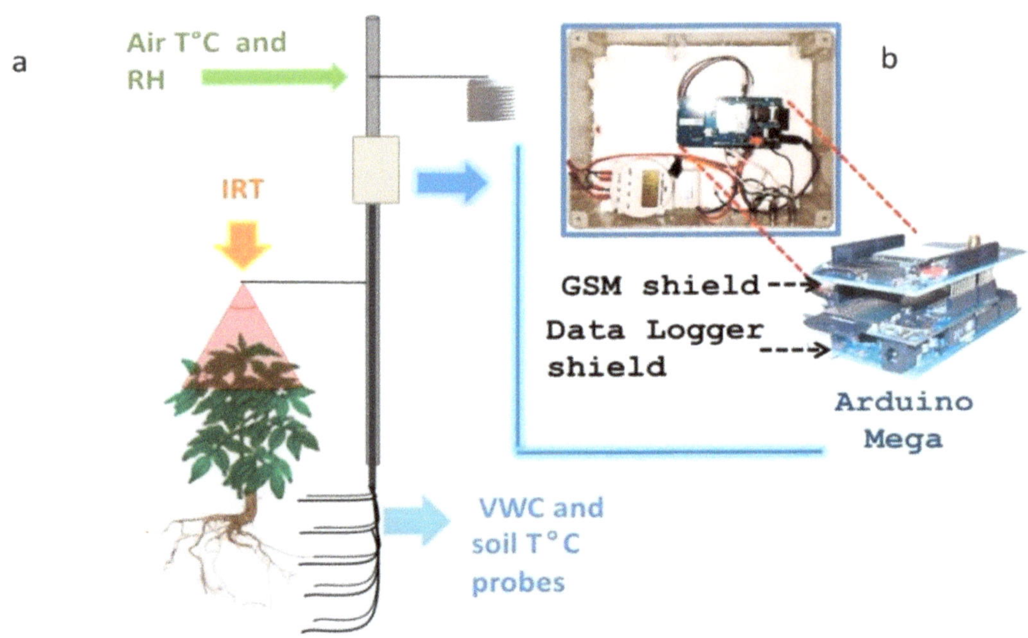

Figure 2. Platform architecture: (**a**) schematic representation of a measurement unit; (**b**) (inset) electronic components inside the indoor enclosure (top-view) and magnification of the micro-controller board and the superimposable expansion shields.

The Microcontroller

The microcontroller is an open source board of the Arduino series [42], handling both analogic and digital sensors, and simply connectable to a computer through a USB cable. The Arduino microcontroller can be programmed through the multi-platform development environment (IDE). All Arduino boards are designed to allow easy connection with other expansion boards (named shields) through a superimposable sequence system. The board used for this project is Arduino Mega 2560 Rev.3.0 featuring an ATMEL microcontroller (ATmega2560) and equipped with 54 digital inputs/outputs and 16 analog inputs (Figure 1).

For field and lab measurements technical specifications of the platform to be constructed around the Arduino microcontroller include:

- Weather proof electrical cabinet IP65 of at least 36 cm × 25 cm × 14 cm for field applications.
- Indoor plastic enclosure to protect electronic components of at least 24 cm × 19 cm × 9 cm for field and lab applications.
- Protection of cables from animal damage through plastic pipes of 4–5 cm diameter.
- External stabilized 12 V–5 V DC/DC power converter.
- Power converter of at least 2 A.
- Capability of data storage of at least 1 GB.
- Capability for data remote transmission with the GPRS protocol at a speed of at least 85.6 kbps.
- In order to ensure flexibility and modularity of sensors it must be capable of supporting the connection to all of the analog ports which means at least 16 water content probes, 5 temperature probes, 1 IR sensor and 1 weather station.

Technical specifications of sensors include:

- Weather sensors needed to operate in the temperature range of −10 to +60 °C and relative humidity from 10% to 90%
- Temperature sensors needed to measure temperatures at least in the range −15 to 60 °C, Humidity sensors needed to Measure relative humidity from 10% to 90%
- Soil water sensors needed to measure soil water content in the range 5% to 50%
- IR sensors need to measure a range from 10 to 40 °C.

Furthermore, for compatibility with the Arduino Mega, sensor signal output must range between 5 mV and 5 V.

The Expansion Boards

(a) Data Logging Shield

The data acquired from the sensors are recorded on a non-volatile and preformatted fat 16–32 memory card type secure digital (SD) through the fully integrated expansion card produced by Adafruit [49]. Data acquisition and timing is completely programmable. The card supports memory cards up to 2 GB and include a real-time clock (RTC) chip ds 1307 [50], it was battery powered independently from the Arduino board. The RTC allows keeping of accurate time, even in the temporary absence of the main power, which is necessary for a correct combination of the time information to the data from the sensor.

(b) GPRS Card for Wireless Transmission

In order to remotely monitor the platform, the system is equipped with a transmission GPRS card connected to the microcontroller through two analog pins at a speed of 115.2 kbps, and able to exchange data remotely to the theoretical maximum speed of 85.6 kbps using a sim phone card. We used the official Arduino GPRS shield developed with Telefonica [51] equipped with a radio modem M10 by Quectel [52] quad-band GSM/GPRS that works at frequencies of GSM850MHz, GSM900MHz, DCS1800MHz and PCS1900MHz. It supports TCP/UDP and HTTP protocols through a GPRS connection. Data recorded in txt/xml format on the SD (see Section 2.1.2 (a)) are transferred to a remote server (see Section 2.1.4) on a daily basis for easy access and to allow periodic monitoring of the proper functioning of the acquisition system.

(c) The Graphic Display

The graphic display used is the mod. 12,864 ZW commercialized by www.robot-domestici.it [53]. It is a graphical liquid crystal display (LCD) 64 × 128 with LED backlight. The display is connected to the microcontroller by a parallel mode connection. This connection required 8 digital pin and 5 V power.

Sensors

The sensors used for monitoring water status are dielectric probes operating at 80 MHz [54] and require a power supply with a voltage between 3.3 and 20 V DC. They provide a voltage reading between 0 and 3 V as a function of the water content of the medium surrounding the sensor with an average consumption of less than 7 mA. Besides the low cost (Table 2), the probes offer several advantages: As reported by the manufacturer [54] and shown by experimental results [55] the sensors are insensitive to water salinity, the probe is small and rugged and does not corrode over time as conductivity-based probes usually do. Communication with the data-logger is analog and occurs through the analog to digital converter at 10-bit resolution, which is included in the Arduino Mega board and converts voltage measurements in the range 0–1023 V. When a voltage of 5 V is provided to the probes, a resolution of about 0.005 V (5/1023) is obtained [45]. Temperature and humidity sensors were selected among those with a digital interface in order to leave the largest possible number of free analog ports on the card to allow for the additional installation of a great number of soil water content sensors. The soil temperature sensor is a Dallas DS18B20 digital thermometer [56] capable of detecting temperatures between −55 and 125 °C with an accuracy of ±0.5 °C from −10 °C to +85 °C. In our system all of the sensors (except the IR probe that required 3 V) were connected to a stabilized 12–5 V DC/DC converter with a power of 3 A that was sufficient to supply the maximum number of sensors connectable to the micro-controller. The thermometer, incorporated in a protective stainless steel capsule, was connected to the microcontroller using a 4.7 pull up resistor (see Scheme II) and a 3.5 mm stainless jack stereo connection. The air temperature and humidity sensor is a DHT22, commercialized by Exp-tech.Inc [57], it allows a simultaneous measurement of temperature and relative humidity (RH) with a range between −40 and 125 °C and an accuracy of ±0.5 °C for temperature and between 0% and 100% RH with an accuracy of 2%–5%. The temperature of vegetation was acquired by a small size non-contact infrared thermometer MLX90614ESF-BAA-00TU-ND that is marketed factory-calibrated by Melexis Microelectronic Systems [58]. This sensor is able to measure object temperature between −70 °C and +380 °C with a resolution of 0.01 °C and an accuracy of 0.5 °C. The field of view is 90. The sensor, encapsulated in a plastic cable gland, was fixed to an adjustable angle support in order to manipulate distance and position to optimize the target view. The interface used for the connection with the microcontroller board was I²C, which is a type of serial communication.

Table 1. Prices of the main components of the platform (source [53])

Platform Electronic Components	Price in €, Vat Excluded
Adafruit datalogging shield	20
Arduino mega	39
Arduino GSM Shield	69
Optional display	18
DC/DC converter 12 V to 5 V 3 A	5
SD card 2 GB	2
AC 110B-220 V to DC 12 V 3 A regulated transformer power supply	10
Additional electronic materials (connectors, resistors, capacitors)	20
Electronic enclosure	10
Total	193

Prices of platform components and sensors are reported in Tables 1 and 2 respectively, shipping costs were not included. For field applications other additional consumable materials might be needed such as plastic pipes, metal poles, cable glands, weather proof electronic cabinet and clamps. These costs (amounting all together less than 100 € on average) were not reported in details since different materials could be purchased depending on size requirements and expected deployment time.

Table 2. Prices for each sensor (source [53,54]), type of measurement and possible applications

Sensors	Type of Measurement	Applications	Price in € (Vat Excluded)
Vegetronix VH400	Soil water content	hydrology, agronomy, soil physics, irrigation, plant physiology, phenotyping, root ecology	28.2
DS18B20	Soil temperature	agronomy, soil physics, irrigation, plant physiology, phenotyping, shoot and root growth and development	7.24
MLX90614	Air temperature and Relative Humidity	agronomy, irrigation, plant physiology, meteorology, crop water requirements	9.80
melexis IR Sensor	Canopy temperature	agronomy, irrigation, plant physiology, phenotyping, crop ecology, plant water stress response	19.95

Data Management

Data management of the platform involves the acquisition, storage and access/transmission of temporal or spatio-temporal data of different categories. The measurement system we propose involves multiple soil water content and soil temperature sensors and therefore a spatio-temporal array of water content and soil temperature data and a temporal dataset for canopy temperature, air temperature and humidity (one sensor each), for each measurement stations. The number and type of sensors to be read as well as sensors acquisition pace is programmable. Data from all sensors are acquired simultaneously.

Data are stored in the data logger to a maximum capacity of 2GB in the txt/xml format. There are three options for access to stored data:

- Periodical data retrieval from the data-logger by direct access to the SD card.
- Remote access to data sent to a server through the GPRS transmission. This allows managing high data throughput as described in Section 2.1.2 (b). The desired pace of data transmission, though, is set through Arduino code, in our example we set it to once a day in order to save energy and traffic fares.
- Real-time access through the LCD display for quick checks, especially in case of problems with the transmission protocol.

All three are available with the platform as described, but options 2 and 3 may be excluded if a display or a transmission unit is not desired. Since there are many possible causes of failure of data retrieval systems, the use of multiple options, though, is strongly advised in order to reduce the risk of data loss or temporary lack of accessibility.

Also, data format allows integrating the platform with existing applications for data visualization, storage and processing on database servers.

This data management architecture allows acquiring and transmitting/accessing information from multiple measurement points throughout field or lab settings, and to use information after processing or as such for real-time decisions such as those related to irrigation events when thresholds are reached in one or multiple sites.

The Platform Power Supply

To power the platform 7–12 V direct current (DC) is recommended. For testing we used an electronic adapter AC 110–220 V to DC 12 V 3 A connected to the electric grid. For open field application, power supply can be provided through a common lead-acid battery to 12 V and the environmental sensors are powered directly by a 12–5 V 3 A high-efficiency digital switch, bypassing the on board linear converter. The battery state-of-charge is monitored simultaneously with the environmental sensors, using a voltage divider, depicted in the schematic diagram of the system. The voltage divider, connected directly to the battery (12 V), produces an output voltage (2.4 V) readable by the Arduino board through its analogical ports.

The overall electronics are protected in a weatherproof case and receive signals from the environmental sensors through connections with quick mounting of the 3.5 mm stereo jack type fixed on the wall of the casing.

The software code is freely available directly from the authors on request.

Soil Water Content Sensors Calibration

The platform was calibrated for the detection of soil water content in non-saline soils (ECe1: 2.5 < 0.1 mS/cm) with three different types of grain size characteristics: a sandy soil (sand 76.48%, silt 16.82%, clay 6.7%), a silty clay soil (sand 30.66%, silt 45.06%, clay 24.28%), and a sand. Each soil was air-dried and sieved to 2 mm, and used to fill 300 mL glass containers. Samples of different water content were obtained by adding different amounts of tap water at the temperature of 20 °C, covering the containers with a screw-lid and equilibrating at 20 °C for 24 h. Containers were then opened and water content was measured with VH400 rev.6.0 probes (Vegetronix™). All samples were weighted fresh and after oven drying at 105 °C for gravimetric water content determination and converted to volumetric water content by multiplying for the bulk density. Stability of dielectric probe readings to temperature was tested by keeping the probes fully immersed in deionized water and recording probe output sensitivity to temperature changes from 3 °C to 50 °C.

Statistical Analysis

The relationship between volumetric water content (VWC) and sensor reading was non-linear following a sigmoidal growth. Data were modelled using a three parameters logistic model in the form:

$$y = \frac{asym}{1 + e^{\frac{(xmid - x)}{scal}}} \quad (1)$$

with: y = volumetric water content ($Kg{\cdot}m^{-3}$); x = sensor output reading (V).

The model was fitted to the pooled data set (*overall model*) and then to each soil texture (*single texture models*). Parameters significance and goodness of fit statistics were calculated. In order to test if soil-specific calibration curves were required, significant differences between single-texture models parameters were assessed using the likelihood-ratio test. Models were validated by a leave-one-out cross validation (LOOC). All data analysis was accomplished in the "R software environment for statistics" [59].

RESULTS AND DISCUSSION

Output Data and Calibration

An example of the log file with data collected from the fully-equipped platform is shown in Figure 3.

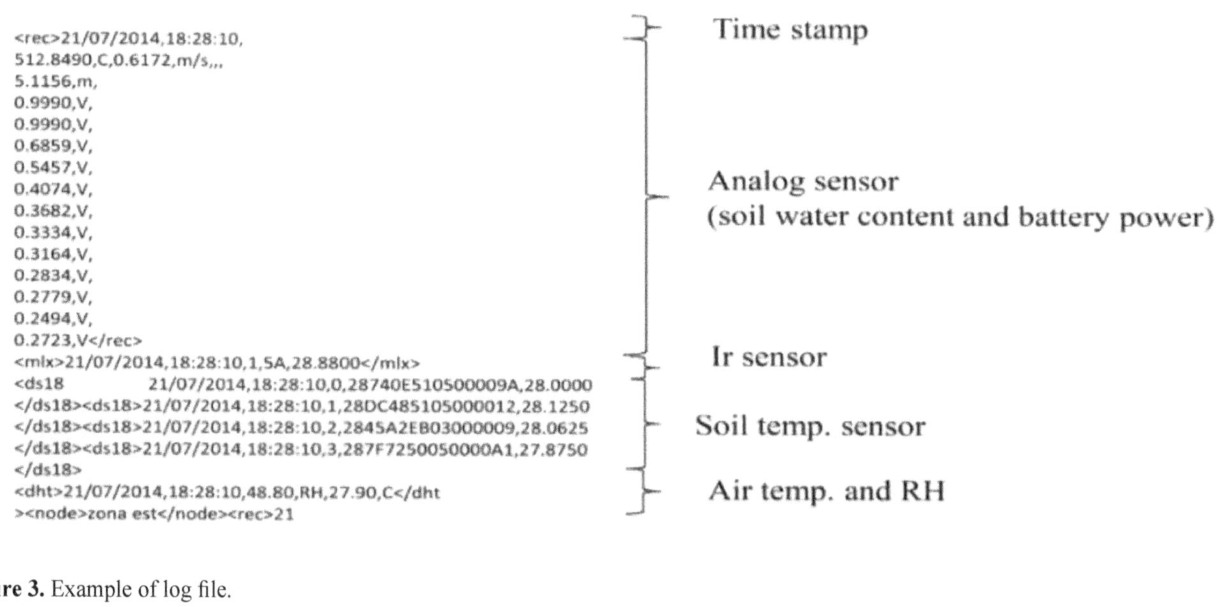

Figure 3. Example of log file.

Output from the analog soil water content sensor in Volts *versus* direct measurements used for calibration are reported in Figure 4 and statistics of the calibration models are shown in Table 3.

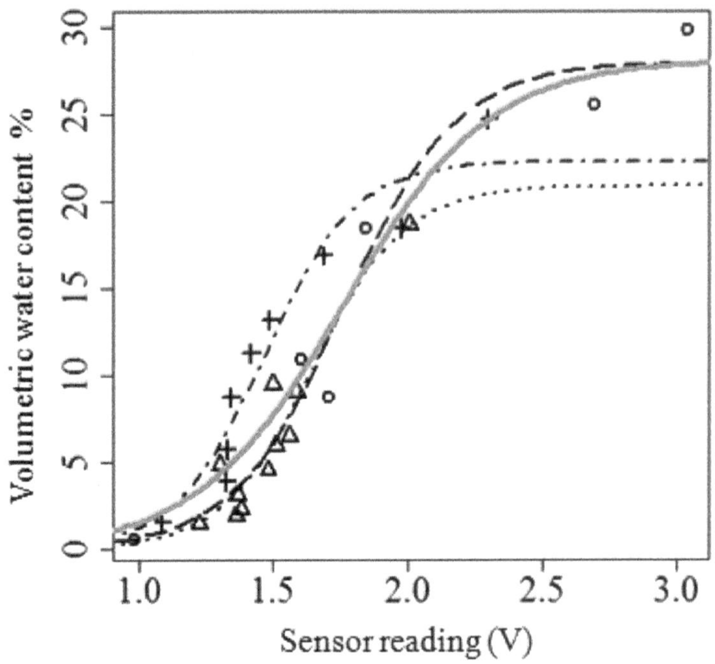

Figure 4. Calibration of soil water content dielectric probes on three soil media. Symbols = data points: Open triangles = sand; open circles = loamy-clay soil; crosses = sandy soil. Lines = model results: Dotted line = sand; dashed line = loamy clay soil; dotted-dashed line = sandy soil; solid grey line = overall model.

Table 3. Summary statistics of calibration models for soil water content sensors. For each soil texture and for the overall model the logistic function parameters estimates were reported together with their standard error and significance, normalized root mean square error (NRMSE) as a measure of goodness of fit. The asterisks "*"indicate the coefficients significance levels. The symbols are displayed as follows: Any *p*-value < 0.001 was designated with three (***) asterisks; *p*-values >= 0.001 and <= 0.01 are shown with two (**) asterisks, *p*-values > 0.01 and < 0.05 are shown with one (*) asterisk

Texture	Parameters	Estimate	Std. Error	t-Value	Pr(>|t|)		NRMSE
Sandy-Loam	xmid	1.470	0.054	27.242	1.62E-07 ***	Residual standard error: 2.307 on 6 d.f. Number of iterations to convergence: 16 Achieved convergence tolerance: 6.816e-06	0.081
	scal	0.166	0.052	3.199	1.86E-02 *		
	Asym	22.263	2.032	10.954	3.44E-05 ***		
Loamy-Clay	xmid	1.765	0.069	25.500	1.32E-04 ***	Residual standard error: 2.951 on 3 d.f. Number of iterations to convergence: 8 Achieved convergence tolerance: 3.206e-06	0.071
	scal	0.206	0.110	1.870	1.58E-01		
	Asym	27.932	2.233	12.510	1.10E-03 **		
Sand	xmid	1.650	0.096	17.172	1.35E-07 ***	Residual standard error: 1.734 on 8 d.f. Number of iterations to convergence: 5 Achieved convergence tolerance: 2.201e-07	0.086
	scal	0.168	0.057	2.969	1.79E-02 *		
	Asym	20.840	3.775	5.520	5.60E-04 ***		
Overall	xmid	1.762	0.065	26.980	2.00E-16 ***	Residual standard error: 2.802 on 23 d.f. Number of iterations to convergence: 4 Achieved convergence tolerance: 2.366e-06	0.09
	scal	0.268	0.044	6.070	3.43E-06 ***		
	Asym	28.034	2.146	13.070	3.97E-12 ***		

Following data behavior the relationship between VWC and sensor reading was modelled by a 3-parameters logistic function. The logistic or autocatalytic model is a nonlinear function defined by three parameters referred to as asymptote, xmid and scal in Table 3 (models summary). The asymptote is the final limiting value attained by the response variable (VWC), xmid is the input value (sensor output reading (V)) at which the inflection point is reached and the "scal" is the inherent growth rate. The purpose of this empirical model was to provide a framework to convert sensor output to water content values, In Table 3, for each texture and for the pooled data (overall model), parameters estimates are reported together with their standard error and significance. The table shows that regardless of the difference in soil texture, the relationship between sensor reading and soil water content is non-linear and always follows a sigmoid behavior. In both pooled and single-texture models the parameters are always significant (*p*-value < 0.05) with the sole exception of the scale parameter in the loamy-clay model. In our data the sandy-loam model shows poor predictions at extremes values (water content below 5% and above 20%). The sand model showed scattered values between 5% and 10% VWC). This might be due to the low cohesion of sand at low water content that causes particles displacement and the possible creation of voids at the interface with the probe. This might be due to the low cohesion of sand at low water content that causes particles displacement and the possible creation of voids at the interface with the probe. Geesing and co-authors [60] discuss the effects of bulk density and particle displacement on capacitive probe readings, highlighting the importance of developing distinct calibration curves for coarse and fine textured soils. In our analysis all of the single-texture model parameters show very similar values, the loamy-clay soil has the highest asymptote and the slowest "growth rate" compared to the other textures. Nevertheless, the single-texture models were found not significantly different (*p*-values > 0.05) at the likelihood ratio test [61] therefore, as is also suggested by manufacturer, at least for the ranges of texture and water content that we tested soil-specific calibration equations are not required.

The overall model fits the data very well ($R^2 = 0.89$) showing a high stability, being able of generating very similar RMSEs during training and validation ($RMSE_{training} = 2.63$; $RMSE_{validation} = 2.61$).

Despite advantages of dielectric-based capacitance probe (CP) leading soil-monitoring sensors, these methods are prone to temperature and salinity drifts. Analytical models for salinity correction and temperature dependent scaled voltage techniques have been validated for different soil types [62,19]. Laboratory calibration of commercial dielectric soil water sensors have shown both linear and non-linear responses to increasing water content depending on soil type, therefore soil-specific calibration equations have been recommended [62]. In field application a calibration method based on the information-sharing in wireless sensor network incorporated spatially variable temperature and salinity data [63].

Figure 5 reports the variation of temperature and the soil water content probe readings during the temperature-drift experiment in water. A difference of about 0.07 V between readings at 3 and 50 °C was found, with a significant drift of 0.0009 $V \cdot °C^{-1}$ ($y = 0.0009x + 2.733$; $R^2 = 0.54$) consistent with previous findings [62].

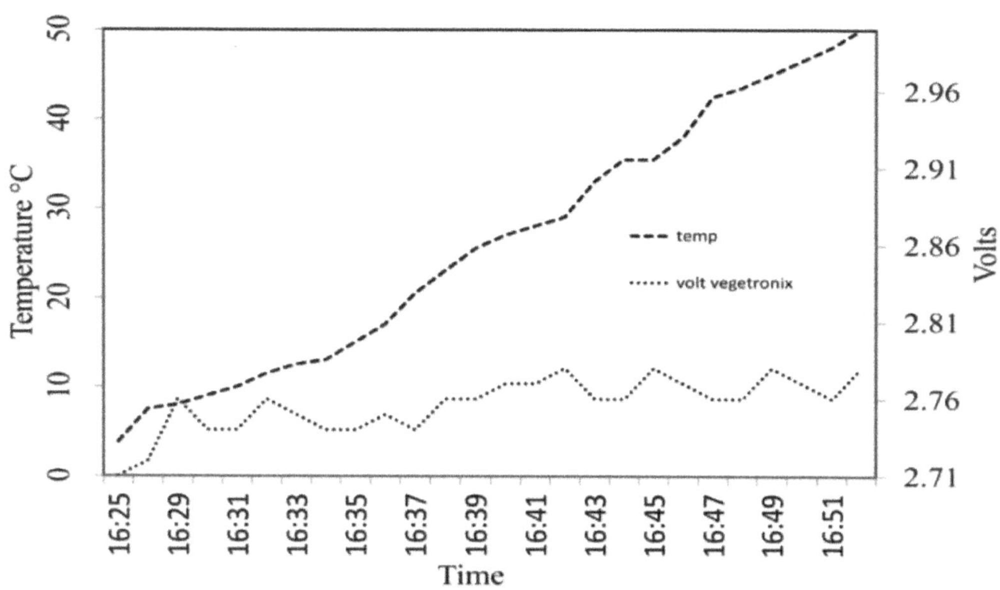

Figure 5. Variation of temperature and dielectric probe readings in water during the temperature-drift experiment.

Mcmichael and Lascano [62] tested the temperature drift of dielectric probes (Model EC20 probe, Decagon Devices, Pullman, WA) at different water content, and found that sensitivity to temperature increases with volumetric water content, ranging from 0.5 $mV \cdot °C^{-1}$ in dry soil to 5.5 $mV \cdot °C^{-1}$ at 0.20 $Kg \cdot m^{-3}$ VWC. This indicates that temperature effects on capacitive probes cannot be ignored especially in the wetter range, and hence temperature data would always be required for correction. Since in field calibration changes in bulk density as well as large temperature fluctuations can be expected, the authors [62] suggest to reduce artifacts linked to temperature sensitivity of CP, place the probes beneath the first soil layers thus avoiding steep gradients of both temperature and water content. In our case the Vegetronix VH400 probe proved to be sensitive to temperature changes in wet conditions (in water) thus temperature data would always be required. A positive linear trend exists. The determination coefficient though is relatively small, the observations are scattered around the regression line and the range of the response variable is narrow (between 2.71 and 2.78 V). The lack of fit of the linear model is partially explained by the fact that between 30 and 50 °C the largest fluctuation of the response variable occur. Within this interval the regression line levels off. Indeed, when the full range of data was modeled by a power relationship (data not shown), which approaches a plateau with increasing temperatures, a better fit is achieved; the determination coefficient rises to 0.65. When the dataset were split in two subsets (from 3 to 30 °C and from 30 to 50 °C) and each one was modeled by a linear regression for the first subset the linear trend was significant ($y = 2.7209 + 0.0017x$; $R^2 = 0.60$ p-value < 0.001) while the dependency disappeared at large values of temperature. For temperature corrections the linear model fit up to values of 30 °C could be used at least below the soil surface, since in most agricultural settings, this temperature threshold value would be plausible in field soils with a water content near saturation (wetter range). The fluctuations of sensor readings as well as the non-linear sensitivity to temperature at high water content are consistent with other studies [18,62], regarding sensor output fluctuations a feasible option to reduce data noise during calibration would be increasing the sensor sampling frequency and averaging out multiple measurements [64]. To develop an absolute temperature correction procedure, however, temperature drift in soil remains to be tested along with possible hysteresis phenomena [62]. Albeit this preliminary fast calibration procedure confirmed the need of equipping the platform with a number of soil temperature sensors conforming to the number of water content probes. Kizito and co-authors [18] demonstrated that increasing dielectric probes measurement frequency up to 70 MHz considerably reduces the sensitivity to conductivity and texture, while sensitivity to temperature was unaltered by the frequency but was found consistently low across soil/solutions and conductivity ranges. The authors argued that even if temperature effects are correctable, the temperature dampening effect in soil would probably reduce this need in many applications. Our probe operates with a higher frequency of measurements (80 MHz) and this makes it even less sensitive to texture and salinity effects while temperature drift, if necessary, can be easily corrected by data processing. The sensor is sensitive to even small changes in water content. This sensitivity is very useful in predicting changes in matric potentials and its implications in variations of soil strength and plant available water.

For this prototype we chose data transmission via GPRS to allow automated data retrieval and remote control of sensors functioning and battery state-of-charge even from remote locations where Wi-Fi is not available. Nevertheless this system can be easily upgraded by adding a low-power Wi-Fi expansion board (*i.e.*, xbee transmission [65]) that allows wirelessly transmission and multipoint communications [46]. For this platform some of the selected sensors are inexpensive to the point of being considered disposable, this is advantageous when probes are buried at considerable depths and thus cannot easily be retrieved. In the case of malfunctions, all sensors can be easily replaced in the field with the aid of a screwdriver. The choice of sensor stems from of a combination of factors such as the objective of measurements, sensor accuracy within the relevant range of measurable values, ease of use and cost. Opting for low-cost equipment, as argued by Mahan and Yeater [34] is not necessarily aimed at reducing the cost of measurements itself but rather on having, at the same cost of a high quality sensor, an increased data density and thus better encompass the spatio-temporal variability of crop attributes and processes.

CONCLUSIONS

An Arduino microcontroller board provided the basis for a multi-sensor platform for the measurement and remote transmission of soil water content alone or coupled with other relevant soil, vegetation and atmosphere parameters. High-frequency dielectric sensors mounted on board proved to be subjected to temperature drifts; hence data collection must be integrated by soil temperature measurements. Meanwhile, the probes are sensitive to even small changes in water content and a single calibration equation can be used, at least in sandy, sandy-loam and loamy-clayey soils.

This work provides a framework for the replication and upgrading of a customized low cost platform, consistent with the open source approach whereby sharing information on equipment design and software facilitates the adoption and continuous improvement of existing technologies. The low cost of all components allows the envisaging of applications where a high number of sensors are required and/or sensors need to be placed in locations where they cannot be easily recovered.

ACKNOWLEDGMENTS

This work was partly funded by PSR 124 "IRRISOL integrated system for differential irrigation management through high resolution differential mapping". Regione Campania. DRD n. 215 07/06/2013. Codice CUP B55C11003040007.

AUTHOR CONTRIBUTIONS

Giovanni Bitella: Platform project and implementation, sensor calibration and writing of the paper; Roberta Rossi: Platform and calibration planning, data analysis, writing and corresponding of the paper; Rocco Bochicchio: Calibration; Michele Perniola: Data discussion; Mariana Amato: Project and calibration planning, data discussion, writing of the paper.

REFERENCES

1. Hedley, C.B.; Yule, I.J. Soil water status mapping and two variable-rate irrigation scenarios. *Precis. Agric.* 2009, *10*, 342–355.
2. Goumopoulos, C.; O'Flynn, B.; Kameas, A. Automated zone-specific irrigation with wireless sensor/actuator network and adaptable decision support. *Comput. Electron. Agric.* 2014, *105*, 20–33.
3. Wendroth, O.; Koszinski, S.; Pena-Yewtukhiv, E. Spatial Association among Soil Hydraulic Properties, Soil Texture, and Geoelectrical Resistivity. *Vadose Zone J.* 2006, *5*, 341–355.
4. Wasson, A.P.; Richards, R.A.; Chatrath, R.; Misra, S.C.; Prasad, S.V.; Rebetzke, G.J.; Kirkegaard, J.A.; Christopher, J.; Watt, M. Traits and selection strategies

to improve root systems and water uptake in water-limited wheat crops. *J. Exp. Bot.* 2012, *63*, 3485–3498.

5. Tuberosa, R. Phenotyping for drought tolerance of crops in the genomics era. *Front. Physiol.* 2012, *3*, 1–26.
6. Passioura, J.B. Phenotyping for drought tolerance in grain crops: When is it useful to breeders? *Funct. Plant Biol.* 2012,*39*, 851–859.
7. International Plant Phenotyping Network-IPPN. Available online: http://www.plant-phenotyping.org accessed on 18 October 2014.
8. Araus, J.L.; Cairns, J.E. Field high-throughput phenotyping: The new crop breeding frontier. *Trends Plant Sci.* 2014, *19*, 52–61.
9. Dardanelli, J.L.; Bachmeier, O.A.; Sereno, R.; Gil, R. Rooting depth and soil water extraction patterns of different crops in a silty loam Haplustoll. *Field Crop. Res.* 1997, *54*, 29–38.
10. Hedley, C.B.B.; Roudier, P.; Yule, I.J.J.; Ekanayake, J.; Bradbury, S. Soil water status and water table depth modelling using electromagnetic surveys for precision irrigation scheduling. *Geoderma* 2013, *199*, 22–29.
11. Bittelli, M. Measuring soil water content: A review. *HortTechnology* 2011, *21*, 293–300.
12. Topp, G.C.; Davis, J.L. Time-Domain Reflectometry (TDR) and Its Application to Irrigation Scheduling. *Adv. Irrig.*1985, *3*, 107–127.
13. Fares, A.; Polyakov, V. Advances in Crop Water Management Using Capacitive Water Sensors. *Adv. Agron.* 2006, *90*, 43–77.
14. Gardner, C.M.K.; Dean, T.J.; Cooper, J.D. Soil water content measurement with a high-frequency capacitance sensor. *J. Agric. Eng.* 1998, *71*, 395–403.
15. Robinson, D.A.; Gardner, C.M.K.; Cooper, J.D. Measurement of relative permittivity in sandy soils using TDR, capacitance and theta probes: Comparison, including the effects of bulk soil electrical conductivity. *J. Hydrol.* 1999,*223*, 198–211.
16. Seyfried, M.S.; Murdock, M.D. Measurement of Soil Water Content with a 50-MHz Soil Dielectric Sensor. *Soil Sci. Soc. Am. J.* 2004, *68*, 394–403.
17. Bittelli, M.; Salvatorelli, F.; Pisa, P.R. Correction of TDR-based soil water content measurements in conductive soils.*Geoderma* 2008, *143*, 133–142.
18. Kizito, F.; Campbell, C.S.; Campbell, G.S.; Cobos, D.R.; Teare, B.L.; Carter, B.; Hopmans, J.W. Frequency, electrical conductivity and temperature analysis of a low-cost capacitance soil moisture sensor. *J. Hydrol.* 2008, *352*, 367–378.
19. Scudiero, E.; Berti, A.; Teatini, P.; Morari, F. Simultaneous monitoring of soil water content and salinity with a low-cost capacitance-resistance probe. *Sensors.* 2012, *12*, 17588–17607.
20. Nakamoto, T. Gravitropic Reaction of Primary Seminal Roots of Zea mays L. Influenced by Temperature and Soil Water Potential. *J. Plant Res.* 1995, *108*, 71–75.
21. Christopher, J.; Christopher, M.; Jennings, R.; Jones, S.; Fletcher, S.; Borrell, A.; Manschadi, A.M.; Jordan, D.; Mace, E.; Hammer, G. QTL for root angle and number in a population developed from bread wheats (Triticum aestivum) with contrasting adaptation to water-limited environments. *Theor. Appl. Genet.* 2013, *126*, 1563–1574.
22. Sanguineti, M.C.; Li, S.; Maccaferri, M.; Corneti, S.; Rotondo, F.; Chiari, T.; Tuberosa, R. Genetic dissection of seminal root architecture in elite durum wheat germplasm. *Ann. Appl. Biol.* 2007, *151*, 291–305.
23. Singh, V.; van Oosterom, E.J.; Jordan, D.R.; Hammer, G.L. Genetic control of nodal root angle in sorghum and its implications on water extraction. *Eur. J.Agron.* 2012, *42*, 3–10.
24. HE, Y.; Yang, J.; Zhu, B.; Zhu, Z. Low Root Zone Temperature Exacerbates the Ion Imbalance and Photosynthesis Inhibition and Induces Antioxidant Responses in Tomato Plants Under Salinity. *J. Integr. Agric.* 2014, *13*, 89–99.
25. Watt, M.; Silk, W.K.; Passioura, J.B. Rates of root and organism growth, soil conditions, and temporal and spatial development of the rhizosphere. *Ann. Bot.* 2006, *97*, 839–855.
26. Blum, A.; Mayer, J.; Gozlan, G. Infrared thermal sensing of plant canopies as a screening technique for dehydration avoidance in wheat. *F. Crop. Res.* 1982, *5*, 137–146.
27. Winterhalter, L.; Mistele, B.; Jampatong, S.; Schmidhalter, U. High throughput phenotyping of canopy water mass and canopy temperature in well-watered and drought stressed tropical maize hybrids in the vegetative stage. *Eur. J. Agron.* 2011, *35*, 22–32.
28. Ndjiondjop, M.N.; Manneh, B.; Cissoko, M.; Drame, N.K.; Kakai, R.G.; Bocco, R.; Baimey, H.; Wopereis, M. Drought resistance in an interspecific backcross population of rice (Oryza spp.) derived from the cross WAB56-104 (O. sativa) × CG14 (O. glaberrima). *Plant Sci.* 2010, *179*, 364–373.
29. Ingram, K.; Real, J.; Maguling, M. Comparison of selection indices to screen lowland rice for drought resistance.*Euphytica* 1990, *48*, 253–260.
30. Vijaya Kumar, P.; Ramakrishna, Y.; Ramana Rao, B.; Khandgonda, I.; Victor, U.; Srivastava, N.; Rao, G.G.S. Assessment of plant-extractable soil water in castor beans (Ricinus communis L.) using infrared thermometry. *Agric. Water Manag.* 1999, *39*, 69–83.
31. Mahan, J.R.; Conaty, W.; Neilsen, J.; Payton, P.; Cox, S.B. Field performance in agricultural settings of a wireless temperature monitoring system based on a low-cost infrared sensor. *Comput. Electron. Agric.* 2010, *71*, 176–181.
32. Guimarães, C.M.; Stone, L.F.; Lorieux, M.; de Oliveira, J.P.; de O'Alencar, G.C.; Dias, R.A.A. Infrared thermometry for drought phenotyping of inter and intra specific upland rice lines. *Rev. Bras. Eng. Agríc. Ambient* 2010, *14*, 148–154.
33. Sezen, S.M.; Yazar, A.; Daşgan, Y.; Yucel, S.; Akyıldız, A.; Tekin, S.; Akhoundnejad, Y. Evaluation of crop water stress index (CWSI) for red pepper with drip and furrow

irrigation under varying irrigation regimes. *Agric. Water Manag.*2014, *143*, 59–70.

34. Mahan, J.R.; Yeater, K.M. Agricultural applications of a low-cost infrared thermometer. *Comput. Electron. Agric.* 2008,*64*, 262–267.

35. O'Shaughnessy, S.A.; Hebel, M.A.; Evett, S.R.; Colaizzi, P.D. Evaluation of a wireless infrared thermometer with a narrow field of view. *Comput. Electron. Agric.* 2011, *76*, 9–68.

36. Hubbart, J.; Link, T.; Campbell, C.; Cobos, D. Evaluation of a low-cost temperature measurement system for environmental applications. *Hydrol. Process.* 2005, *19*, 1517–1523.

37. Garcia-sanchez, F.; Garcia-haro, J. Wireless sensor network deployment for integrating video-surveillance and data-monitoring in precision agriculture over distributed crops. *Comput. Electron. Agric.* 2011, *75*, 288–303.

38. Ruiz-Garcia, L.; Lunadei, L.; Barreiro, P.; Robla, I. A review of wireless sensor technologies and applications in agriculture and food industry: State of the art and current trends. *Sensors* 2009, *9*, 4728–4750.

39. Pierce, F.J.; Elliott, T.V. Regional and on-farm wireless sensor networks for agricultural systems in Eastern Washington. *Comput. Electron. Agric.* 2008, *61*, 32–43.

40. Agudo, J.E.; Pardo, P.J.; Sánchez, H.; Pérez, A.L.; Suero, M.I. A low-cost real color picker based on arduino. *Sensors*2014, *14*, 11943–11956.

41. Fisher, D.K. Open-Source Hardware Is a Low-Cost Alternative for Scientific Instrumentation and Research. *Mod. Instrum.* 2012, *1*, 8–20.

42. Arduino-Home. Available online: http://arduino.cc (accessed on 18 October 2014).

43. Kelley, C.D.; Krolick, A.; Brunner, L.; Burklund, A.; Kahn, D.; Ball, W.P.; Weber-Shirk, M. An Affordable Open-Source Turbidimeter. *Sensors* 2014, *22*, 7142–7155.

44. Koenka, I.J.; Sáiz, J.; Hauser, P.C. Instrumentino: An open-source modular Python framework for controlling Arduino based experimental instruments. *Comput. Phys. Commun.* 2014, *185*, 2724–2729.

45. Leeuw, T.; Boss, E.; Wright, D. *In situ* measurements of phytoplankton fluorescence using low cost electronics. *Sensors*2013, *13*, 7872–7883.

46. Lian, K.Y.; Hsiao, S.J.; Sung, W.T. Intelligent multi-sensor control system based on innovative technology integration*via* ZigBee and Wi-Fi networks. *J. Netw. Comput. Appl.* 2013, *36*, 756–767.

47. Thalheimer, M. A low-cost electronic tensiometer system for continuous monitoring of soil water potential. *J. Agric. Eng.* 2013, *44*. [CrossRef]

48. Tanenbaum, J.; Williams, A. Democratizing technology: Pleasure, utility and expressiveness in DIY and maker practice. Proceedings of the SIGCHI Conference on Human Factors in Compting Systems, Paris, France, 27 April–2 May 2013; pp. 2603–2612.

49. Adafruit Assembled Data Logging shield for Arduino ID: 1141. Available online: http://www.adafruit.com/product/1141 accessed on 18 October 2014).

50. DS1307 64 × 8, Serial, I²C Real-Time Clock. Available online: http://datasheets.maximintegrated.com/en/ds/ds1307.pdf(accessed on 18 october 2014).

51. Telefonica. Available online: http://www.telefonica.com/en/home/jsp/home.jsp (accessed on 18 October 2014).

52. Quectel Wireless Solutions-Dedicated M2M Wireless Module Supplier. Available online: http://www.quectel.com/(accessed on 18 October 2014).

53. Robots-Robot-Domestici. Available online: http://www.robot-domestici.it/joomla/index.php (accessed on 18 October 2014).

54. VH400 Soil Moisture Sensor. Available online: http://vegetronix.com/Products/VG400/ (accessed on 18 October 2014).

55. Sudha, M.N.; Valarmathi, M.L.; Babu, A.S. Energy efficient data transmission in automatic irrigation system using wireless sensor networks. *Comput. Electron. Agric.* 2011, *78*, 215–221.

56. DS18B20 Programmable Resolution 1-Wire Digital Thermometer. Available online: http://datasheets.maximintegrated.com/en/ds/DS18B20.pdf (accessed on 18 October 2014).

57. EXP TECH. Available online: http://www.exp-tech.de/Sensoren/Temperatur/DHT22-temperature-humidity-sensor-extras.html (accessed on 18 October 2014).

58. MLX90614 Digital, Plug & Play, Infrared Thermometer in a TO-can. Available online: http://www.melexis.com/Infrared-Thermometer-Sensors/Infrared-Thermometer-Sensors/MLX90614-615.aspx (accessed on 18 October 2014).

59. The Comprehensive R Archive Network. Available online: http://cran.r-project.org/ (accessed on 18 October 2014).

60. Geesing, D.; Bachmaier, M.; Schmidhalter, U. Field calibration of a capacitance soil water probe in heterogeneous fields. *Aust. J. Soil Res.* 2004, *42*, 289–299.

61. Nonlinear Regression and Nonlinear Least Squares An Appendix to An R Companion to Applied Regression, Second Edition. Available online: http://socserv.mcmaster.ca/jfox/Books/Companion/appendix/Appendix-NonlinearRegression.pdf (accessed on 1 august 2014).

62. Mcmichael, B.; Lascano, R.J. Laboratory Evaluation of a Commercial Dielectric Soil Water Sensor. *Vadose Zone J.* 2003,*2*, 650–654.

63. Wang, N.; Zhang, N.; Wang, M. Wireless sensors in agriculture and food industry—Recent development and future perspective. *Comput. Electron. Agric.* 2006, *50*, 1–14.

64. Van Iersel, M. Sensors for improved efficiency of irrigation in greenhouse and nursery production. *Hort Technol.* 2013,*23*, 735–746.

65. Arduino-ArduinoXbeeShield. http://arduino.cc/en/Main/ArduinoXbeeShield (accessed on 18 October 2014).

CHAPTER 11

SOIL AND WATER CONSERVATION STRATEGIES AND THEIR IMPACTS ON LIVELIHOODS

Isaurinda Baptista, Luuk Fleskens, Coen Ritsema, António Querido, Jacques Tavares, António D. Ferreira, Eduardo A. Reis, Samuel Gomes and Anabela Varela

Keywords: Land degradation, desertification, sustainable land management (SLM), livelihoods, poverty, governance, Cape Verde

INTRODUCTION

Cape Verde (Cabo Verde, in Portuguese), is a small island country where a mix of climatic, geomorphologic, pedologic and human factors has resulted in widespread land degradation, with negative consequences for the livelihood of the population and its fragile environment [1,2]. The scarcity of natural resources and the weakness of the productive system make poverty a structural occurrence in the country.

Desertification, associated with the occurrence of extreme drought episodes, has been present throughout the country's history and has led to periods of food production deficit and famine, which have, in the past, caused thousands of deaths [1]. Following the last big famine in the late 1940's, the authorities took measures to eradicate famines from the country. The stabilization of the agricultural landscape with erosion control measures and the maintenance of sustainable yields became absolute priorities, not just for environmental protection, but also for survival. Since the country's independence in 1975, successive governments have focused their rural development policies on soil and water conservation (SWC) strategies to address desertification, water scarcity, and soil erosion, aiming to reconstruct the ecological potential and reduce poverty in rural areas [3]. This effort has completely changed the landscape to a level where SWC techniques can be found everywhere, with the exception of rock outcrops. This arduous task was performed in the interval of two generations, since the last big famine in 1948, and more assertively in 1975, using a set of SWC techniques [1] and a governance strategy involving the communities affected by the drought.

Several types of SWC measures were implemented throughout the country (*i.e.*, terraces, half-moons, live barriers, contour rock walls, contour furrows, micro-catchments, check dams, runoff water collecting dams, water reservoirs, wells and afforestation with drought resistant species), aiming to hold the soil in place, the water in the soil and to combat desertification [1].

Current efforts to reverse and prevent land degradation focus on the concept of sustainable land management (SLM) which refers to the use of land resources to meet present needs without compromising the ability of future generations to meet their own needs [4,5]. SLM technologies include agronomic, vegetative, structural, and management measures to control land degradation and enhance productivity in the field [6]; thus,

---

Citation: Baptista I., Fleskens L., Ritsema C., Querido A., Tavares J., Ferreira A.D., Reis E.A., Gomes S., Varela, Soil and Water Conservation Strategies and their Impacts on Livelihood, doi: 10.3390/land4010022.

SWC measures implemented in Cape Verde promote the SLM concept. The increased efforts and financial resources allocated to the promotion of SLM calls for adequate monitoring and assessment (M&A) of their impact and benefits. Nevertheless, SLM must reflect the human-environment interactions. This requires a system to systematically assess the impacts and benefits from SLM measures that remain a fundamental gap [7].

Investments in SLM measures are enormous at the national level have been huge, yet a clear overview of their extent and combined benefits in terms of agriculture productivity, conservation effectiveness, sustainability, and rural people's well-being, is still lacking. Information on past interventions is scattered and of little influence for new SLM activities. Implementation of the five-year research project, DESIRE (Desertification Mitigation and Remediation of Land: A global approach for local solutions), contributed to fill part of the gap for the Ribeira Seca Watershed (RSW), giving policy makers and implementing institutions a spatial overview of past and ongoing processes, so that they can plan future activities. The DESIRE project developed an approach for establishing SLM strategies in response to desertification, consisting of five steps:

- establishing land degradation and SLM context and sustainability goals with stakeholders;
- identifying, evaluating and selecting SLM strategies with stakeholders;
- trialing and monitoring SLM strategies;
- up-scaling SLM strategies; and
- disseminating the knowledge gathered [8].

Even though consistent biophysical impact and cost-benefit studies of SWC measures are lacking, their contribution to the improvement of environmental conditions of the treated landscape is visible [9], making crop production possible, and consequently improving the livelihood of the rural people [1]. The implementation of SWC measures, particularly afforestation, has contributed to change the "lunar" aspect of the landscape to more pleasant and greener ones [10].

Since most of the rain-fed cropland is on steep slopes, the main concern has been to protect the hillsides from runoff and erosion caused by heavy rain events, while neglecting in-field agronomic measures that lead to sustainable productivity increase, such as soil cover and nutrient management. Rain-fed crop yields remain low, with maize grain yield not exceeding 1 ton·ha^{-1}. The low input farming system, dominated by continuous maize and beans intercropping still faces severe climate conditions, inadequate crop and land management practices, and land degradation.

In this paper, we aim to corroborate the importance of SLM measures for Cape Verde, the strategies undertaken to address desertification, and the impact of those measures on the environment and rural livelihood. More specifically, we analyze the government strategies towards building resistance against the harsh conditions and, based on the DESIRE approach and results within the RSW, we analyze the state of land degradation and its drivers, survey and map the existing SWC measures and assess their effectiveness against land degradation and impact on crop yield and livelihoods. We explore the governance aspects behind the relative success of Cape Verde in tackling desertification and rural poverty, to verify the replication potential in countries with similar adverse conditions.

METHODS

Study Area

Cape Verde Natural Resource Endowment

Rainfall and climate: Rainfall, as the dominant climate factor influencing land degradation, is extremely variable both in space and time and strongly influenced by elevation and topography. It results from the seasonal migration of the Inter-Tropical Convergence Zone (ITCZ) around the equator, which originates a single short wet season from July to October, and a long dry season from November to June [11,12]. The climate is predominantly semiarid. Rainfall typically ranges from 200 to over 650 mm in the upwind façade to less than 100 mm in the downwind side.

Water resources: Cape Verde has no permanent surface stock of fresh water, creating an almost total dependence on ground water supplies for domestic and agricultural use. Ground and superficial water resources are scarce as they depend on the erratic rainfall, susceptible to be lost to the sea. In an average rainfall year, about 20% (180 million m^3) is lost through runoff; more than 50% is lost as evaporation; and only about 13%–17% recharges the aquifers [13]. About 60,000 m^3·day^{-1} are currently being pumped for irrigation, representing a serious overexploitation of available groundwater resources that contributes to soil and water salinization.

Soil resources and land management: Of the 4033 km^2 of land surface that the country comprises, about 10% is cultivated. The soils are mainly of volcanic origin, medium to coarse textured, steep, low in organic matter and generally shallow. Of the arable surface (41,000 ha), >90% is used for rain-fed agriculture while about 5% is used for irrigated agriculture. About 23% of the country's surface is forested. Severe land degradation occurs in the whole country, in different ways and extent according to the physical characteristics of the islands. To prevent land degradation and to enable agriculture production, both structural and biological SWC and water harvesting measures have been implemented at a vast scale. For a more comprehensive review of the SWC and water harvesting measures, please refer to [1].

Ribeira Seca Watershed: The Largest Watershed in Santiago Island

Ribeira Seca Watershed (RSW) is located on the east-central side of the Santiago Island, between latitude 15°07'40"W and longitude 23°32'05"W (Figure 1). With a drainage area of about 72 km^2, it represents 4 agro-climatic zones: semi-arid (49%), arid (20%), sub-humid (20%), and humid (11%) based on altitude, vegetation and relief [14].

The climate is characterized by a dry season that lasts 8–9 months (November to June) and a short, humid season of 3–4 months (July to October), with the humid period coinciding with high temperatures (Figure 2). Rainfall is extremely heterogeneous and its spatial-temporal distribution irregular.

The watershed has a maximum altitude of 1394 m and comprises seven sub-watersheds, characterized by steep slopes and annual precipitation that varies from <200 mm downstream to >650 mm upstream. The maximum average monthly temperature is 28.1 °C and the minimum is 16.6 °C. The dominant land use is rain-fed agriculture, particularly the staple crops (maize and beans) and groundnut, occupying 83% of the area. The remaining area is used for: irrigated crops (sugarcane, fruits, vegetables, cassava and sweet potato) 5% (362 ha) and forest 4% (251 ha). In addition; 1% are rock outcrops and 7% are built environment (Figure 3). Livestock keeping is an important activity in the watershed as most family farmers own animals, such as cows, goats, pigs, and chickens that often graze freely.

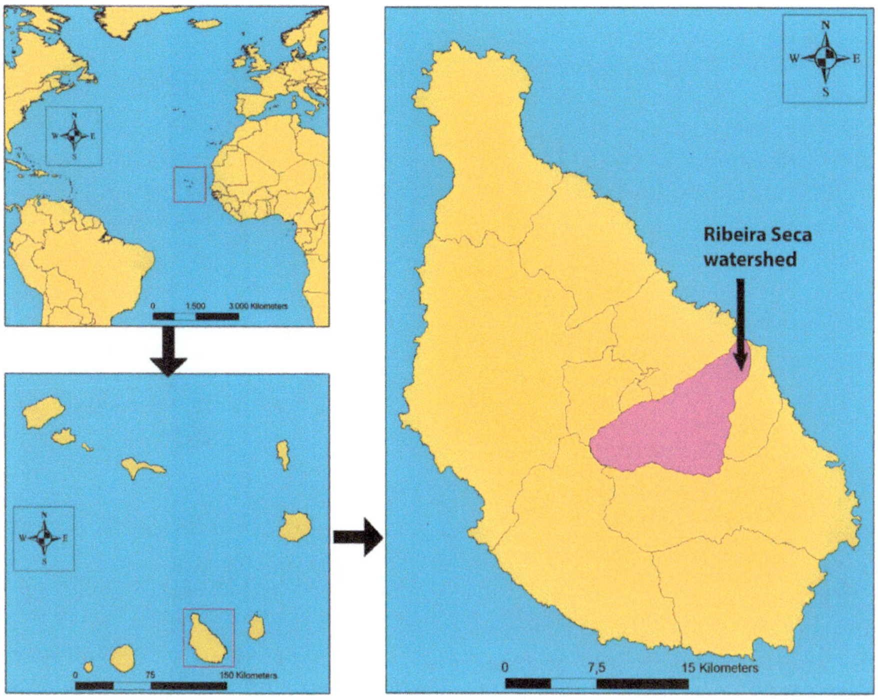

Figure 1. Location of the Ribeira Seca Watershed within Santiago Island and Cape Verde.

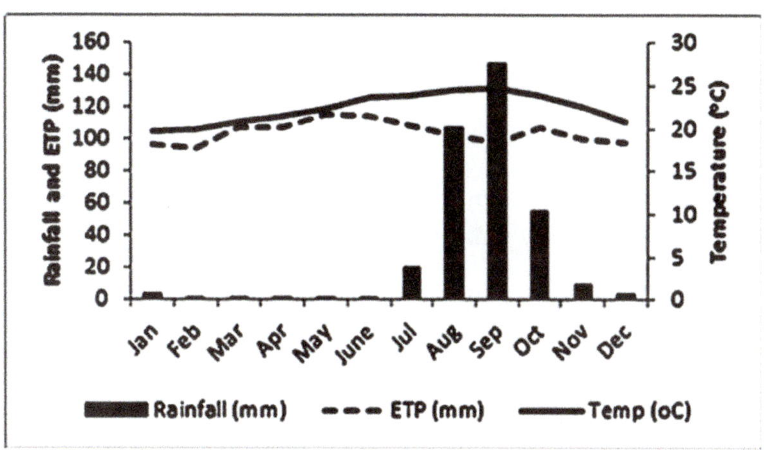

Figure 2. Mean annual rainfall distribution, evapo-transpiration and temperature for São Jorge Station (subhumid zone of Ribeira Seca Watershed): period 1973–2010.

The soils, mainly Regosols and Cambisols developed on basaltic substrate, are shallow and low in organic matter (OM), generally with low to medium fertility [15] and medium to coarse texture, and exhibit marked symptoms of degradation (*i.e.*, rills and gullies). Deeper soils, *i.e.*, Kastanozems with higher OM content and Vertisols, can be found on the plateaus (*achadas*) of less steep slopes. In the valley bottoms or *ribeiras*, Fluvisols are predominant and used for irrigated agriculture.

The Ribeira Seca Watershed has an estimated total population of about 15,000 people, with an even gender distribution. Of the women, 43% are head of household. The population is young, with 77% below 35 years old (of which 48% are below 15) and features an 83% literacy rate [16]. Poverty (people living on <$1.25 a day [17]) affects >60% of the population.

RSW was selected as the study area since it is the largest watershed of Santiago (the main agricultural island) that can be considered a reference for SWC works.

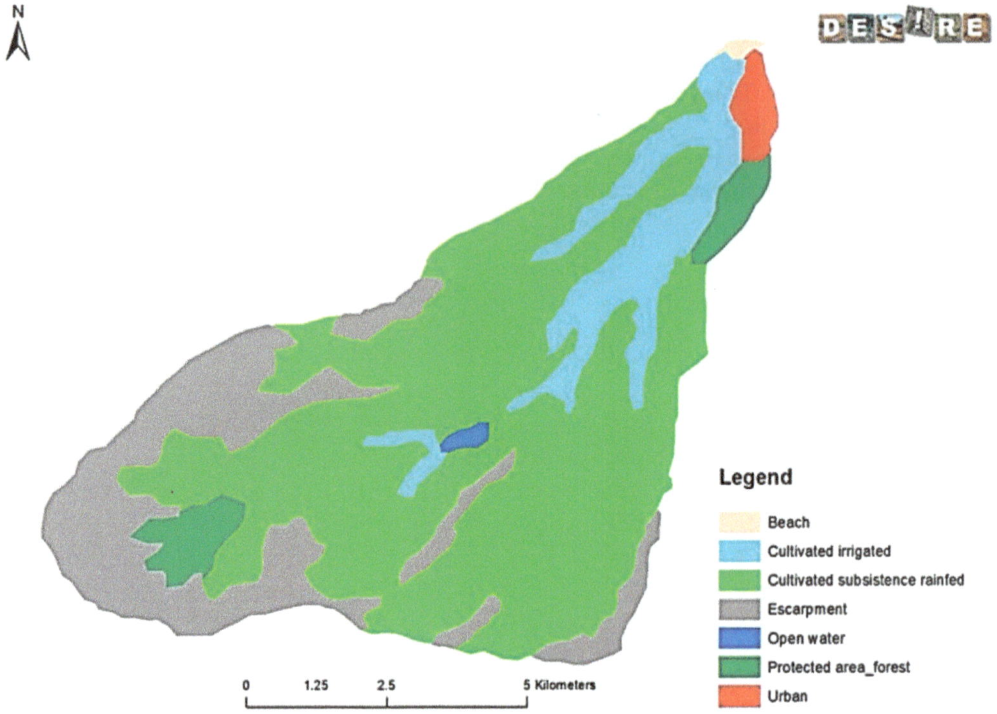

Figure 3. Map of land use types in the Ribeira Seca Watershed showing the dominance of rain-fed farming. Adapted from [18].

Methodological Approach

For this study, we used a mixed methods approach, consisting of literature and policy review, qualitative and quantitative field assessment and analysis of local and scientific knowledge. Tapping into local and expert opinion formed an important part of the methodology, both to triangulate results obtained using other methods, and because measurements and observations were not always possible.

An analyses of Government actions towards building resilience against the harsh drought conditions (Section 3) was conducted through national documents, such as strategic and action plans (*i.e.*, the United Nations Conventions on Desertification—UNCCD, Biological Diversity—CBD, and Climate Change—UNFCC), policy regulations, and project implementation reports.

Assessment and mapping of land degradation status (drivers, types, and extent), existing SLM measures (types and extent), effectiveness of measures, and expert recommendation of conservation measures were carried out during implementation of the DESIRE project (2007–2012) using the "DESIRE approach" [8].

Land degradation and existing SLM measures were mapped using the WOCAT-LADA-DESIRE methodology and evaluated by applying WOCAT (World Overview of Conservation Approaches and Technologies) questionnaires on SLM technologies and approaches at local level [19]. WOCAT [20] offers a systematic methodology to assess both the institutional aspects (approaches) and practical implementation aspects (technologies) of SLM using expert questionnaires.

Detailed mapping coverage of the existing SWC measures was carried out using aerial photography [21]. The extent of land degradation was determined in % on the mapping unit, based on the land use system as the basic unit of evaluation; while the degree of degradation refers to the intensity of the land degradation process, weighted against the extent for each map unit: Degree × Extent (%)/100.

The effectiveness of SLM measures was defined in terms of how much the measures reduce the degree, or prevent land degradation and was weighted against the extent for each mapping unit: Effectiveness × Extent (%)/100.

Expert recommendation concerning interventions on how to address degradation was based on four groups: adaptation, prevention, mitigation and rehabilitation.

Stakeholders involvement in the project was accomplished through workshops following a standard approach developed for sharing of local knowledge [6,10,22] and through WOCAT questionnaires that generated data for the mapping.

To assess the biophysical impact of selected SWC measures (terraces, vegetation barriers and rock walls—*arretos*) on dryland crop yield and soil quality, thirty six plots of 25 m^2 were selected in farmers' fields, in three agroecological zones (arid, semiarid and subhumid), during the 2010 rainy season. In each plot, maize, bean and total biomass yield and soil quality parameters (nutrients, textures, organic matter, depth of top layer, *etc.*) were measured and compared to untreated plots using ANOVA and LSD, assuming other drivers (*i.e.*, yield drivers) being constant. Results of other studies were also used as evidences of biophysical impact.

BUILDING RESILIENCE TO CHALLENGE THE HARSH CONDITIONS: SLM GOVERNANCE FACTORS

Eradicating Famine

Cape Verde's history comprises remarkable events of crop failure and food insecurity that caused widespread starvation from the 16th to the 19th centuries, registering several famines and epidemics [1]. In the first half of the 20th century, six famines occurred, causing the starvation of more than 75,000 people [23]. The famine episodes were closely associated with the occurrence of extended periods of extreme drought. The last famine in 1947–1948 caused an almost 50% reduction of the Cape Verde population, and Amaral [24] points out that in the Santiago Island alone the population decreased by 65%.

Following the 1947–1948 famine, the colonial authorities took measures to mitigate the problem. The first was to employ people locally on the so-called *Frentes de Alta Intensidade de Mão-de-Obra*—FAIMO (High Intensity Labor Fronts) to implement SWC measures, such as terraces, dams, afforestation and irrigation infrastructure, to reduce erosion risk, conserve soil and retain water, thus stabilizing the agricultural landscape. FAIMO was a national program that ensured jobs for thousands of people in rural areas, particularly in dry years. A second measure was to expatriate people to other colonies to work in plantations. A third initiative was the installation of an agriculture technical school to guarantee technical know-how, dissemination and extension. These measures, together with large-scale emigration led to the eradication of famine from the country [1].

Environmental Awareness and Actions

The post-independence governments, facing successive dry years in the late 1970s and early 1980s, oriented their rural environmental actions towards combating desertification and land degradation by managing scarce financial resources, elaborating strategic instruments, creating proper institutional frameworks, setting policies and regulations and adhering to regional and international agreements. The establishment of an environmental restoration strategy was linked to political, cultural and economic objectives as the governments aimed to reduce rural poverty by creating jobs and reducing the rural exodus.

Financial resources management. From 1975 to late 1990s, Cape Verde received food aid which, in agreement with international donors, was internally commercialized to support development, with a large sum of the revenue channeled to SWC works and afforestation programs implemented through FAIMO [1,25,26]. The country also benefited from programs and projects with watershed development components such as: Community-based Project for the Development of Agriculture and Livestock (PRODAP/FIDA, 1993–1997), Watershed and Applied Research Development (WARD, 1993–1997), and the Millennium Challenge Account (MCA, 2007–2011), that contributed to implementing SWC works throughout the islands. Nowadays a large share of the Ministry of Rural Development's (MDR) budget is still being channeled to SWC through watershed planning and management projects, but the funds come from government loans for development.

National Instruments: The successive post-independence Governments defined instruments for the implementation of a national development strategy, aiming to mainstream environmental issues in the planning process to eradicate poverty and to promote sustainable development [27]. The most relevant ones are: The National Development Plans (PND); the Growth and Poverty Reduction Strategy (DECRP I: 2004–2007, II: 2008–2011, III: 2012–2016); the National Environment Action Plan (PANA II: 2004–2014); the National Program to Combat Poverty (PNLP 1996–2008); the National Adaptation Program of Action (NAPA 2008–2012); the Agricultural Development Strategic Plan (PEDA: 2004–2015); and the National Action Plan to Combat Desertification (PAN-LCD: 1998). PANA II, which promotes natural resources management, the use of efficient techniques, local participation in the sustainable use of natural resources, sustainable management of biodiversity and defines policies for food security, incorporates the objectives of PAN-LCD. For more than 10 years the UNCCD offered the outline of a global framework in the fight against desertification and Cape Verde has been positively evaluated for its implementation. Cape Verde is currently updating its PAN-LCD, aligning it with the UNCCD 10-year strategic plan (2008–2018).

Regional and international agreements: At regional level, Cape Verde is a member of CILSS (Permanent Interstate Committee for Drought Control), which is the regional organization for the formulation, analysis, coordination and harmonization of strategies and policies on food security and combating the effects of drought and desertification in the Sahelian countries. At the international level, the country is a member of the three United Nations Conventions (CBD, UNFCC, and UNCCD). The fact that Cape Verde was the first African country and the second in the world to ratify the UNCCD in 1995 [3] shows the strong commitment towards combating desertification, protecting the natural resources and fighting poverty in the country.

Institutional context. Combating land degradation through SWC measures constitutes the main pillar of the sustainable rural development strategy in Cape Verde. An institutional framework has been set up, adapted and strengthened through the years, to support both the current activities of rural development and UNCCD implementation. Currently, MDR is the coordinating institution and an umbrella organization of key departments such as the Direction General of Agriculture and Rural Development (DGADR); the National Institute of Agriculture Research and Development (INIDA), the National Institute of Rural Engineering and Forest (INERF/SONERF) and the National Institute of Water Resource Management (INGRH/ANAS). MDR collaborates with the Ministry of Environment, Housing and Territory Management (MAHOT) for crosscutting environmental issues.

Participatory approach. Stakeholders' participation in the post-independence implementation of SWC measures went through three major phases [3]:

1. From 1975 to 1985, participation of the population was achieved through FAIMO, which were institutionalized employment schemes that provided people with temporary paid work to alleviate their immediate household needs. The employment scheme and the top-down approach created their own dynamics: while successful in mobilizing the participation of land users, they led to a lack of perceived responsibility by landowners to maintain the structures [25], with consequences for the effectiveness of environmental conservation [26].
2. From 1985 to 1995, in an attempt to change the concept of FAIMO by diversifying rural livelihood sources and introducing more participation of local people, there was some involvement of the stakeholders,

particularly land users. However, it was not an effective participation since the approach did not emphasize participation as a process. Stakeholders' participation, as a process, needs to be supported by an approach that emphasizes empowerment, fairness, trust, and learning [28]. Thus, this can be considered a transitional phase to an effective stakeholders' participation.

3. From the late 1990's onwards, the ratification of the UNCCD harnessed the participation of local communities in drawing up a national action program to combat desertification (PAN-LCD). Since then, a full partnership with the population has been established, with community-based organizations playing a major role not only in implementing and maintaining SWC works, but also in contributing local knowledge and choosing the best local measures.

There is general consensus that success in combating land degradation and natural resource management can best be achieved with active involvement of affected populations in developing solutions, leveraging their knowledge and experiences and combining them with the knowledge of researchers and managers in all planning steps [10,29,30,31,32,33,34].

Also, taking the watershed as the planning unit, with full involvement of community associations and farmers, has enabled the capacity of local people to implement major rural engineering SWC works. Thus, SWC activities acquired new dynamics, moving from essentially work provision schemes to effective participation in both building and maintenance of SWC measures.

Stakeholders' awareness, training and capacity building. Since the 1990s an increasing attention was paid to awareness, training and capacity building of stakeholders on environmental conservation, which has positively influenced the success of SWC in the country. Stakeholders have become more conscious about the need for SWC and maintenance. Technical capacity building was crucial for executing engineering works. Currently, all environmental projects include awareness, training and capacity building activities for stakeholders.

As a result of these governance aspects, Cape Verde is currently recognized as a success case on transforming a harsh environment into a less hostile living environment and building resilience to absorb shocks from extreme drought periods, which no longer result in deaths due to famine or even starvation. From being one of the poorest countries in the world at independence in 1975 [23], Cape Verde graduated to a medium human development country in 2007 [35], and is currently pursuing a new development paradigm. New solutions have to be sought to provide food for a fast growing population, while intensifying soil productivity on highly degraded land and this should pass through the endorsement of SLM strategies.

DRIVERS OF LAND DEGRADATION AND DESERTIFICATION

Land degradation in Cape Verde, particularly in Ribeira Seca, results from a combination of climate, human and topographic factors.

Climatic drivers. Extended droughts have reduced vegetation cover, exposing bare land to erosion, while heavy rainfall events during the wet season generate high rates of runoff transporting enormous quantities of soil. Erosion by water (*i.e.*, 43 and 23 $t \cdot ha^{-1} \cdot yr^{-1}$, at watershed and plot levels, respectively) seriously affects the quality of the environment, food security, sustainability and longevity of the limited arable land in the RSW [26,36].

Human drivers. In RSW, human activity exerts strong pressure on the limited land and water resources, contributing to land degradation in several ways [16]: (a) Inadequate agricultural practices such as intensive cultivation of steep slopes without adequate conservation measures and excessive weeding by hoe; (b) Overexploitation of the aquifers, through over-pumping of wells, which has resulted in sea water intrusion, leading to the salinization of soil and water in the valley bottom. Salinity has affected soil and water quality up to about 2.5 km from the coast into the valley; (c) Construction of the Poilão dam (2/3 of the way to the sea) has blocked the aquifer movement downstream and also contributed to salinization of the lands close to the sea; (d) rural poverty, leading to deforestation due to tree cuttings for domestic use as energy for cooking; and excessive grazing by animals.

Topographic and pedological factors. Rainfall is strongly influenced by elevation, with highest erosivity values at high elevations, coinciding with steep slopes and shallow soils, which make these areas susceptible to erosion [12]. Despite the natural fertility of the soils due to their volcanic nature, intensive cultivation without replenishment of soil nutrients has resulted in the decline of fertility in the dryland soils. Since farmers do not apply any fertilizers on dryland soils, the only source of nutrients for the maize crop comes from the intercropped beans, which is not sustainable.

These driving forces place pressures on the land and have particular impacts at different levels. Cape Verde society and authorities have already produced some responses to those impacts and also seek better knowledge on how to respond. Decisions on responses are governed by a range of international, national and local policy regulations and agreements (seeSection 3). Nongovernmental Organizations (NGOs), governments and civil society are working together in natural resource management to combat environmental degradation through reduction of poverty, as it is believed that, by reducing poverty, the pressure on natural resources will also decrease environmental degradation.

STATUS OF LAND RESOURCES AND MANAGEMENT

Monitoring and Assessment in the Ribeira Seca Watershed

Monitoring and assessment (M&A) of SLM measures in Cape Verde are at an incipient stage, with few environmental projects containing these important components. Most projects lack a scientifically robust and consistent baseline, which is a key constraint to identifying priorities for action and monitoring their impact. This paper focuses on the implementation of the DESIRE project.

The bottom-up Desire approach to combat desertification advocated SLM based on inventories of local knowledge using an integrative participatory approach in close collaboration with local stakeholders, but with a sound scientific basis for effectiveness at various scales [8]. With the completion of the project, the data from RSW is part of a web-based harmonized information system [18] featuring the: socio-cultural, economic and environmental context; stakeholders goals and drivers of changes; status of land degradation and land management; identification, documentation and evaluation of SLM options; prioritization and selection of remediation options; field trials of SLM techniques with scientific and stakeholders' monitoring; biophysical and economic effects at field and regional scale; evaluation of options and recommendations; and strategy for disseminating recommendations at various levels. Given the DESIRE project's limited duration of five years, its methods, experiences, and know-how should be integrated into long-term programs (*i.e.*, UNCCD) to ensure their continued use and ultimate effectiveness.

Type, Degree, Extent and Trend of Land Degradation and Desertification Risk

The RSW presents physical land degradation in various forms, degree, extent and rate (Figure 4a–d). The degree of degradation is light to moderate, possibly due to the large extent of conservation measures, covering 80%–100% of the cultivated rain-fed land. The main type of land degradation is water erosion, estimated to affect 26%–50% of the watershed area under subsistence rain-fed agriculture, manifesting first as loss of top soil, second as mass movement and third as gullies. Chemical degradation due to salinization in cultivated land under irrigation is also significant, affecting 11%–25% of the surface area. Wind erosion and biological degradation, associated to the loss of top soil and soil life, are less significant and affect 0%–10% of the RSW.

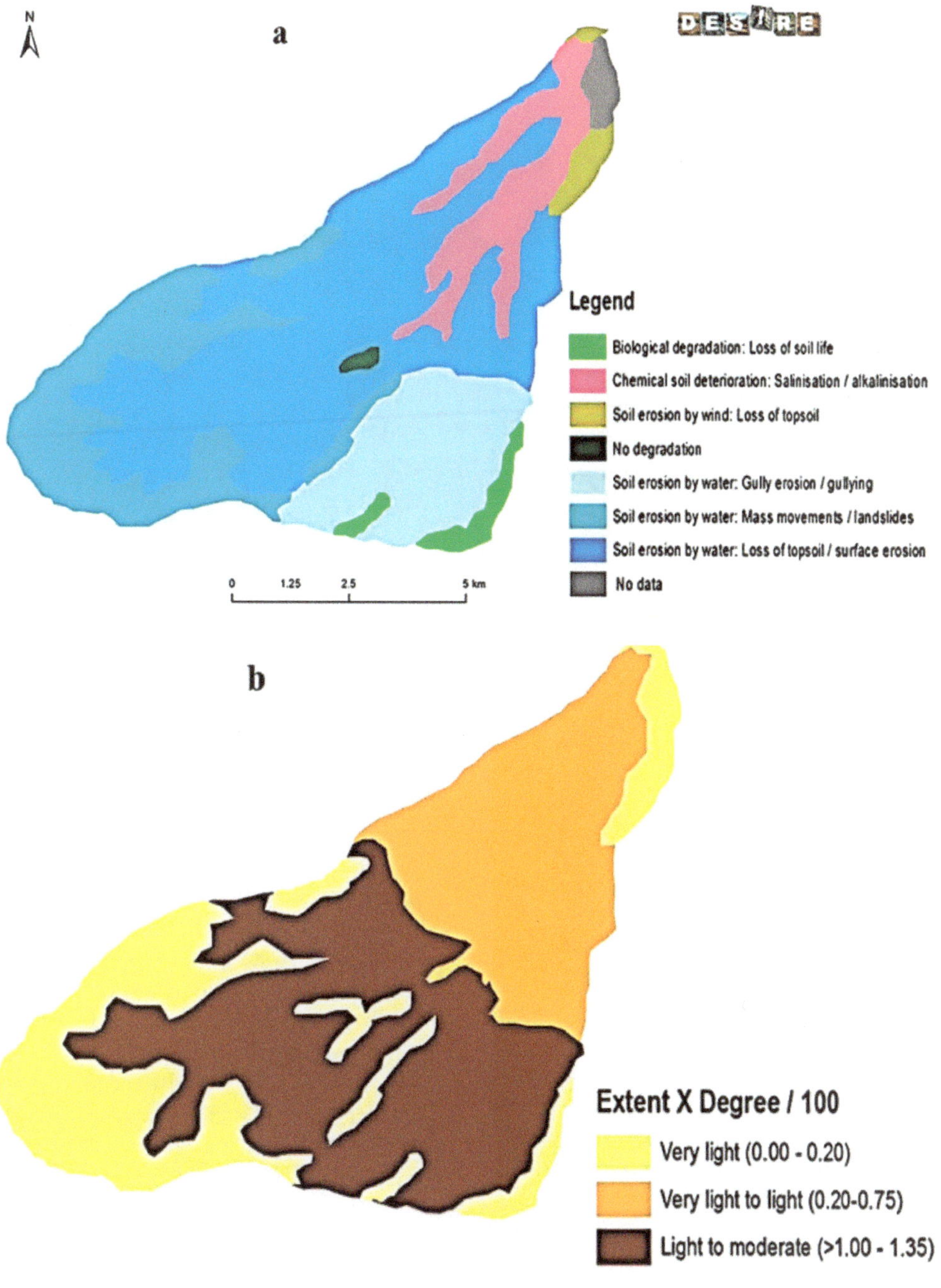

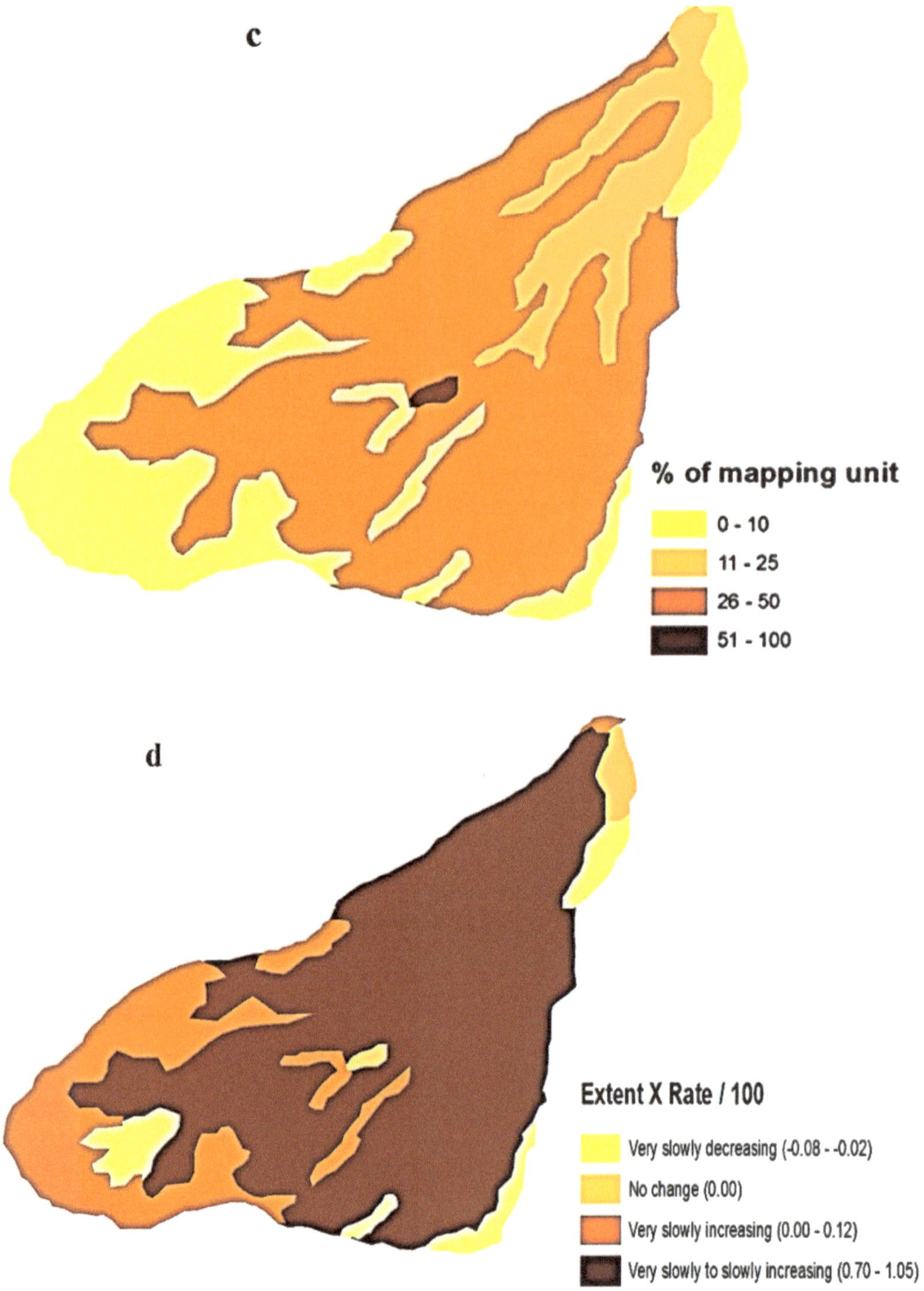

Figure 4. (**a**) Type, (**b**) degree, (**c**) extent and (**d**) rate of land degradation in the Ribeira Seca Watershed (RSW) on as the dominant type, moderate degree, and slight to moderate rate of degradation. Adapted from [18].

The rate of land degradation on the hillsides, as indicated by the trend over the recent past 10-year period, was estimated as very slowly to slowly increasing.

Magnitude of Soil Erosion

Several attempts have been made to quantify erosion in the RSW both at plot [2,26,37,38] and sub-watershed levels [39,40,41,42,43], with results indicating great spatial and temporal variability depending on slope, land-use, rainfall amount and intensity. Mean erosion rates ($t{\cdot}ha^{-1}{\cdot}yr^{-1}$) for traditional farming vary from 0.2 to 23 at plot level and from 0.1 to 43 at sub-watershed level under natural rainfall (Table 1). The smaller rates at plot level correspond to low slope areas and the larger to the steeper slopes. Given the large variability within the results, the high rates of erosion and the shallow soils on steep terrain, there is a need to carry out longer term assessments and to establish standard tolerable rates for Cape Verde hillsides, allowing policy makers to better plan SWC interventions.

Table 1. Mean rates of soil erosion in the Ribeira Seca Watershed at different scales and periods

Scale/Location/% Slope	Period of Measurements	Land Use/Conservation Measure	Magnitude ($t{\cdot}ha^{-1}{\cdot}yr^{-1}$)	Reference
Sub-watershed: Seca (variable slopes)	1985–1989	Mixed land-use/Check dams	4.66	[39]
Sub-watershed: Longueira (variable slopes)	2005–2009	Mixed land-use and conservation measures	42.7	[42]
Sub-watershed/Grande/variable slopes			1.57	
Sub-watershed: Godim/variable slopes			0.10	
Plot/O. Pequeno/≤37	2011–2012	Maize and beans	9.10 (16.6) *	[37]
Plot/Serrado/≤23			2.13 (3.21) *	
Plot/S. Jorge/≤8			0.30 (0.4) *	
Plot/Godim)/50	1994–1996	Maize and beans	16.0 (48) **	[2]
Plot/Godim/35–60		Maize and beans	23 (69) **	[26]

* Maximum rates; ** total for period.

Existing Land Management Measures

Ribeira Seca is one of the watersheds in the country where most SWC measures, both structural and biological (vegetation), were implemented. Structural techniques comprise check dams, contour rock walls, contour furrows, micro catchments, terraces, water reservoirs and retaining dams (Figure 5a). Biological techniques consist of vegetation barriers with different species (*i.e.*, *Aloe vera*, *L. leucocephala*, *F. gigantean*), pigeon pea cultivation and reforestation with drought-tolerant species. Vegetative measures, including tree/shrub cover, implemented to protect the steep hillsides, are most widespread and cover more than 80% of the watershed area (Figure 5b,c). Along the valley bottom and the steep water lines, about 200 check dams are present, contributing to decrease runoff rate and increase infiltration. The implementation of SWC techniques modifies landscape functions at different spatial scales [44] and in Cape Verde they have produced dramatic changes both at plot and watershed scales. Figure 6a–h) illustrates some relevant SWC measures in the watershed, some of which (*i.e.*, afforestation and *Aloe vera* hedges) were documented as successful and promising conservation measures [8,45].

Recently, the government constructed large dams to collect and store runoff from the watersheds, with a total of five dams already built in the two largest islands. Ribeira Seca encloses one of the largest rainwater collecting dams built in the country, with a capacity of 1.2 million m^3. The construction of this large piece of infrastructure (see Figure 6c) completely changed the surrounding landscape, allowing 188 ha of irrigated land. Drip irrigation was introduced in the 1990's as a more efficient and water saving irrigation system, and has been promoted since. In rain-fed areas in the downstream section of the watershed, large concrete water reservoirs were built to collect and stock rain water for supplementary irrigation, human and animal consumption [1,3,16].

Afforestation (Figure 6h) with numerous drought-tolerant species is one of the key techniques to address land degradation and has increased from <3000 ha in 1975 to 92,000 ha in 2012. It provides soil protection against erosion, fuel wood for cooking and animal feed, and enables better use of rainfall for increased crop productivity. Moreover, the large afforested area has conveyed more green color to the "*lunar*" aspect of the landscape. The total biomass production has reached 800,000 ton in Cape Verde and the total carbon sequestered in the biomass, 402,500 ton with Santiago Island contributing with 60% of the total [46].

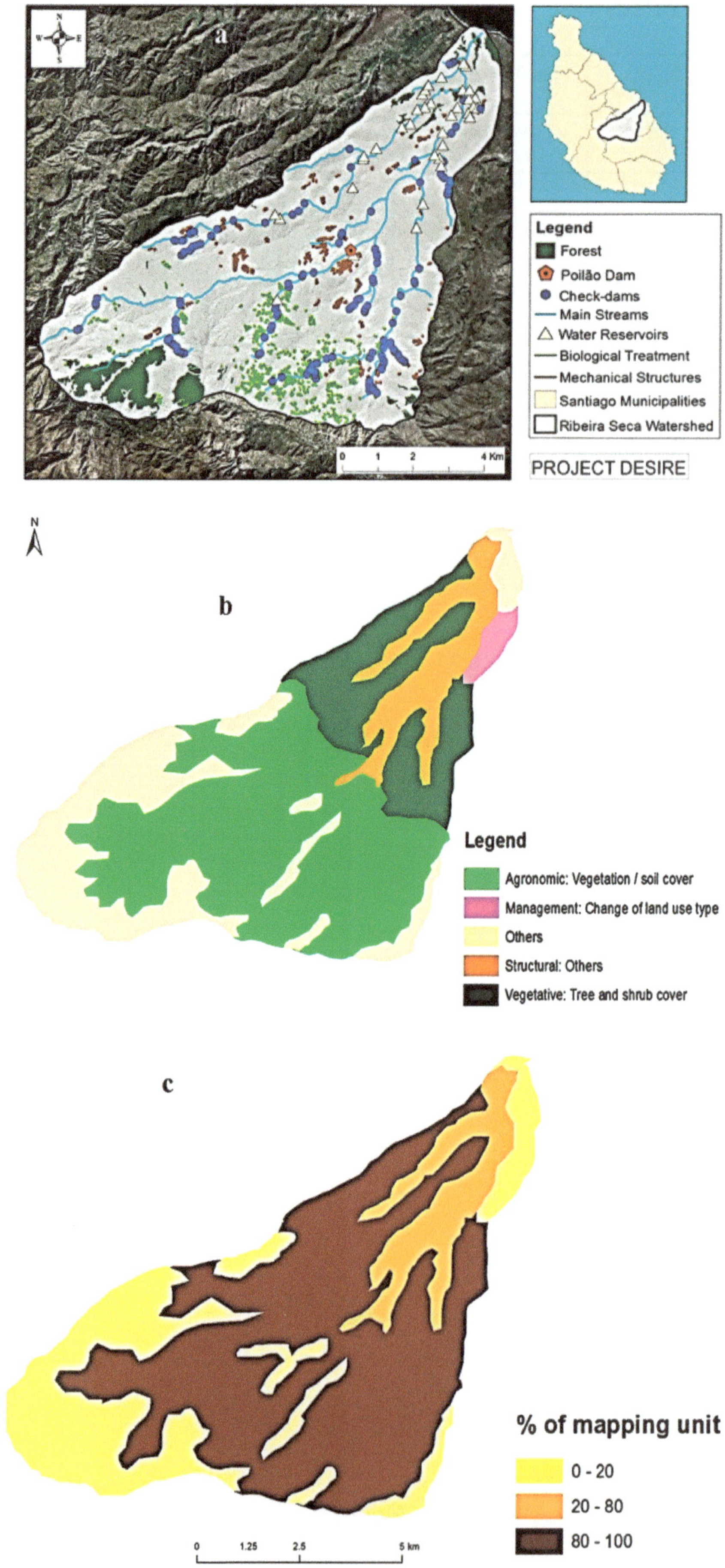

Figure 5. (**a**) Existing soil and water conservation measures (**b**) land conservation group measures and (**c**) extent as % of mapping unit showing the importance of vegetation measures as the major conservation group at hillslope level. (b) and (c) adapted from [18].

Figure 6. Vegetation and structural soil and water conservation measures in the Ribeira Seca Watershed: (**a**) contour rock walls; (**b**) terraces (São Jorge); (**c**) Water collecting dam (Poilão) during rainy season, d. check dams on water ways (Godim); (**e**) Aloe Vera barriers (São Jorge), (**f**) *L. Leucocephala* hedges (Godim); (**g**) Pigeon-pea on terraced fields; and (**h**) Afforestation with different species (*Longueira Baixo*).

EFFECTIVENESS AND IMPACT OF SWC

Effectiveness for Slope Protection and Soil Cover

Though the few studies carried out to evaluate the impact of SWC techniques on slope protection are not consistent, there is a consensus that the techniques implemented in the last half century played a positive role in the fight against land degradation [1,10,26]. The effectiveness of SWC measures refers to indications of how much or how well the measures reduce or prevent degradation [8]. The area of the watershed under rain-fed agriculture shows moderate to high effectiveness of the implemented measures (Figure 7a) with an increasing efficiency trend (Figure 7b). A small area, corresponding to the irrigated land at the valley bottom, shows low to very low conservation effectiveness with a slightly increasing trend. Although >80% of the watershed benefits from conservation measures, their effectiveness trend shows the need for additional SLM measures, either as mitigation, prevention, rehabilitation or a combination, according to the land use type [8,47]. For the valley bottom affected by salinization, mitigation of the problem and prevention of further degradation are recommended; for the biological degradation with loss of soil life at high elevation, mitigation measures are recommended; for the severe loss of top soil in the upstream part of the watershed, a combination of mitigation, rehabilitation and rehabilitation measures is recommended; while downstream rehabilitation measures are recommended (Figure 7c).

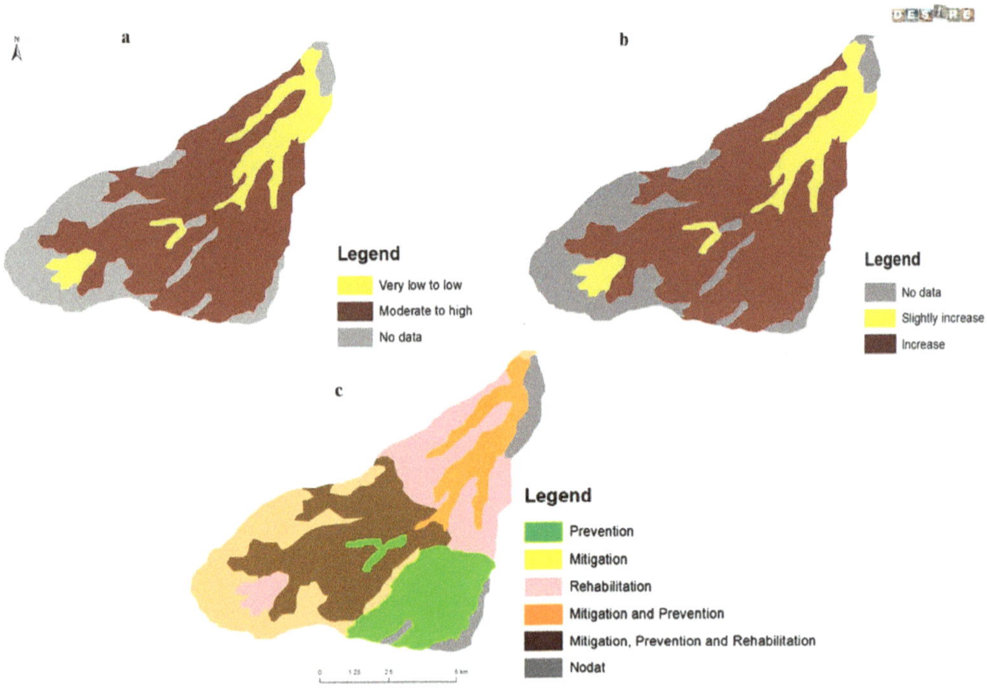

Figure 7. Effectiveness of conservation measures in the Ribeira Seca Watershed, weighted by area, evidencing: (**a**) the moderate to high effectiveness of the implemented measures at slope level, (**b**) the increasing conservation efficiency trend on the slopes and (**c**) expert recommendation on measures. Adapted from [18].

At hillslope level, Ferreira [1] and colleagues reported a positive effect of different SWC techniques (*i.e.*, vegetation strips, afforestation, terraces, contour stone walls and vegetation strips) on parameters like soil compaction, erosion index, rock outcrops and stone, organic matter, vegetation cover and soil accumulation. Smolikowski [2] and colleagues showed the effectiveness and the importance of vegetation barriers and mulch against water erosion when compared with traditional cropping practices (maize and beans), while Querido [26] reported that the use of 1.5 ton/ha of mulch reduced soil erosion by up to 95% and significantly increased infiltration. Baptista [37] and colleagues also reported up to 100% reduction in soil loss from steep dryland hillsides with a combination of crop residue mulching, pigeon-pea hedges and organic amendments.

Agronomic trials of pigeon-pea runoff barriers under the DESIRE project [48] resulted in positive ecological impact such as 28% reduction in area with erosion rates >10 $t{\cdot}ha^{-1}{\cdot}yr^{-1}$ and >20% increase in both water availability and ground water recharge. As off-site impact, >20% reduction in flooding and 20% less stream discharge were also reported. However, these are experimental results that have not been applied at larger scale.

Impact on Crop Yield

Impact studies of techniques on dryland crop productivity carried out by INIDA have turned out inconsistent through the years, depending on yearly rainfall and agro-ecological zone. In 2005, a year considered of deficient rainfall both in amount and distribution, a significant increase of 87% was registered for maize grain and biomass yield for contour stone walls when compared to non-treated plots, assuming other yield drivers constant [49]. However, this significant increase was only verified in the semiarid agro-ecological zones and not in humid, sub-humid or arid zones.

Table 2. Some indicators of biophysical impact of pigeon-pea technology in the Ribeira Seca Watershed (RSW) at field scale. Source: [10,48]

Parameter	Before Technology	Year 1	Year 2
Soil cover (%)	<1	35	48–87
Dry organic mulch (%)	<5	31	60
Grain yield ($t{\cdot}ha^{-1}$)	<0.2	1.9	1.8
Biomass (% increase)			115

On the other hand, in 2010, a year of good rainfall amount and distribution, none of the assessed SWC measures (terraces, vegetation barriers and contour rock walls) resulted in a significant increase in maize, bean or biomass yield, in any of the three agro- ecological zones, when compared with untreated field plots. Likewise, no significant effect on soil quality parameters such as nutrients availability, depth of topsoil and organic matter was registered. The absence of an increase in yield could be explained by the positive water balance observed during the growing period, indicating that soil moisture was not limiting crop yield. However, since overall mean crop yields were low (0.6 and 0.3 $t{\cdot}ha^{-1}$ for maize and bean, respectively), other factors such as low soil fertility (*i.e.*, N and OM content) might also have influenced crop development and yield [50].

Other researchers [10,48] reported increased biophysical impact of the implemented pigeon-pea (*Cajanus cajan*) technology in RSW, at field scale trials, compared to without implementation (Table 2). Soil cover increased up to 87%, dry mulch by 60%, grain yield by 89%, and biomass production by 115% after the second year of implementation.

To improve the quality of biophysical impact assessment of SWC measures on crop yield, there is a need for more systematic and longer-term evaluation, allowing the establishment of a more consistent trend based on rainfall amount and distribution, agro-ecological zone and soils characteristics.

SUSTAINABLE RURAL LIVELIHOOD

Livelihood Approach to Soil and Water Conservation

The SWC measures implemented contributed to the improvement of the economic and environmental conditions of the treated lands in Cape Verde by making crop production possible, and consequently improving the livelihood of the people. The SLM measures implemented in the RSW have, to a great extent, imparted resilience to the agro-ecosystems, making them able to resist droughts and land degradation as evidenced by the high effectiveness of the measures, greener landscape, more water available, slight to moderate degree of degradation and moderate risk of desertification. These positive aspects contribute to people's well-being as evidenced by farmers' testimonies showing satisfaction, for example, with the pigeon-pea technology that lead to high yield and better soil cover, with positive impacts on soil conservation, human and animal feed [10]. Evaluation of the global scenario for food production [48] shows that the pigeon-pea technology can achieve very significant yield increases, both per area (2568 $kg{\cdot}ha^{-1}$) and per capita (1218 kg), thus, reducing poverty level. Pigeon-pea technology also resulted in 50% increase in production area and 50% reduction in production risk in RSW.

In addition, surface runoff water stored in the Poilão dam (1.714 m^3) is used for irrigation of 188 ha of land, allowing farmers to increase yield and diversity their livelihood sources of income [51].

Though benefits of SWC measures are in line with the sustainable rural livelihoods framework [52], dryland agriculture sustainability as a whole may be questioned as unsuitable conditions such as high rates of soil erosion in many parts of the country, low rain-fed productivity, inadequate agricultural practices and high incidence of rural poverty (>60%), still occur.

Cultural Value of Rain-Fed Crops

Maize was introduced in Cape Verde from Brazil in the XVI century and evolved as the predominant dryland crop and the preferred staple food for the population [11]. Despite the low yields (300–700 kg maize grain per ha) and frequent crop failure, farmers routinely plant maize every year, regardless of the economic loss and environmental degradation arising from this agricultural activity. Land degradation has been widely associated with dryland farming practices [10,37] as the conventional maize cultivation is practiced on steep lands without SWC measures, is excessively hoe-weeded, has low soil cover and thus leads to high soil erosion rates.

Maize is at the basis of local people diet and farmers are reluctant to cultivate other cereals. Several attempts were made to introduce crops like sorghum and quinoa that research proved to be better suited for the short rainy season, but the crops were not accepted by local people. Maize and beans production in dryland does not contribute more than 5% to rural household revenues, and some families incur net costs for their cultivation [53]. Moreover, the market price of imported maize and beans is lower than local production cost. Thus, the importance of dryland crops like maize and beans is more due to their cultural than economic value, making them unlikely to be substituted for crops potentially more economically valuable or that promote SLM.

Importance of Dryland Production for Rural Family Food Security

Rain-fed maize production in Cape Verde does not cover more than 10%–15% of consumption in a normal rainfall year [54]; nonetheless, it plays a crucial role in guaranteeing food security of the rural families. Firstly, it provides food grain (maize and beans) for family consumption so that in years of "good" rainfall families do not have to buy these food items for the whole year. Secondly, it provides crop biomass and stubble grazing, so the families keeping livestock do not spend money buying animal feed. Animals like goats, cattle, and pigs contribute up to 47% to the rural household income as well as food for the families [53]. Thirdly, rain-fed production guarantees maize and bean seeds for subsequent years, so that farmers do not have to buy them.

POLICY ORIENTED RECOMMENDATIONS

Cape Verde's government should reinforce its policy regarding SLM practices, as they have numerous environmental, economic, and social benefits, reaching far beyond their potential to reduce land degradation and desertification. SLM also addresses global concerns such as water scarcity, resource-use efficiency, food security, poverty relief, climate change, and biodiversity conservation [55]. These multiple benefits justify investments in SLM and may require funding schemes from different sources, especially when involving small-holder farmers and poor people.

The huge annual spending on implementation of SLM calls for concerted efforts to standardize documentation and evaluation. New efforts towards SLM should build on existing knowledge either from within the country or from similar environments elsewhere. Identifying and assessing existent knowledge on SLM and making it widely available needs a harmonized methodology for comprehensive data collection, knowledge management, and dissemination, such as the DESIRE approach applied in the RSW.

The DESIRE experience in Cape Verde emphasized the importance of a close partnership between researchers and local stakeholders in developing and evaluating SLM options to meet local needs and priorities. Stakeholders' active involvement in the project has resulted in the application of the SLM measures tested. It is important to consider local knowledge and traditional approaches to land management together with the latest research tools (*i.e.*, WOCAT-LADA-DESIRE mapping, DESIRE expert system, decision support tool), and to combine perceptions from both sources. Such tools can be used in reporting to UNCCD, thus contributing to the implementation of the convention in the country.

CONCLUSIONS

(Cabo Verde in Portuguese) has long recognized land degradation and desertification as major threats to its environment with direct impact on human well-being and social welfare. The country moved from experiencing regular crop failure and starvation to being a success case on managing a harsh environment and scarce natural resources to provide food security and improve the livelihoods of local inhabitants. Quoting the Minister of Rural Development (Mrs. Eva Ortet), allusive to the Desertification day, 2014, "*Cape Verde has successfully faced its limitations*".

The dual orientation of rural development policies towards combating desertification and reducing poverty based on massive implementation of SWC measures has resulted in building resilience to extreme drought periods.

The country's relative success in managing harsh conditions can be attributed to "good governance" in combating land degradation and desertification as evidenced by the positive evaluation of the UNCCD and DESIRE project implementation. Good governance here refers to: a responsible handling of funds, an adequate framework and institutional setup, an appropriate set of instruments, the inclusion of civil society, and capacity building at national and international levels.

Although the positive impact of the implemented SWC measures on the environment, economy, food production, livelihood and combating land degradation in the country is generally acclaimed, monitoring and assessment of biophysical and socioeconomic impacts of SLM remain incipient, poorly quantified and documented, being restricted to a few isolated studies with limited scope.

There is a need for better integration of biophysical and socioeconomic aspects of SLM through a robust scientific framework. As an example of a standardized and harmonized knowledge management system, the DESIRE methodology provided a key pillar for informed decision-making at different scales by: giving SLM measures a sound scientific basis; improving indicators; assessing and developing promising SLM strategies with stakeholders; evaluating SLM measures on regional scale; and disseminating results, guidance and decision support tools suitable for all relevant stakeholders. DESIRE's methods, experiences, and know-how should be integrated into long-term programs (*i.e.*, UNCCD) to ensure their continued use and ultimate effectiveness. Joint commitment of all institutions, projects, and actors involved in SLM is needed to build a harmonized knowledge system for scaling up SLM.

ACKNOWLEDGMENTS

We are grateful to the Netherlands Universities Foundation for International Cooperation (NUFFIC) for funding the PhD project (through the NFP-PhD-10/132), the Instituto Nacional de Investigação e Desenvolvimento Agrário of Cape Verde (INIDA) for providing logistic facilities for field work, and the DESIRE project Desertification Mitigation and Remediation of Land—A Global Approach for Local Solutions (under EF6—contract # 037046 GOCE) for providing assessment data for the study and partial funds for the field work. We are particularly thankful to the Cape Verde DESIRE team for carrying out most of the field assessment and ISRIC (International Soil Reference and Water Conservation) team for producing the land degradation and conservation maps. We are grateful to the five anonymous reviewers for their valuable comments to the initial version of the manuscript.

AUTHOR CONTRIBUTIONS

Isaurinda Baptista, the lead author of the manuscript, conducted the majority of the literature review, developed the concept and structure of the paper and wrote the first drafts using various input and editing suggestions from co-authors. Luuk Fleskens revised the first draft of the manuscript and provided a better logical structure and fundamental concepts. Coen Ritsema revised the second and third draft of the paper and provided supporting evidences. António Querido contributed to the conception of the manuscript and through the editorial process. Jacques Tavares and Eduardo Reis carried out most of the field assessments for the elaboration of the various land degradation and conservation maps. António Ferreira provided supporting information and contributed as well through the editorial process. Samuel Gomes provided literature review data for Section 3; Anabela Varela carried out biophysical field assessment related to impact of SWC techniques on crop yield and soil quality.

REFERENCES

1. Ferreira, A.D.F.; Tavares, J.; Baptista, I.; Coelho, C.O.A.; Reis, A.; Varela, L.; Bentub, J. Efficiency of overland and erosion mitigation techniques at Ribeira Seca, Santiago Island, Cape Verde. In *Overland Flow and Surface Runoff*; Hydrological Science and Engineering Book Series; Wong, T.S.W., Ed.; Nova Science Publishers, Inc.: Singapore, 2013; pp. 113–135.
2. Smolikowski, B.; Puig, H.; Roose, E. Influence of soil protection techniques on runoff, erosion and plant production on semi-arid hillsides of Cabo Verde. *Agric. Ecosyst. Environ.* **2001**, *87*, 67–80.
3. Direcção Geral do Ambiente (DGA). *PANAII—Segundo Plano de Acção Nacional para o Ambiente (2004–2014)*; Ministério do Ambiente, Agricultura e Pescas: Praia, Cape Verde, 2004.
4. Cowie, A.L.; Penman, T.D.; Gorissen, L.; Winslow, M.D.; Lehmann, J.; Tyrrell, T.D.; Twomlow, S.; Wilkes, A.; Lal, R.; Jones, J.W.; *et al.* Towards sustainable land management in the drylands: Scientific connections in monitoring and assessing dryland degradation, climate change and biodiversity. *Land Degrad. Dev.* **2011**, *22*, 248–260.
5. World Overview of Conservation Approaches and Technologies (WOCAT). *Where the Land is Greener—Case Studies and Analysis of Soil and Water Conservation*

Initiatives Worldwide; Liniger, H.P., Critchley, W., Eds.; Technical Centre for Agriculture and Rural Cooperation (CTA): Wageningen, The Netherlands; Food and Agriculture Organization (FAO): Rome, Italy; United Nations Environment Programme (UNEP): Nairobi, Kenya; Centre for Development and Environment (CDE), University of Berne: Berne, Switzerland, 2007.

6. Schwilch, G.; Bachmann, F.; Liniger, H.P. Appraising and selecting conservation measures to mitigate desertification and land degradation based on stakeholder participation and global best practices. *Land Degrad. Dev.* **2009**, *20*, 308–326.
7. United Nations University (UNU-INWEH). *Guidelines for the Preparations and Reporting on Globally Relevant SLM Impact Indicators for Project Level Monitoring*; UNU-INWEH: Hamilton, ON, Canada, 2011.
8. Schwilch, G., Hessel, R., Verzandvoort, S., Eds.; *Desire for Greener Land. Options for Sustainable Land Management in Drylands*; Centre for Development and Environment, University of Bern (CDE): Bern, Switzerland; ALTERRA-Wageningen UR, World Soil Information (ISRIC) and Technical Centre for Agriculture and Rural Cooperation (CTA): Wageningen, The Netherlands, 2012.
9. Moreno, A.; Pereira, L.S. *As obras de engenharia rural e os seus impactos nos recursos hídricos em Cabo Verde*; 7ª Silusba da APRH: Evora, Portugal, 2005.
10. Tavares, J.; Ferreira, A.J.D.; Reis, E.A.; Baptista, I.; Amoros, R.; Costa, L.; Furtado, A.M.; Coelho, C. Appraising and selecting strategies to combat and mitigate desertification based on stakeholder knowledge and global best practices in Cape Verde Archipelago. *Land Degrad. Dev.* **2013**, *25*, 45–57.
11. Langworthy, M.; Finan, T. *Waiting for Rain: Agriculture and Ecological Imbalance in Cape Verde*; Technology and Engineer Book; Lynne Rienner Publishers: Boulder, CO, USA, 1997; p. 212.
12. Sanchez-Moreno, J.F.; Mannaerts, C.M.; Jetten, V. Influence of topography on rainfall variability in Santiago Island, Cape Verde. *Int. J. Climatol.* **2013**.
13. Instituto Nacional de Gestão e Desenvolvimento dos Recursos Hídricos (INGRH). *Plano de Acção Nacional para Gestão Integrada dos Recursos Hídricos—PAGIRE*; INGRH Serie No 45, 2; República de Cabo Verde: Praia, Cape Verde, 2010.
14. Diniz, A.C.; Matos, G.S. Carta de Zonagem Agro-ecológica e de Vegetação de Cabo Verde: I-Ilha de Santiago. *Garcia da Orta; Serie de Botânica* **1986**, *8*, 39–82.
15. Instituto Nacional de Investigação e Desenvolvimento Agrário (INIDA). *Manual de Fertilidade do solo e fertilização das culturas*; Sociedade Astória, Lda: S. Jorge dos Órgãos, Cape Verde, 1997.
16. Instituto Nacional de Investigação e Desenvolvimento Agrário-Desertification Mitigation and Remediation of Land (INIDA-DESIRE). *Characterization of the Ribeira Seca Watershed—Santiago Island—Cape Verde Hotspot*; Field Guide for the DESIRE 2nd Plenary Meeting; INIDA-DESIRE: Praia, Cape Verde, 2008.
17. World Bank. *World Development Report 2000/2001: Attacking Poverty*; World Bank: Washington, DC, USA, 2000.
18. DESIRE Project—Harmonised Information System: Providing Local Solutions to Global Sustainable Land Management Problems. Available online: http://www.desire-his.eu (accessed on 1 June 2014).
19. World Overview of Conservation Approaches and Technologies (WOCAT). Questionnaires on SLM technologies and approaches (basic). In *A Framework for the Evaluation of Sustainable Land Management (Revised)*; Liniger, H.P., Schwilch, G., Meckdachi Studer, R., Hauert, C., van Lynden, G., Critchley, W., Eds.; Centre for Development and Environment, Institute of Geography, University of Berne: Berne, Switzerland, 2008.
20. World Overview of Conservation Approaches and Technologies (WOCAT). Available online: http://www.wocat.net(accessed on 1 June 2014).
21. *Aerial Photography and Imagery, Ortho-Corrected–2010 Digital Orthophotos–Lafayette County*; Aerial Data Service, Inc.: Tulsa, OK, USA, 2010.
22. Schwilch, G.; Bestelmeyer, B.; Bunning, S.; Critchley, W.; Herrick, J.; Kellner, K.; Liniger, H.P.; Nachtergaele, F.; Ritsema, C.J.; Schuster, B.; *et al.* Experiences in monitoring and assessment of sustainable land management. *Land Degrad. Dev.* **2011**, *22*, 214–225.
23. Lesourd, M. *État et société aux îles du Cap-Vert. Collection «Hommes et Sociétés»*; Éditions Karthala: Paris, France, 1995; p. 524.
24. Amaral, I. *Santiago de Cabo Verde: A terra e os Homens*; Memórias da Junta de Investigação do Ultramar, segunda série, n°48; Universidade do Algarve: Faro, Portugal, 2007; p. 444.
25. Haagsma, B.; Reij, C. Frentes de trabalho: Potentials and limitations of large scale labour employment for soil and water conservation in Cape Verde. *Land Degrad. Dev.* **1993**, *4*, 73–85.
26. Querido, A.L. *Watershed System Analysis for Evaluating the Efficiency of Soil and Water Conservation Works: A Case Study in Ribeira Seca, Santiago Island, Cape Verde*; Faculty of Geo-Information Science and Earth Observation, University of Twente (ITC): Enschede, The Netherlands, 1999.
27. *National Adaptation Programme of Action on Climate Change (NAPA 2008–2012)*; Global Environmental Fund/United Nations Development Programme (GEF/UNDP): Praia, Cape Verde, 2007.
28. Pretty, J. Participatory learning for sustainable agriculture. *World Dev.* **1995**, *23*, 1247–1263.
29. Reed, M.S. Stakeholder participation for environmental management: A literature review. *Biol. Conserv.* **2008**, *141*, 2417–2431.
30. Mutekanga, F.P.; Kessler, A.; Leber, K.; Visser, S. The use of stakeholders' analysis in integrated watershed management. *Mt. Res. Dev.* **2013**, *33*, 122–131.
31. Reed, M.S.; Fraser, E.D.G.; Morse, S.; Dougill, A.J. Integrating methods for developing sustainability

indicators that can facilitate learning and action. *Ecol. Soc.* **2005**, *10*, r3.

32. Reed, M.S.; Buenemann, M.; Atlhopheng, J.; Akhtar-Schuster, M.; Bachmann, F.; Bastin, G.; Bigas, H.; Chanda, R.; Dougill, A.J.; Essahli, W.; *et al.* Cross-scale monitoring and assessment of land degradation and sustainable land management: A methodological framework for knowledge management. *Land Degrad. Dev.* **2011**, *22*, 261–271.
33. Stringer, L.C.; Reed, M.S. Land degradation assessment in Southern Africa: Integrating local and scientific knowledge bases. *Land Degrad. Dev.* **2007**, *18*, 99–116.
34. Schwilch, G.; Bachmann, F.; de Graaff, J. Decision support for selecting SLM technologies with stakeholders. *Appl. Geogr.* **2012**, *34*, 86–98.
35. The Committee for Development Policy, Development Policy and Analysis Division, United Nations. Monitoring the progress of graduated countries—Cape Verde. In Proceedings of Expert Group Meeting for Review of the List of the Least Developed Countries, New York, NY, USA, 16–17 January 2011; 2011; CDP2012/PLEN/11.
36. Tavares, J.; Amiotte-Suchet, P. Rainfall erosion risk mapping in volcanic soils of Santiago Island, Cape Verde Archipelago. *Afr. Geosci. Rev.* **2007**, *14*, 399–414.
37. Baptista, I.; Ritsema, C.J.; Querido, A.; Ferreira, A.D.F.; Geissen, V. Improving rainwater-use in Cape Verde drylands by reducing runoff and erosion. *Geoderma* **2015**, *237–238*, 283–297.
38. Barry, O.; Smolikowski, B.; Roose, E. Bilan de la lutte anti-érosive dans les zones d'agriculture pluviale au Cap-Vert; un projet innovant, le PRODAP. *Agric. Dév. CIRAD-CA* **1995**, *5*, 57–68.
39. Lopes, V.; Mayer, J. Watershed management program on Santiago Island, Cape Verde. *Environ. Manag.* **1993**, *17*, 51–57.
40. Mannaerts, C.M. *Contribution à l'évaluation de l'érosion des sols au Cap Vert*; Rapport Inédit; GCP/CVI/002/BEL.FAO; Food and Agriculture Organization (FAO): Rome, Italy, 1986; p. 43.
41. Mannaerts, C.M. Assessment of the Transferability of Laboratory Rainfall—Runoff and Rainfall—Soil Loss Relationship to Field and Catchment Scales: A Study in the Cape Verde Islands. Ph.D. Thesis, University of Twente, Enschede, The Netherlands, 1993.
42. Tavares, J. TavaresSoil Erosion in Cape Verde: A Study of Processes and Quantification at the Scale of Three Watersheds of the Santiago Island. Ph.D. Thesis, Bourgogne University, Dijon, France, 2010.
43. Sabino, A. As estruturas de conservação do solo e água em Cabo Verde e a quantificação dos impactos na área do Watershed Development Project (Ilha de Santiago). In Proceedings of the 1as Jornadas sobre Agricultura de Cabo Verde, Comunicações do Instituto de Investigação Científica Tropical, Série de Ciências Agrárias, Lisboa, Portugal, 23–25 October 1990; 1990; pp. 91–124.
44. Wakindiki, C.; Ben-Hur, M. Indigenous soil and water conservation techniques: Effects on runoff, erosion, and crop yields under semi-arid conditions. *Aust. J. Soil Res.* **2004**, *40*, 367–379.
45. Liniger, H.P.; Mekdaschi Studer, R.; Hauert, C.; Gurtner, M. *Sustainable Land Management in Practice—Guidelines and Best Practices for Sub-Saharan Africa*; TerrAfrica: Halfway House Midrand, South Africa; World Overview of Conservation Approaches and Technologies (WOCAT): Berne, Switzerland; Food and Agriculture Organization of the United Nations (FAO): Rome, Italy, 2011.
46. Direcção Geral de Agricultura, Silvicultura e Pecuária (DGASP). *ventario Florestal Nacional*; DGASP: Praia, Cape Verde, 2013.
47. World Overview of Conservation Approaches and Technologies (WOCAT)/Land Degradation Assessment in Drylands (LADA)/Desertification Mitigation and Remediation of Land (DESIRE). *A Questionnaire for Mapping Land Degradation and Sustainable Land Management*; Liniger, H.P., van Lynden, G., Nachtergaele, F., Schwilch, G., Eds.; Centre for Development and Environment, Institute of Geography, University of Berne: Berne, Switzerland, 2008.
48. Desertification Mitigation and Remediation of Land (DESIRE). *Pesera Model Simulation for Ribeira Seca Watershed*; DESIRE Scientific Report Series (WP 5.3); University of Leeds: Leeds, UK, 2011.
49. Instituto Nacional de Investigação e Desenvolvimento Agrário (INIDA). *Avaliação da produtividade das culturas de sequeiro*; Relatório do estudo realizado pelo INIDA para o Agriculture Cooperative Development International ACDI/VOCA; INIDA: São Jorge, Cape Verde, 2006.
50. Varela, A.C. Avaliação do impacto das técnicas de conservação de solo e água na produção agrícola pluvial e na qualidade do solo: Estudo de caso da bacia hidrográfica da Ribeira Seca, Santiago. Master's Thesis, Universidade de Cabo Verde/Universidade Técnica de Lisboa (UniCV/ISA-UTL), Praia, Cape Verde, 2012.
51. Ministério de Desenvolvimento Rural (MDR). *Diagnóstico da implementação dos instrumentos*; Plano Estratégico de Desenvolvimento Agrícola-Plano Nacional de Investimento Agrícola (PEDA-PNIA): Praia, Cape Verde, 2012.
52. Department for International Development (DFID). *Sustainable Livelihood Guidance Sheets*, 1999. Available online: http://www.livelihoods.org (accessed on 1 June 2014).
53. Instituto Nacional de Investigação e Desenvolvimento Agrário (INIDA). *Avaliação económica de 4 explorações agrícolas em Santiago*; Relatório de actividades; INIDA: São Jorge dos Órgãos, Cape Verde, 2013.
54. Food and Agriculture Organization of the United Nations (FAO). *The State of Food and Agriculture: Paying Farmers for Environmental Services*; FAO: Rome, Italy, 2007.
55. Food and Agriculture Organization of the United Nations (FAO). *The State of the World's Land and Water Resources for Food and Agriculture: Managing Systems at Risk*; FAO: Rome, Italy; Earthscan: London, UK, 2011.

CHAPTER 12 SOIL DEGRADATION: CHALLENGES AND SOLUTIONS

Ranjan Bhattacharyya, Birendra Nath Ghosh, Prasanta Kumar Mishra, Biswapati Mandal, Cherukumalli Srinivasa Rao, Dibyendu Sarkar, Krishnendu Das, Kokkuvayil Sankaranarayanan Anil, Manickam Lalitha, Kuntal Mouli Hati and Alan Joseph Franzluebbers

Keywords: Land degradation; soil erosion; conservation agriculture; agroforestry; nutrient management; sustainable crop intensification

SOURCES OF LAND DEGRADATION

Land degradation is not being adequately addressed, but is of vital importance to raise awareness so that future land management decisions can lead to more sustainable and resilient agricultural systems. Of India's total geographical area (328.7 Mha), 304.9 Mha comprise the reporting area with 264.5 Mha being used for agriculture, forestry, pasture and other biomass production. The severity and extent of soil degradation in the country has been previously assessed by many agencies (Table 1). According to the National Bureau of Soil Survey and Land Use Planning [1] ~146.8 Mha is degraded. Water erosion is the most serious degradation problem in India, resulting in loss of topsoil and terrain deformation. Based on first approximation analysis of existing soil loss data, the average soil erosion rate was ~16.4 ton $ha^{-1}year^{-1}$, resulting in an annual total soil loss of 5.3 billion tons throughout the country [2]. Nearly 29% of total eroded soil is permanently lost to the sea, while 61% is simply transferred from one place to another and the remaining 10% is deposited in reservoirs.

Soil degradation has become a serious problem in both rainfed and irrigated areas of India. India is losing a huge amount of money from degraded lands (Table 2). This cost is documented by declining crop productivity, land use intensity, changing cropping patterns, high input use and declining profit [13,14,15,16]. Reddy [17] valued the loss of production in India at Rupees (Rs) 68 billion in 1988–1989 using the National Remote Sensing Agency (NRSA) dataset. Additional losses resulting from salinization, alkalinization and waterlogging were estimated as Rs 8 billion. Of late, in a comprehensive study made on the impact of water erosion on crop productivity, it was revealed that soil erosion due to water resulted in an annual crop production loss of 13.4 Mt in cereal, oil seeds and pulse crops equivalent to ~US$162 billion [18].

Citation: Bhattacharyya, R.; Ghosh, B.N.; Mishra, P.K.; Mandal, B.; Rao, C.S.; Sarkar, D.; Das, K.; Anil, K.S.; Lalitha, M.; Hati, K.M.; Franzluebbers, A.J Soil Degradation: Challenges and Solutions, Sustainability 2015, 7, 3528-3570. doi:10.3390/su7043528.

Table 1. Extent of land degradation in India, as assessed by different organizations

Organizations	Assessment Year	Reference	Degraded Area (Mha)
National Commission on Agriculture	1976	[3]	148.1
Ministry of Agriculture-Soil and Water Conservation Division	1978	[4]	175.0
Department of Environment	1980	[5]	95.0
National Wasteland Development Board	1985	[6]	123.0
Society for Promotion of Wastelands Development	1984	[7]	129.6
National Remote Sensing Agency	1985	[8]	53.3
Ministry of Agriculture	1985	[9]	173.6
Ministry of Agriculture	1994	[10]	107.4
NBSS&LUP	1994	[11]	187.7
NBSS&LUP (revised)	2004	[12]	146.8

Table 2. Estimates on the annual direct cost of land degradation in India

Parameters	NRSA [19]	ARPU [20]	Sehgal and Abrol [21]
Area affected by soil erosion (Mha)	31.5	58.0	166.1
Area affected by salinization, alkalinization and waterlogging (Mha)	3.2	-	21.7
Total area affected by land degradation (Mha)	34.7	58.0	187.7
Cost of soil erosion in lost nutrients (Rs billion)	18.0	33.3	98.3
Cost of soil erosion in lost production (Rs billion)	67.6	124.0	361.0
Cost of salinization, alkalinization and waterloggingin lost production (Rs billion)	7.6	-	87.6
Total direct cost of land degradation (Rs billion)	75.2	-	448.6

Apart from faulty agricultural activities that led to soil degradation (discussed in the next Section), other human-induced land degradation activities include: land clearing and careless management of forests, deforestation, over-grazing, improper management of industrial effluents and wastes, surface mining, and industrial development. Each of these factors are discussed briefly, but offering greater detail is beyond the scope of this review.

Overgrazing, Deforestation and Careless Forest Management

Overgrazing and deforestation have caused degradation in eight Indian states which now have >20% wasteland (Source: Wasteland atlas of India by national remote sensing agency; NRSA). Loss of vegetation occurs due to cutting beyond the silviculturally permissible limit, unsustainable fuelwood and fodder extraction, encroachment by agriculture into forest lands, forest fires and overgrazing, all of which subject the land to degradation forces. A cattle population of 467 million grazes on 11 Mha of pastures, implying an average of 42 head per hectare of land compared to a sustainable threshold level of 5 animals per hectare [22]. High livestock density in arid regions causes overgrazing, resulting in decreased infiltration and accelerated runoff and soil erosion. Due to overgrazing, soil loss is 5 to 41 times greater than normal at the mesoscale and 3 to 18 times greater at the macroscale [23]. Tendency of cultivation on slopes in the 1990s led to deforestation and land degradation [24]. Impoverishment of the natural woody cover of trees and shrubs is a major factor responsible for wind and water erosion. This occurs because the per capita forest land in the country is only 0.08 ha compared to a requirement of 0.47 ha to meet basic needs, thus creating excessive pressure on forest lands.

Urban Growth, Industrialization and Mining

An increase in industrialization, urbanization and infrastructure development is progressively taking away considerable areas of land from agriculture, forestry, grassland and pasture, and unused lands with wild vegetation. Opencast mining is of particular focus because it disturbs the physical, chemical, and biological features of the soil and alters the socioeconomic features of a region. Negative effects of mining are water scarcity due to lowering of water table, soil contamination, part or total loss of flora and fauna, air and water pollution and acid mine drainage. Overburden removal from mine area results in significant loss of vegetation and rich topsoil [25]. Overburden removal is normally done by blasting or using excavators, resulting in generation of large volume of waste (soil, debris and other material). Open-pit mines produce 8 to 10 times as much waste as underground mines [26]. The magnitude and significance of impact on the environment due to mining varies from mineral to mineral and also on the potential of the surrounding environment to absorb the negative effects associated with geographical disposition of mineral deposits and size of the mining operations. Mineral production generates enormous quantities of waste/overburden and tailings/slimes and a huge land area is degraded (Table 3).

Table 3. Mineral Production, waste generation and land affected in 2005-06 (Data source: Sahu [25])

Mineral	Production (Mt)	Overburden/Waste (Mt)	Estimated Land Affected (ha)
Coal	407	1493	10,175
Limestone	170	178	1704
Bauxite	12	8	123
Iron ore	154	144	1544
Others	9	19	-

Natural and Social Sources of land Degradation

Natural causes of land degradation include earthquakes, tsunamis, droughts, avalanches, landslides, volcanic eruptions, floods, tornadoes, and wildfires (discussed in more detail in Section 3). Some underlying social causes of soil degradation are land shortage, decline in per capita land availability, economic pressure on land, land tenancy, poverty, and population increase.

Land Shortage, Land Fragmentation and Poor Economy

In India, small land holdings are a prominent feature, particularly in rainfed regions. Some 80% of farmers' holdings are ≤2 ha, accounting for >50% of agricultural output. Average size of land holding declined from 2.3 ha to 1.3 ha during 1970–2000 with per capita land of 0.32 hectare in 2001 [27]. Small land holdings lead to severe economic pressures on farmers. Because of such pressure, labor, land and capital resources limit the use of green manuring or soil conservation structures. Therefore, land shortage and poverty, taken together, lead to non-sustainable land management practices as a direct source of degradation. This is also the underlying reason for two other direct causes of land degradation, improper crop rotations and unbalanced fertilizer use [28].

Despite several interventions by the Indian Government, land degradation is still a serious problem. Some programs have included Integrated Watershed Management in the catchment of flood prone areas-1980–1981; National Land Use and Development Council-1985; National Wasteland Development Board–1985; National Watershed Development Projects for Rainfed Areas-1985–1986; Reclamation and Development of Alkali and Acid soil-1985–1986; National Land Use Policy-1988; Integrated Wasteland Development Project-1989–1990; Constitution (74th amendment) Act-1992 (Regulation of Land Use) and National Rainfed Area Authority-2006. The United Nations Environmental Programme (UNEP) indicated that over the preceding 20 years the problem of land degradation had continued to worsen due to human activities and climate change causing prolonged or frequent droughts that aggravated land degradation. Other underlying causes included increasing population:land ratio (Agriculture share in GDP fell from 35% in 1981 to 13% in 2012); market and institutional failures; externality and tenurial system–insecure property rights.

Population Increase

India's land area is about 2.5% of the global land area, where as it supports more than 16% of the global human population and ~20% of the world's livestock population. Steady increases in human population, as well as livestock population, and the widespread incidence of poverty, are exerting heavy pressures on India's limited land resources. Urban sprawl is a consequence of increasing urban population. As urban population increases, infrastructure requirements including transportation, water and sewage facilities, housing, schools, commerce, health, and recreation all contribute to urban sprawl [29].

AGRICULTURAL ACTIVITIES LEADING TO LAND DEGRADATION IN INDIA

"Most of the area under cultivation in India has been under cultivation for hundreds of years, and had reached its state of maximum impoverishment many years ago ... In this connection it must be remembered that deficiency of combined nitrogen is the limiting factor throughout the greater part of India" (The Royal Commission on Agriculture in India Report, [30] (p. 76)). The Green Revolution brought about a technological breakthrough, leading to the use of short duration high yielding varieties that helped intensify land use within a year by increasing the area under irrigation and greatly increasing the use of chemicals such as fertilizers and pesticides. Agricultural production of India increased from 50 Mt to over 250 Mt, over the last five decades. This, however, had further consequences, including loss of plant biodiversity and environmental pollution. Widespread land degradation caused by inappropriate agricultural practices has a direct and adverse impact on the food and livelihood security of farmers. Basically, degradation is caused by erosion, which results in the loss of topsoil through the action of water and wind, or waterlogging, which results in soil salinization. Maheswarappa *et al.* [31] observed that (i) the C-sustainability index was high in 1960, and was indicative of the minimum usage of inputs prior to the onset of the Green Revolution and (ii) thereafter, the C-sustainability index decreased because of greater C-based inputs, in which a linear relationship exists between C inputs and C outputs.

Agricultural activities and practices can cause land degradation in a number of ways depending on land use, crops grown and management practices adopted. Some of the common causes of land degradation by agriculture include cultivation in fragile deserts and marginal sloping lands without any conservation measures, land clearing through clear cutting and deforestation, agricultural depletion of soil nutrients through poor farming practices, overgrazing, excessive irrigation, overdrafting (the process of extracting groundwater beyond the safe yield of the aquifer), urban sprawl and commercial development, and land pollution including industrial waste disposal to arable lands.

Low and Imbalanced Fertilization

Intensive farming practices, particularly with wheat (*Triticum aestivum* L.) and rice (*Oryza sativa* L.) in India, have virtually mined nutrients from the soil. The already imbalanced consumption ratio of 6.2:4:1 (N:P:K) in 1990–1991 has widened to 7:2.7:1 in 2000–2001 and 5:2:1 in 2009–2010 compared with a target ratio of 4:2:1. As food grain production increased with time, the number of elements deficient in Indian soils increased from one (N) in 1950 to nine (N, P, K, S, B, Cu, Fe, Mn, and Zn) in 2005–2006. Although the use of fertilizers has increased several fold, the overall consumption continues to be low in most parts of the country. Wide spread Zn deficiency, followed by S, Fe, Cu, Mn and B in are common throughout the country. Every year, ~20 Mt of the three major nutrients are removed by growing crops [32], but the corresponding addition through inorganic fertilizers and organic manures falls short of this harvest. Another estimate suggests that for the past 50 years, the gap between removals and additions of nutrients has been 8 to 10 Mt N + P_2O_5 + K_2O per year [33]. In addition nutrient loss through soil erosion is another reason for soil fertility depletion, accounting for an annual loss of 8 Mt of plant nutrients through 5.3 billion tons of soil loss [34].

Excessive Tillage and Use of Heavy Machinery

Excessive tillage coupled with use of heavy machinery for harvesting and lack of adequate soil conservation measures causes a multitude of soil and environmental problems. Decline in soil organic matter (SOM) leads to limited soil life and the poor soil structure. Puddling of soil for paddy rice degrades soil physical properties and has negative impacts on soil biology [35]. Poor physical condition of soil leads to poor crop establishment and waterlogging after irrigation. Intensive agriculture has also led to doubling of irrigated cropland over the past four decades, from 19% to 38% of the cropped area. Much of this water has been extracted from limited ground water resources. Improper use and maintenance of canal irrigation has contributed significantly to soil degradation problems like waterlogging and salinization. Excess nitrate has leached into groundwater due to heavy N fertilizer use. Unnecessary tillage for land preparation and planting, indiscriminate irrigation, and excessive fertilizer applications are the main sources of greenhouse gas (GHG) emission from agricultural systems.

Crop Residue Burning and Inadequate Organic Matter Inputs

The NBSS&LUP data [21] show that nearly 3.7 Mha suffer from nutrient loss and/or depletion of SOM. Burning of crop residues for cooking, heating or simply disposal is a pervasive problem in India and contributes to SOM loss. According to the Ministry of New and Renewable Energy [36], ~500 Mt of crop residues are generated every year and ~125 Mt are burned. Crop residue generation is greatest in Uttar Pradesh (60 Mt) followed by Punjab (51 Mt) and Maharashtra (46 Mt). Among different crops, cereals generate 352 Mt of residues followed by fibre crops (66 Mt), oilseeds (29 Mt), pulses (13 Mt) and sugarcane (*Saccharum officinarum*) (12 Mt). Rice (34%) and wheat (22%) are the dominant cereals contributing to crop residue generation [37].

Poor Irrigation and Water Management

Improper planning and management of irrigation systems and extraction of ground water in excess of the recharge capacity have resulted in a rise of the water table in most canal command areas. Specific issues of concern are inefficient use of irrigation water, poor land development, seepage from unlined water courses, non-conjunctive use of surface and ground water resources and poor drainage. Expansion of canal irrigation (like the Indira Gandhi Nahar Project, for instance) has been associated with widespread waterlogging and salinity problems in areas, such as in the Indo-Gangetic Plains (IGP). In arid, semi-arid and sub-humid regions, large areas have been rendered barren due to the development of saline-sodic soils because of poor irrigation and drainage management. Cracking of soil from poor irrigation management leads to bypass flow of water and subsequent nitrate leaching [29]. Cracks not closing properly leave a U-shaped trace, and upon drying these cracks can expand and cause soil shrinkage.

Poor Crop Rotations

Improper crop rotation coupled with lack of proper soil and water conservation measures are important reasons contributing to soil erosion in lands under cultivation. In addition, cultivation of marginal lands on steep slopes, in shallow or sandy soils, with laterite crusts, and in arid or semi-arid regions bordering deserts has resulted in land degradation. Agricultural production in marginal areas with low SOM due to unsuitable cropping patterns has been the major cause of accelerated wind and water erosion. Wind erosion is a serious problem in arid and semi-arid regions, in coastal areas with sandy soils, and in the cold desert regions of Leh in the extreme north of India.

Pesticide Overuse and Soil Pollution

Indiscriminate use of pesticides together with sewage sludge and composted municipal wastes leads to contamination of soil and water with toxic substances and heavy metals. Heavy metal pollution is due to improper disposal of industrial effluents and use of domestic and municipal wastes and pesticides. Some commercial fertilizers also contain appreciable quantities of heavy metals, which have undesirable effects on the environment. Indiscriminate use of agro-chemicals, such as fertilizers and pesticides, is often responsible for land degradation.

EXTENT AND CAUSES OF SOIL DEGRADATION BY REGION

The extent of land degradation in India, as estimated by NBSS&LUP and Indian Council of Agricultural Research (ICAR) is given in Table 4.

Table 4. State-wise extent of various kinds of land degradation in India (Mha). Data source: NBSS&LUP-ICAR [12] on 1:250,000 scale. TGA is total ground area

State	Water Erosion	Wind Erosion	Water Logging	Salinity/ Alkalinity	Soil Acidity	Complex Problem	Total Degraded Area	% of Degraded Area to TGA
Andhra Pradesh + Telengana	11.5	0	1.9	0.5	0.9	0.2	15.0	54.5
Goa	0.1	0	0.1	0	0	0	0.2	43.9
Karnataka	5.8	0	0.9	0.1	0.1	0.7	7.6	39.8
Kerala	0.1	0	2.1	0	0.1	0.3	2.6	67.1
Tamil Nadu	4.9	0	0.1	0.1	0.1	0.1	5.3	41.0
Manipur	0.1	0	0	0	1.1	0.7	1.9	42.6
Mizorum	0.1	0	0	0	1.1	0.7	1.9	89.2
Meghalaya	0.1	0	0	0	1.0	0	1.2	53.9
Assam	0.7	0	0	0	0.6	0.9	2.2	28.2
Arunachal Pradesh	2.4	0	0.2	0	2.0	0	0	53.8
Nagaland	0.4	0	0	0	0.1	0.5	1,0	60.0
Sikkim	0.2	0	0	0	0.1	0	0.2	33.0
Tripura	0.1	0	0.2	0	0.2	0.1	0.6	59.9
Himachal Pradesh	2.8	0	1.3	0	0.2	0	4.2	75.0
Jammu and Kashmir	5.5	0.1	0.2	0	0	0	7.0	31.6
Uttar Pradesh + Uttarakhand	11.4	0.2	2.4	1.4	0	0	15.3	52.0
Delhi	0.1	0	0	0	0	0.0	0.1	55.4
Haryana	0.3	0.5	0.1	0.3	0	0.2	1.5	33.2
Punjab	0.4	0.3	0.3	0.3	0	0	1.3	25.4
Bihar + Jharkhand	3.0	0	2.0	0.2	1.0	0	6.3	36.1
West Bengal	1.2	0	0.7	0.2	0.6	0.1	2.8	31.0
Union Territories	0.2	0	0	0	0	0.0	0.2	24.8
Gujarat	5.2	0.4	0.5	0.3	0	1.7	8.1	41.5
Rajasthan	3.2	6.7	0	1.4	0	0.1	11.4	33.2
Madhya Pradesh + Chhattisgarh	17.9	0	0.4	0	7.0	1.1	26.2	59.1
Maharashtra	11.2	0	0	1.1	0.6	0.3	13.1	42.4
Orissa	5.0	0	0.7	0	0.3	0.1	6.1	39.3
Grand Total (Mha)	93.7	9.5	14.3	5.9	16.0	7.4	146.8	-

The Planning Commission of India has delineated 15 agro-climatic regions to form the basis for agricultural planning in the Eighth Plan. These are: 1. Western Himalayan Region, 2. Eastern Himalayan Region, 3. Lower Gangetic Plains Region, 4. Middle Gangetic Plains Region, 5. Upper Gangetic Plains Region, 6. Trans-Gangetic Plains Region, 7. Eastern Plateau & Hills Region, 8. Central Plateau & Hills Region, 9. Western Plateau & Hills Region, 10. Southern Plateau & Hills Region, 11. East Coast Plains & Hills Region, 12. West Coast Plains & Ghats Region, 13. Gujrat Plains and Hills Region, 14. Western Dry Region, 15. The Island Region. Similar agro-climatic regions have been combined to form six major regions. Region-specific causes and extent of degradation are described in the online Supplementary Information.

STRATEGIES TO MITIGATE LAND DEGRADATION

The salient mitigation techniques for reversing land degradation in India and their applicability in major agro-climates are given in Table 5.

Table 5. Major land degradation mitigation techniques in the agro-climatic zones of India

Mitigation Technologies	Hilly Areas	Indo-Gangetic Plains	Dryland and Desert Areas	Southern Peninsular India	Central India	Coastal Areas
	Applicability					
Soil Erosion Control	√	√	√	√	√	√
Water Harvesting, Terracing and Other Engineering Structures	√	√	√	√	√	√
Landslide and Minespoil Rehabilitation and River Bank Erosion Control	√	√	√	√	√	√
Intercropping and Contour Farming	√	√	√	√	√	√
Subsoiling					√	
Watershed Approach	√	√	√	√	√	√
Participatory Resource Conservation and Management	√	√	√	√	√	√
Integrated Nutrient Management and Organic Manuring	√	√	√	√	√	√
Reclamation of Acid and Salt Affected Soils and Drainage (Desalinization)	√	√	√	√	√	√
Remediation of As contamination		√				√
Water Management and Pollution Control	√	√	√	√	√	√
Irrigation Management for Improving Input Use Efficiency	√	√	√	√	√	√
Judicious Use of Distillery Effluent	√	√	√	√	√	√
Reforestation, Grassland and Horticulture Development	√	√	√	√	√	√
Vegetative Barriers and Using Natural Geotextiles, Mulching and Diversified Cropping	√	√	√	√	√	√
Agroforestry	√	√	√	√	√	√
Conservation Agriculture (CA)	√	√	√	√	√	√
Intensive Cropping and Integrated Farming Systems (IFS)	√	√	√	√	√	√
Disaster (Tsunami) Management						√

Soil Erosion Control

Tolerance to soil loss (T) is defined as the upper threshold limit of soil erosion that can be allowed without degrading long term productivity of a particular soil. If soil erosion rates are greater than T, mitigation measures are needed to achieve sustainable productivity. T-values of the hilly regions of India, as estimated by Mandal *et al.* [38], are given in Table 6 and Table 7. It is projected that ~59% of land within the hilly region requires some form of erosion management to achieve T [38].

Table 6. Area under different erosion rates and soil loss tolerance limits in the northwestern Hills

Erosion Categories Based on Soil Erosion (ton ha^{-1}year^{-1})	**Very Low (<5)**	**Low (5 to 10)**	**Moderate (10 to 20)**	**Severe (20–40)**	**Very Severe (>40)**	**Others**
Area (Mha) under each category	1.7	2.5	3.3	1.9	4.5	19.2
	(5.2) *	(7.5)	(9.8)	(5.8)	(13.7)	(58.0)
T-value (ton ha^{-1}year^{-1})	2.5	5.0	7.5	10.0	12.5	Rocks/ unreported
Area (Mha) under each T value	0.4	0.3	3.5	9.0	1.3	18.7
	(1.2)	(0.8)	(10.6)	(27.2)	(3.9)	(56.3)

* Values in the parentheses are percentages of total area. Data source: Mandal *et al.* [38].

Table 7. Area under different potential erosion rates and soil loss tolerance limits in the northeastern Hills (Source: Mandal *et al.* 38). * Values in the parentheses are percentages of area

Erosion Categories Based on Soil Erosion (ton ha^{-1}year^{-1})	**Very Low (<5)**	**Low (5 to 10)**	**Moderate (10 to 20)**	**Severe (20-40)**	**Very Severe (>40)**	**Others**
Area (Mha) under each category	1.2	5.8	4.6	3.6	8.2	4.1
	(4.5) *	(21.2)	(16.8)	(13.0)	(29.8)	(14.8)
T-value (ton ha^{-1}year^{-1})	2.5	5.0	7.5	10.0	12.5	Rocks /unreported
Area (Mha) under each T value	-	0.1	4.7	13.1	5.8	3.8
		(0.3)	(17.1)	(47.7)	(21.0)	(13.9)

* Values in the parentheses are percentages of total area.

Soil conservation measures, such as contour ploughing, bunding, use of strips and terraces, can decrease erosion and slow runoff water. Mechanical measures, e.g., physical barriers such as embankments and wind breaks, or vegetation cover (and use of vegetative buffer strips and geotextiles) and soil husbandry are important measures to control soil erosion [39]. In addition, conservation agriculture (CA), agro-forestry, integrated nutrient management (INM) and diversified cropping also conserve soil and water. These are discussed sequentially as physical, chemical and biological means of soil conservation and land degradation mitigation in the following sections.

Water Harvesting, Terracing and Other Engineering Structures

Mechanical soil and water conservation measures are required for controlling soil erosion, retaining maximum rainfall within the slope and safe disposal of excess runoff from the top to the foot hills of India. These structures are often used in case of extreme soil degradation. The measures are: Bunding-small earthen barriers built on agricultural lands with slopes ranging from 1%–6% slope. Bunds are used in agriculture to collect surface *run-off*, increase water infiltration and prevent soil erosion. Graded bunds-constructed in medium to high rainfall areas of ~600 mm year^{-1}. Contour bunds- either mechanical or vegetative barrier created across the slope. A study conducted at Doon valleys in the northwestern hills region indicted that contour bunds decreased runoff 25%–30% compared to field bunds [40]. Bench terrace and half moon terrace-adopted where soil depth is >1.0 m. Half-moon terraces are level circular beds having 1 to 1.5 m diameter cut into half-moon shape on the hill slopes. Beds are used for planting and maintaining saplings of fruit and fodder trees in horticulture/agro-forestry land uses. Grassed waterways-channels laid out preferably on natural drainage lines in the watershed. Water harvesting ponds-dug-out embankment type of water harvesting structure used for creating seasonal and perennial ponds at the foot of a micro-watershed for irrigation and fish farming purposes.

In vertisols (of central India), graded broad bed and furrow system of land configuration improves surface drainage and allows better water infiltration. It also facilitates drainage of excess water through grassed waterways. However, the broad bed and furrow system is not as effective for shallower Vertic soils, as it encourages runoff. Runoff and soil loss were lower from broad bed and furrow land surface management practices than from a flat on grade system (Table 8). The broad bed and furrow system decreased soil loss to a greater extent (31% to 55%) than its effect on runoff volume (24% to 32%) compared with that of flat on grade system.

Table 8. Seasonal rainfall, runoff and soil loss from different land configuration, broad-bed and furrow (BBF) and flat on grade (FOG) (Data source: Mandal *et al.* [41])

Year	Rainfall (mm)	Runoff (mm)		Soil Loss (ton ha^{-1})	
		BBF	FOG	BBF	FOG
2003	1058.0	163.0 (15.4%)	214.9 (20.3%)	2.0	2.9
2004	798.2	124.0 (15.5%)	183.3 (23.0%)	0.7	1.5
2005	946.0	177 (18.7%)	246 (26.1%)	1.4	3.1
2006	1513.0	502 (33.2%)	873 (57.7%)	3.5	6.4

Values within parentheses indicate the percent of total rainfall.

Landslide and Minespoil Rehabilitation and River Bank Erosion Control

High soil erosion rates were checked and brought within permissible limits (Table 9) by using bioengineering treatments on landslide affected (Nalotanala watershed; area ~60 ha) and minespoil affected (Sahastradhara watershed; area ~64 ha) areas. Restoration of limestone minespoil areas resulted in improved water quality through a reduction in Ca content (Table 10). For river bank erosion control, bio-engineering technologies such as spurs, retaining walls and earthen embankments may be used in conjunction with suitable vegetation such as giant cane (*Arundo donax)*, five-leaf chaste trees (*Vitex negundo)*, morning glory (*Ipomoea sp.)*, Bamboo (*Bambusa vulgaris*), napiergrass (*Pennisetum purpureum*) or munja (*Saccharum munja)*[40].

Table 9. Effect of bioengineering measures on landslide (1964–1994) and minespoil rehabilitation (1984–1996) project [40]

Particulars	Landslide Project		Minespoil Project	
	Before Treatment	After Treatment	Before Treatment	After Treatment
Sediment load (ton ha^{-1} $year^{-1}$)	320	6	550	8
Vegetative cover (%)	<5	>95	10	80

Table 10. Water quality parameters (mg L^{-1}) for treated and untreated minespoils (Data source: CSWCR&TI Vision [40])

	Ca	Mg	SO_4
Treated mine	74	34	138
Untreated mine	188	39	240

Intercropping and Contour Farming

Agronomical practices like use of cover crops, mixed/inter/strip cropping, crop rotation, green manuring and mulch farming are vital practices associated with integrated nutrient management. Growing soybean (*Glycine max*)/groundnut (*Arachis hypogoea*)/cowpea (*Vigna radiata*) with maize (*Zea mays*)/jowar (*Sorghum bicolor*)/bajra (*Pennisetum glaucum*) is a common example of intercropping in the drylands [39]. Strip cropping is a combination of contouring and crop rotation in which alternate strips of row crops and soil conserving crops are grown on the same slope, perpendicular to the wind or water flow in drylands and hilly regions, respectively. Intercropping cowpea with maize (2 rows of cowpea with 1 row of maize) decreased runoff by 10% and soil loss by 28% compared to pure maize. Minimum runoff (36% of rainfall) was recorded under barnyard millet (*Echinochloa frumentacea* L.) followed by black soybean (*Glycine max* L.) and maize which was 37% and 42%, respectively. Black soybean and maize alone had maximum soil loss of 7.1 and 6.7 ton ha^{-1}, respectively, followed by barnyard millet (4.8 ton ha^{-1}). The practice of line sowing of wheat and mustard (*Brassica juncea* L.) crops and maintaining row ratio of 8:1 ensured optimum use of space and soil moisture, increased wheat equivalent yield by 14% and net returns by 30% compared to mixed sowing (Table 11) [42,43].

Table 11. Water use efficiency, yield and net return as affected by different technologies and crop rotation in farmers' fields of Uttarakhnad, Jammu and Kashmir and Himachal Pradesh

Intercropping	Crops	Water Use Efficiency (kg $ha^{-1}mm^{-1}$)			Yield (t ha^{-1})			Net Return (INR ha^{-1})		
		C	T	% Increase	C	T	% Increase	C	T	% Increase
Maize + cowpea (1:2) – wheat	Maize	3.19	5.60	76	2.21 *	3.67 *	66	4448	11,690	163
	Wheat	5.30	8.31	57	1.13	1.64	46	3176	6149	88

Maize – wheat + mustard (9:1)	Maize	3.00	4.34	45	1.94	2.75	42	3248	8658	163
	Wheat	6.33	9.66	50	1.31 *	1.93**	47	4455	9041	105
Maize – potato – onion (irrigated)	Maize	3.09	4.52	46	1.95	2.86	46	3361	9135	172
	Potato	53.70	76.50	42	17.10	23.50	33	9775	19,250	97
	Onion	18.87	25.45	35	12.05	15.10	25	38,700	51,050	32

Source: Ghosh [44], C-Conventional, T-Intercropping/crop rotation *—Maize equivalent yield; ** wheat equivalent yield. 60 INR (Indian Rupees) ~ 1USD (2014).

When crops like maize, sorghum and castor (*Ricinus communis* L.) are cultivated along with legumes such as groundnut, green gram (*Vigna radiata* L.), black gram (*Vigna mungo* L.), soybean and cowpea in inter-row spaces, sufficient cover on the ground is ensured and erosion hazards decreased [45]. Pathak *et al.* [46] reported several soil conservation measures based on rainfall in a particular area (Table 12).

Table 12. Soil and water conservation measures to be taken up based on seasonal rainfall in the Peninsular India (Source: Pathak *et al.* [46])

Seasonal Rainfall (mm)	Soil and Water Conservation Measures	
<500	Contour cultivation with conservation furrows, ridging, Sowing across slope, Mulching, Scoops, Off season tillage, Inter row water harvesting system, Small basins, Field bunds, Khadin	Tied ridges, contour bunds
500–750		Zingg terrace, modified contour bunds and broad bed furrow
750–1000	Broad bed furrow (vertisols), field bunds, vegetative bunds and graded bunds	Conservation furrows, sowing across slope, conservation tillage, Lock and spill drains, small basins, nadizingg terrace
>1000		Level terrace and zingg terrace (conservation bench terrace)

Subsoiling

Low infiltration rate is one of the major problems of black soils (Vertisols) in central India. In Vertisols, improved tillage practices, particularly deep tillage (subsoiling with chisel plough), can improve soil water storage by greater infiltration and minimizing water stress. A study with three tillage treatments consisting of conventional tillage (CT), CT + subsoiling in alternate years, and CT + subsoiling in every year showed that the basic infiltration rate and soil water storage in the 90 cm profile were greater in CT + subsoiling every year than in CT [47].

Watershed Approach

Integrated watershed management, which involves soil and water conservation coupled with suitable crop management, is another excellent strategy for mitigating soil erosion. Development and management of watershed resources to achieve optimum production without causing deterioration in the resources base is integrated watershed management. It involves construction of check dams along gullies, bench terracing, contour bunding, land leveling and planting of grasses. These strategies will increase percolation of water, decrease runoff and improve water availability. Several reviews are available on the performance of watershed development projects [48,49], as well as their limitations. An operational research project on watershed management at Fakot by the CSWCR&TI during 1975–1986 is a successful example of participatory integrated watershed management approach [50]. Conservation agriculture along with above-said practices has great potential to reverse soil loss.

Participatory Resource Conservation and Management

A case study in Netranahalli Watershed (Karnataka) in the Southern Peninsular India stressed the importance of involvement of communities for conservation of natural resources (mainly soil and water) and their management. Improvement in ground water levels, soil and moisture conservation, development of irrigation facilities, water regeneration capacity, forestry and horticulture development, change in land use pattern and cropping pattern, improvement in animal health, employment and income generation were noticed by Adhikari *et al.* [51]. In a joint programme by Bangalore Regional Centre of NBSS&LUP, Nagpur and Tamil Nadu State Department of Agriculture at Shivagangai district, Tamil Nadu, water harvesting (in better maintained existing tanks), recharging (in aquifers) and providing drainage facilities (in lowlands) prevented water erosion and decreased incidence of salinity and sodicity [52].

Integrated Nutrient Management and Organic Manuring

Integrated nutrient management, *i.e.*, the application of NPK mineral fertilizers along with organic manure, increases crop productivity, improves SOC content, and decreases soil loss. In the northwestern hill region, integrated nutrient management improved soil health and SOC storage in all cropping systems. Kundu *et al.* [53] and Bhattacharyya *et al.* [54] observed that about 19% and 25% of gross C input contributed to greater SOC content after 30 years of rainfed or after nine years of irrigated soybean-wheat production, respectively. Annual farmyard manure addition improved labile (movable; short-lived) and long-lived C pools [55]. Nearly 16% (mean of all treatments) of the estimated added C was stabilized into SOC both in the labile and recalcitrant pools, preferentially in the 0–30 cm soil layer (Figure 1). However, the labile:recalcitrant SOC ratios of applied C stabilized was largest in the 15–30 cm soil layer (Figure 2). The labile pool constituted about 62% of the total SOC in the 0–45 cm soil layer and about 50% of the applied C stabilized in the labile pool (Figure 3). The integrated nutrient management approach of 5 ton ha^{-1} of farmyard manure +50% recommended fertilizer led to an additional grain yield of 2.65 ton ha^{-1} in paddy-maize cropping system (reference). Under rainfed conditions, C retention rate varied from 0.61 to 1.8 ton ha^{-1} $year^{-1}$ in different crop rotations, which also enhanced crop yield (Table 13). However, with green manuring, wheat had greater water use (289 mm) than wheat in a wheat-fallow system (273 mm) or wheat (270 mm) rotated with maize [56].

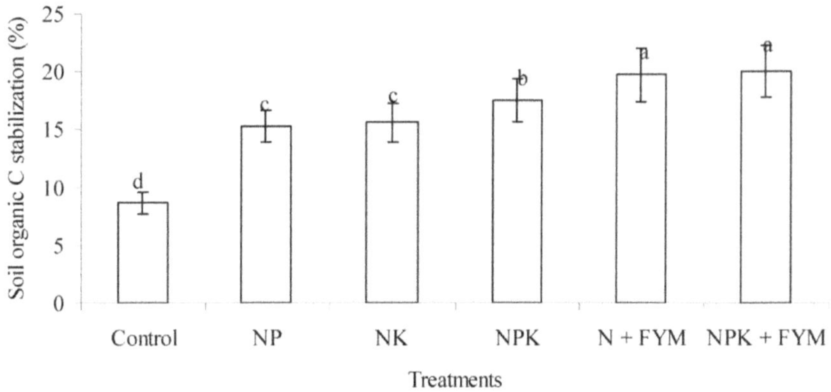

Figure 1. Soil organic C (SOC) stabilization in the 0 to 45 cm soil layer as affected by 32 years of continuous annual fertilization under soybean-wheat cropping in a sandy clay loam soil of the Indian Himalayas. Bars with the same lowercase latter indicate that the values are not significantly different (at $P < 0.05$ according to Tukey's HSD tests). Error bars indicate standard errors. (Source: Bhattacharyya *et al.* [55]).

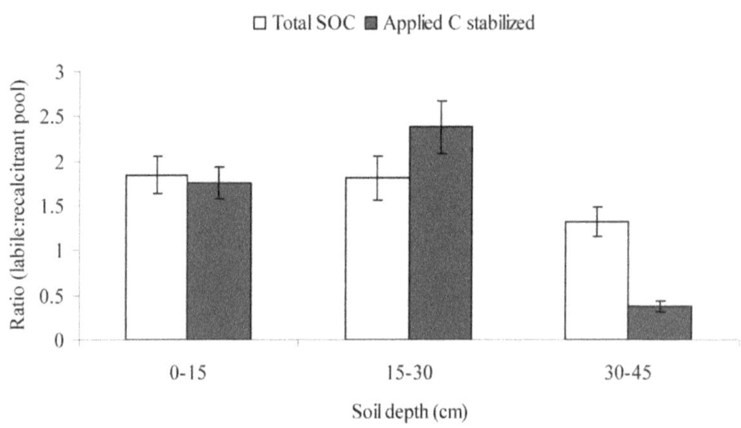

Figure 2. Ratios of labile and recalcitrant pools of total SOC and applied C stabilized in soils by depth after 32 years of cropping with different fertilization (error bars indicate standard error of mean).

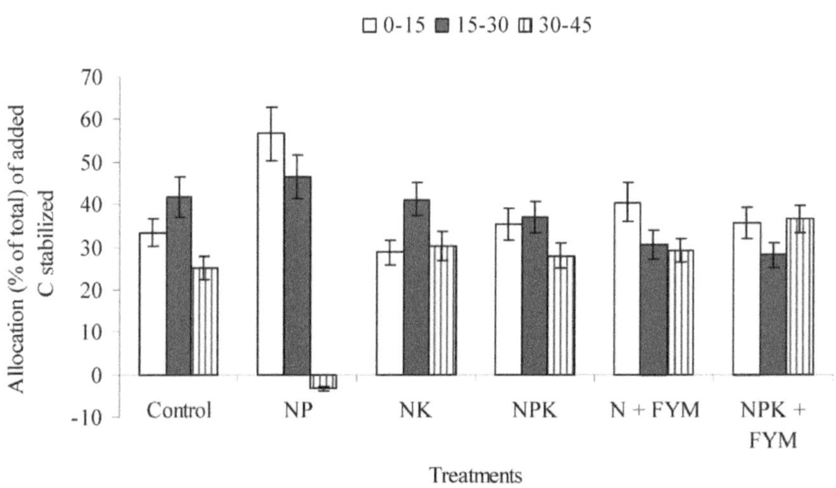

Figure 3. Depth (cm) distribution of total estimated added C stabilized after 32 years of fertilization (error bars indicate the standard error of mean; Source: Bhattacharyya *et al.*[55]).

Table 13. Fertilization impacts on carbon retention in the 0–15 cm layer and crop yield change in the Indian Himalayas (Data Source: Bhattacharyya *et al.* [54,57,58,59])

Rainfed Management Practices	Duration of Adoption (year)	Carbon Retention Over Control (Mg $ha^{-1}year^{-1}$)	Yield Change Over	Yield Change over Unfertilized Control/Two Irrigations (Mg $ha^{-1}year^{-1}$)
NPK + FYM application-rainfed	32	0.87	Unfertilized control	2.31 (S) & 1.17 (W)
NPK + FYM application-irrigated	9	1.28	Unfertilized control	0.80 (S) & 1.74 (W)
FYM at 15 Mg ha^{-1}	3	1.63	Unfertilized control	6.2 (GP) & 7.1 (FB) & 0.55 (BC)
FYM at 10 Mg ha^{-1}	3	1.80	Unfertilized control	3.5 (GP) & 1.3 (R)
Four irrigations in wheat	4	0.35	Two irrigations	0.17 * (R) & 0.44 (W)

S—soybean, W—wheat, SEY—soybean equivalent yield, FB—French bean, GP—Garden pea, BC—baby corn. * indicates not significant.

In the lower Indo-Gangetic Plains, Mandal *et al.* [60] observed 25%–38% greater accumulation of total SOC with NPK + FYM/compost than the control (no fertilizer). The order of such accumulation under different cropping systems was rice-mustard-sesame > rice-fallow-rice > rice-wheat-fallow > rice-wheat-jute (*Corchorus sp.*) > rice-fallow-barseem (*Trifolium alexandrinum*), over the control (Table 14). Mandal *et al.* [60] further observed that the amount of residue C inputs in the rice-wheat-fallow system (3.33 Mg ha^{-1} $year^{-1}$) was similar to that in the rice-fallow-berseem (3.17 Mg ha^{-1} $year^{-1}$), but the rate of annual C accumulation in the former was more than double (0.27 Mg C ha^{-1} $year^{-1}$) than that in the latter (0.13 Mg C $ha^{-1}year^{-1}$). Higher N content in crop residues of berseem (2.6%) and jute (1.8%) but lower lipids and lignin in rice-fallow-berseem and rice-wheat-jute systems may have accelerated decomposition and thus hastened loss of C. Crop residues from rice and wheat, which have low N content, are likely to be more efficient in C sequestration than the residue of crops like berseem and jute, which have higher N content. Likewise, in the drylands, Srinivasarao *et al.* [61] found that integrated nutrient management could improve C accumulation rates up to 0.45 ton $ha^{-1}year^{-1}$ in a groundnut based cropping system (Table 15).

Table 14. Effects of balanced fertilization (NPK and NPK + FYM or compost) on C build up in soils under different cropping systems (Data source: Mandal *et al.* [60])

Cropping System	C Build-Up (%) in Treatments over the Control Plots		C Build-Up Rate (Mg C $ha^{-1}year^{-1}$) over the Control Plots	
	NPK	NPK + FYM	NPK	NPK+FYM
R-M-S	51.8 a	55.7 a	1.91 a	2.05 a
R-W-F	16.8 c	23.4 c	0.27 b	0.37 c
R-F-B	9.3 d	24.7 c	0.13 c	0.36 c
R-W-J	14.9 c	32.3 b	0.11 c	0.25 d
R-F-R	33.5 b	54.8 a	0.28 b	0.45 b

Build-up = [(NPK//NPK + FYM – Control)/Control] × 100; Build-up rate = [(NPK//NPK + FYM – Control)/year]; R-M-S, rice-mustard-sesame; R-W-F, rice-wheat-fallow; R-F-B, rice-fallow-berseem; R-W-J, rice-wheat-jute; R-F-R, rice-fallow-rice, FYM, farmyard manure.

Table 15. Carbon accumulation rate in soil (0–20 cm) and potential carbon emission reduction (CER) under different INM practices (Data source: Srinivasarao *et al.* [61])

Production Systems	Suggested INM Practice	C Accumulation (ton $ha^{-1}year^{-1}$)		Potential CER from the Suggested Practice	
		Farmers' Practice	Suggested Practice	ton ha^{-1}	Value (US $)
Groundnut-based (in Alfisols)	50% RDF + 4 ton groundnut shell ha^{-1}	0.08	0.45	0.370	1.85
Groundnut–finger millet (in Alfisols)	FYM 10 ton + 100% RDF (NPK)	−0.138	0.241	0.379	1.90

Finger millet–finger millet (in Alfisols)	FYM 10 ton + 100% RDF (NPK)	0.046	0.378	0.332	1.66
Sorghum-based (in Vertisols)	25 kg N ha^{-1} (through FYM) + 25 kg N ha^{-1}(through urea)	0.101	0.288	0.187	0.94
Soybean-based (in Vertisols)	6 ton FYM ha^{-1} + 20 kg N + 13 kg P	−0.219	0.338	0.557	2.79
Rice-based (in Inceptisols)	100% organic (FYM)	−0.014	0.128	0.142	0.71
Pearl millet-based (in Aridisols)	50% N (inorganic fertilizer) + 50% N (FYM)	−0.252	−0.110	0.142	0.71

CER at US$ 5 ton^{-1} C (prevailing market price of CER for agroforestry and other related practices); RDF, Recommended dose of fertilizer; FYM, Farmyard manure.

Integrated nutrient management also decreases soil loss. Runoff and soil loss increased with increase in slope from 0.5% to 2.0% at Bellary (Table 16). However, in the treatments with application of recommended rate of fertilizer along with farm-yard manure, it was comparatively low. Plots under coir pith compost and integrated nutrient management improved maize yield and rainwater use efficiency from 4.95 to 5.79 kg ha^{-1} mm at Ayalur watershed, Tamil Nadu (Table 17).

Table 16. Runoff and soil loss under different crops on varying slopes at research farm, Bellary (Source: CSWCR&TI Annual Report [62])

Treatments	**Runoff (mm)**						**Soil Loss (ton ha^{-1})**					
	Sorghum			**Chickpea**			**Sorghum**			**Chickpea**		
	0.5	**1.0**	**2.0**	**0.5**	**1.0**	**2.0**	**0.5**	**1.0**	**2.0**	**0.5**	**1.0**	**2.0**
	Slope (%)											
With fertilizer	52.3	66.78	94.8	48.71	64.45	84.56	2.45	4.04	5.67	2.01	2.72	4.79
Without fertilizer	63.16	66.85	101.79	49.06	65.64	92.99	2.72	4.79	6.08	2.19	3.31	5.35

Table 17. Effect of coir pith compost and integrated nutrient management (INM) on maize —a Case study in Ayalur watershed, Tamil Nadu (Source: Kannan *et al.* [63])

Particulars	**Farmers' Practice**	**INM**	**Control**	**Coir Pith**
Yield (ton ha^{-1})	4.5	5.5	4.2	4.9
Additional yield (ton ha^{-1})	-	1.0	-	0.7
Additional cost (Rs)	-	2747	-	-
Additional benefits(Rs)	-	8000	-	-
Rain WUE (kg $ha^{-1}mm^{-1}$)	10	12.2	4.95	5.79

Reclamation of Acid and Salt Affected Soils and Drainage (Desalinization)

Liming is the most desirable practice for amelioration of acid soils. Lime raises soil pH, thereby increasing the availability of plant nutrients and reducing toxicity of Fe and Al [64,65,66]. Sharma and Sarkar [64] and Bhat *et al.* [66] recommended low dose of lime (*i.e.*, one-tenth to one-fifth of lime requirement) applied along with fertilizers in furrows at the time of sowing. Bhat *et al.* [67] also tested low-cost locally available basic slag, a by-product of a steel factory as an ameliorant for acidic red and lateritic soils of West Bengal under mustard-rice.

Management of saline soils involves tillage, irrigation and leaching. Inversion tillage can decrease potential soluble salt accumulation in the root zone compared to zero tillage [68]. However, deep tillage may bring more salts to the soil surface and root zone. The most efficient method is through application of high quality irrigation water (low electrical conductivity) and growing of salinity tolerant crops. Tolerant crops also support formation of stable soil aggregates, which help to improve soil tilth. Rice is the potential crop for reclamation of sodic soils. Salt affected soils are reclaimed by leaching followed by application of green manures. Gypsum is the major chemical used for reclamation of alkali soils. Other amendments used are: phosphogypsum or acid formers like pyrites, sulphuric acid, aluminium sulphate and sulphur. The treated field should be kept submerged with good quality water to facilitate reaction and subsequent leaching. In addition, proper drainage through deep and open drains can be adopted wherever problems persist. Restoration of salt-affected soils can also lead to a significant increase in SOC pool. Garg [69] observed a dramatic increase in SOC of a sodic soil planted with perennials (e.g., mesquite) after 8 years. Bhojvaid and Timmer [70] also reported an increase in SOC pool by restoration of salt-affected soils and a similar potential exists [71].

To reclaim non-saline sodic soil, incorporation of relatively soluble Ca salt like gypsum, phosphogypsum, iron salt like pyrite, $CaCl_2$, sulfuric acid (H_2SO_4), or other acid formers like sulphur (S), lime-sulphur (9% Ca + 25% S), ferric sulfate and aluminium sulfate to replace exchangeable Na from the clay complex, along with recommended water and crop management practices, have been researched by many [72,73,74]. Reclaiming acid sulfate soils may follow approaches like: (i) pyrite and soil acidity can be removed by leaching after drying and aeration; (ii) pyrite oxidation can be limited or stopped and existing acidity inactivated by maintaining a high water table, with or without (iii) additional liming and fertilization with phosphorus, though liming may often be uneconomical in practical use. For coastal acid sulfate soils of Sundarbans, application of lime, superphosphate and rock phosphate have been found useful [75].

Since 1970 in India, there has been significant commercial development using various desalination technologies, including distillation, reverse osmosis and electrolysis [76]. Desalination mostly uses fossil fuels. Many facilities in coastal region are using reverse osmosis for desalinization. For example, at Kalpakkam reactor, Tamil Nadu, 1.8 million liters of water is being produced per day. Installation of one-way sluice gates on the river banks or any other suitable location to drain out excess water from the land during low tides in river, use of subsurface tile drains combined with moling perpendicular to the former [77], and open surface drains along with moling perpendicular to it [78] are some of the practices.

Remediation of As Contamination

Mitigation of As contamination could be achieved by replacing *boro* rice requiring more ground water with summer legumes and pulses, decreased irrigation coupled with addition of zinc sulfate, greater use of organic/green manures that moderate As toxicity in soils and plants [79,80], and phytoremediation employing hyper-accumulating plants like brake fern (*Pteris vittata*) and water hyacinth (*Eichornia crassipes*). Blue-green algae also have ability to decontaminate as of paddy soils through accumulation in its biomass and subsequent removal.

Water Management and Pollution Control

Promoting water conservation and efficient water management along with expansion of irrigation facilities, drip irrigation and sub-surface irrigation in some areas holds promise. Domestic and municipal wastes, sludges, pesticides, industrial wastes, *etc.* need to be used with utmost caution to avoid the possibility of pollution of soil. Mined land can be better reclaimed by proper back filling and spreading topsoil over the surface [81]. Reclaimed land after mining can be used for planting trees. The use of geo-textiles, permeable fabrics which separate, filter, reinforce, protect or drain the soil, will help the re-vegetation process [57,82].

Sen and Oosterbaan [83] presented a practical working method on integrated water management for Sundarbans through surface gravity induced drainage during the rainy season (through land shaping)-cum-excess rainwater storage for irrigation during dry season. Ambast and Sen [84] developed a user-friendly software 'RAINSIM' primarily for small holdings in the Sundarbans region based on hydrological processes, as well as in different agro-climatic regions. The software may be used for (i) computation of soil water balance; (ii) optimal design of water storage in the "on-farm reservoir" concept for converting up to 20% of the watershed; (iii) design of surface drainage in deep waterlogged areas to decrease water congestion in 75% of the area; and (iv) design of a simple linear program to propose optimal land allocation.

Irrigation Management for Improving Input Use Efficiency

Scheduling of irrigation based on critical stages of crops, or atmospheric demand stimulates optimum plant growth and increases the transpiration component of evapo-transpiration loss of water, thereby improves crop productivity and decreases soil degradation. Compared to flooding, sound surface irrigation methods like sprinkler or drip give better input efficiency. It not only improves the yield levels and input use efficiencies but also saves considerable volume of water. Besides this, in Vertisols, a considerable volume of irrigation water is lost beyond the root zones through bypass flow when irrigation water is applied through flooding. Loss of water through bypass flow in Vertisols could also be decreased by adopting irrigation application based on the atmospheric demand of water, *i.e.*, by adopting (irrigation water/cumulative pan evaporation) based scheduling. Bandyopadhyay *et al.* [85] reported that irrigation scheduling at 0.8 irrigation water/cumulative pan evaporation significantly improved the soil water extraction, root length density and grain yield of wheat over irrigation at 0.6 IW/CPE. Besides, integrated use of 75% of the recommended doses of NPK to wheat with farm-yard manure at 5 Mg ha^{-1} or poultry manure at 1.5 Mg ha^{-1} or phosphocompost at 5 Mg ha^{-1} to the rainy season crops (like soybean or sorghum) significantly improved the root length density, yield and water use efficiency of wheat over application of 100% NPK to both crops, leading to a saving of 25% fertilizer NPK in both the seasons and improvement of the use efficiency of the fertilizer nutrients.

Judicious Use of Distillery Effluent

In the vicinity of distillery industries, irrigation water as distillery effluent can be applied judiciously as a waste by-product and this technique has a considerable impact on mitigating land degradation. Both spent wash and post-methanated effluent were tested in a field experiment on soybean-wheat system for five years in a Vertisol. The SOC of the surface (0–15 cm) layer and aggregate stability were greater with application of both techniques. Proportion of macroaggregates was greater with spent wash than with post-methanated effluent, as well as compared with no distillery effluents and NPK+farm-yard manure treatments. Macroaggregate-associated C was also greater in spent water treated plots. Plots receiving waste by-produts had greater SOC, mean-weight diameter of aggregates, and percent macro- and macroaggregate-associated C than farmers' typical practice [86].

Reforestation, Grassland and Horticulture Development

In the hills, the majority of the upper slope is covered with horticultural crops using half-moon terraces and contour bunds and the remaining one-third of the lower section is used for cultivation of cereals, or oil crops with bench terraces. The following crops may be grown: (1) Fruit trees in half-moon terraces (triangular system of planting) on contour; (2) Pineapple (*Ananas comosus* L.) in two rows planted closer together in contour bunds; (3) Vegetables like bean (*Phaseolus vulgaris* L.), cowpea (*Vigna sinensis* L.), guar or clusterbean (*Cyamopsis tetragonoloba* L.), pea (*Pisum sativum* L.) and good cover crops like sweet potato in the interspaces of the contour and (4) Ginger (*Zinziber officinale* L.) and turmeric (*Curcuma longa* L.) grown in the inter-space area of contours. Grewal [87] found that under Eucalyptus-Bhabar grass (*Eulaliopsis binata*) system, soil loss was negligible (0.07 ton ha^{-1}) (Table 18). Likewise, reforestation and grassland development in wastelands have great potential to decrease land degradation.

Table 18. Soil loss under different land use systems in Shivaliks (Source: Grewal [87])

Land Use Systems	No. of Years of Observations	Soil Loss (ton ha^{-1})	Runoff (% of Total Rainfall)	N Loss (kg ha^{-1})	K Loss (kg ha^{-1})
Eucalyptus-Bhabar grass	6	0.1	0.1	0.5	0.9
Acacia catechu–forage grass	3	0.2	2.0	7.0	0.5
Leucaena-Napier grass (*Pennisetum purpureum* L.)	3	0.3	4.4	6.6	1.2
Teak (*Tectona grandis* L.)-*Leucaena*-Bhabar	3	0.4	3.3	2.1	0.6
Eucalyptus-Leucaena-Turmeric	5	0.6	2.6	2.5	0.7
Poplar (*Liriodendron tulipifera*)-*Leucaena*-Bhabar	5	1.5	4.8	5.9	1.1
Sesamum (*Sesamum indicum*)-Rapeseed (*Brassica napus*)	3	2.7	20.5	42.5	3.0
Cultivated fallow	3	5.6	23.0	51.3	5.0

Vegetative Barriers and Using Natural Geotextiles, Mulching and Diversified Cropping

In general, results from the Himalaya region indicate that vegetative barriers can decrease runoff by 18%–21% and soil loss by 23%–68% on slopes varying from 2%–8% (Table 19). Vegetative barriers of *Guinea* grass (Plate 1), *Khuskhus* and *Bhabar*were effective (after 3–4 years) in reducing soil loss by 6–8 ton ha^{-1}year^{-1} and runoff by 33%–38% [40]. Maize and wheat yield increased ~32 and 10%, respectively, due to conserved moisture in the hilly region [88]. Pigeonpea (*Cajanas cajan*), because of its very good canopy cover (95%–98%) as a vegetative barrier, was effective in reducing runoff (28%–29%) and soil loss (2.1 to 2.6 ton ha^{-1}) in a finger millet (*Eleusine coracana* L.)/kodo millet (*Paspalum scrobiculatum* L.)-lentil (*Lens esculentus* L.) cropping sequence. Pigeonpea improved SOC along with addition of 22 to 41 kg of N ha^{-1} in the soil. The practice increased maize yield 5%–10% and wheat yield 10%–15% in the hills.

Table 19. Effect of grass barriers on yield, runoff and soil loss in different slopes of the northwestern hill region (Data Source: CSWCR&TI Vision [40])

Particulars	Slope (%)					
	2	4			8	
	Guinea Grass	Guinea Grass	Khus khus	*Bhabar*	Guinea Grass	Khus khus
Runoff (% of total rainfall)	25.8	33.3	35.1	37.9	38.90	40.52
Soil loss (ton ha^{-1}year^{-1})	3.27	6.12	6.72	8.34	9.45	9.87
Maize yield (kg ha^{-1})	2530	2460	2444	2296	2285	2180
Wheat yield after maize (kg ha^{-1})	2852	2693	2555	2362	2415	2385
Dry grass yield (kg ha^{-1}year^{-1})	1675	1540	542	1090	1375	485

Plate 1. Vegetative barrier of Guinea grass and Bhabar grass (Source: CSWCR&TI Vision [40]).

Research results in neighboring countries, like China, Thailand and Vietnam indicated that even short-term use of biological geotextiles (maize stalk mats, bamboo mats, borassus and buriti mats and wheat straw mats) in highlands significantly improved biomass and decreased runoff and soil loss [89,90,91,92]. Borassus mats (geotextile mats manufactured from *Borassus aethiopum* L.) and buriti mats (geotextiles manufactured from *Mauritia flexuosa* L.) were also tested for soil conservation in a loamy sand soils of the UK. The results are very novel and worth mentioning here. Results reveal that: (i) borassus mats significantly decreased soil splash erosion; (ii) complete cover by borassus mats is unnecessary and only 10% mat cover as buffer strips had similar erosion control to completely covered plots; and (iii) borassus mat-covered plots maintained SOM and other selected soil properties [93,94,95,96]. Vegetative barriers are also used to mitigate soil degradation in non-arable areas. Vegetative barriers of tree species are effective in controlling runoff and soil loss on 4% slope [97]. Total sediment deposited along hedgerows (3-year period) and tree rows (9-year period) ranged from 184 to 256 ton ha^{-1}, equivalent to 15 to 20 mm soil depth (Table 20).

Table 20. Sediment deposition along vegetative barriers at Dehradun (Source: Narain*et al.* [97])

Vegetative Barrier	No. of Years	Sediment Deposited (ton ha^{-1}year^{-1})	Average Deposition (ton ha^{-1}year^{-1})	Soil Loss s(ton ha^{-1}year^{-1})
Leucaena hedge in turmeric field	3	47.3	15.8	7.6
Leucaena hedge in maize field	3	184.0	61.3	12.1
Leucaena trees in maize field	9	256.5	28.5	8.8
Eucalyptus trees in maize field	9	185.6	20.6	5.8
Leucaena trees in turmeric field	9	90.1	10.1	6.8
Eucalyptus trees in turmeric field	9	103.7	11.5	7.1

In a novel attempt in Bangladesh, implementation of jute geotextiles aided by native vegetation cover was investigated in 2009. Combined presence of jute geotextiles and vegetation cover decreased erosion rates by ~95% and runoff by ~70% with respect to bare plots (that had ~18 ton ha^{-1} year^{-1} soil loss) on a 20% land slope [98].

Agroforestry

Agroforestry systems are an appropriate management tool for both acid and salt-affected soils, because perennial woody vegetation recycles nutrients, maintains soil organic matter, and protects soil from surface erosion and runoff [99]. Four multipurpose tree species were compared with a control plot (without tree plantation) for soil fertility status in an acid soil of India. The presence of trees improved the physico-chemical and microbial biomass parameters by storing greater SOC [100]. Tree vegetation in an agroforestry system serves two major purposes: (i) the fine root system holds soil in place, reducing susceptibility to erosion; and (ii) plant stems decrease the flow velocity of runoff, enhancing sedimentation.

Nair [101] stated three environmental benefits of agroforestry systems: water-quality enhancement, C sequestration, and soil improvement. These benefits are based on the perceived ability of (i) vegetative buffer strips to decrease surface transport of agrochemical pollutants; (ii) large volumes of aboveground and belowground biomass of trees to store C deeper in the soil profile; and (iii) trees enhance soil productivity through biological N_2 fixation, efficient nutrient cycling, and deep capture of nutrients. Legume-based agroforestry has the capacity to support biological N fixation to enhance subsequent soil N availability and therefore improve soil fertility and crop yields [102].

Biosaline (agro) forestry is the cultivation of trees and/or crops on salt-affected soils. Some tree species are less susceptible to soil salinity and sodicity than agricultural crops and hence the cultivation of these trees can help regenerate these soils. In alkaline waste lands, mechanical impedance is a major cause of poor root proliferation. This problem could be overcome by planting *Prosopis juliflora*, which has roots to vertically penetrate a hard pan [103]. Mishra *et al.* [104] opined that soil erosion can be decreased in alkaline soils with *Prosopis juliflora* and *Casuarina equisetifolia* due to the formation of stable soil aggregates in the surface layers. Kaur *et al.* [105] analyzed the role of agroforestry systems (*Acacia*, *Eucalyptus* and *Populus* along with rice–berseem (*Trifolium alexandrinum* L.)) to improve soil organic matter, microbial activity and N availability and observed that: (i) microbial biomass C and N were greater by 42% and 13%, respectively, in tree-based systems than mono-cropping; (ii) soil organic C increased by 11%–52% due to integration of trees along with crops after 6–7 years.

In India, many tolerant species for saline soils have been tried since long (Table 21), like: *Prosopis juliflora*, *Salvadorapersica*,*S. oleoides*, *Tamarixericoides*, *T. troupii*, *Salsolabaryosma etc.*, successful on sites with ECe > 35 dS m^{-1}, *Tamarixarticulata*, *Acacia farnesiana*, *Parkinsonia aculeate* on sites with moderate salinity (ECe 25–35 dS m^{-1}), *Casuarina (glauca, obesa, equiselifolia)*, *Acacia tortilis*, *A. nilotica*, *Callistemon lanceolata*, *Pongamia pinnata*, *Eucalyptus camaldulensis*, *Albizia lebbeck* on sites with moderate salinity (ECe 15–25 dS m^{-1}), trees like *Casuarina cunninghamiana*, *Eucalyptus tereticornis*, *Acacia catcechu*, *A. ampliceps*, *A. eburnea*, *A. leucocephala*, *Dalbergia sissoo*, *etc.* on sites with lower salinity (ECe 10–15 dS m^{-1}).

Table 21. Ameliorative effects of tree plantation on salt affected soils of India

Region	Tree Species	Soil Depth (cm)	Original		After		References
			pH	EC (dS m^{-1})	pH	EC (dS m^{-1})	
Karnataka	*Acacia nilotica* (age 10 years)	0–15	9.2	3.73	7.9	2.05	Basavaraja *et al.* [106]
Karnal	*Eucalyptus tereticornis* (age 9 years)	0–10	10.06	1.90	8.02	0.63	Mishra *et al.* [107]
Lucknow and Bahraich in north India.	*Terminaliaarjuna*	0–15	9.60 ± 0.42	1.47 ± 0.45	8.40 ± 0.27	0.31 ± 0.07	Singh and Kaur [108]
	Prosopisjuliflora				8.70 ± 0.33	0.42 ± 0.06	
	Tectonagrandis				6.15 ± 0.23	0.06 ± 0.006	

Conservation Agriculture (CA)

Conservation agriculture refers to a set of principles, grounded in sound science that is gradually being adopted globally. The concept includes: (1) causing minimum disturbance to the soil surface by using no- or minimum-tillage; (2) keeping the soil surface covered all the time through practices such as retention of crop residue, mulching, or growing cover crops; (3) adopting crop sequences or rotations that include agroforestry in spatial and temporal scales; and (4) controlled traffic [109]. Collectively these practices lead to an increase in water stable aggregates, greater SOC concentrations, and protection from wind and water erosion. Conservation agriculture-based crop management technologies include zero tillage (ZT) with residue recycling; laser assisted precision land levelling, direct drilling into the residues and direct seeding.

In the Himalayan region, year-round ZT under irrigated rice-wheat system with two irrigations at critical growth stages [110], year-round ZT with integrated nutrient management under an irrigated rice-wheat system [58], and 10-cm stubble retention (under CA) of rice and wheat crops for maximum yield and fodder production [111] are novel technologies (Table 22). Zero tillage enhanced macroaggregate-associated SOC and intra-aggregate particulate organic C under a rainfed finger millet-lentil system (Figure 4), but only in the topsoil [57,58,59]. Plots with minimum tillage (MT; a 50% tillage reduction) improved SOC stock in the 0–15 cm layer, as well as soybean yield. Under direct-seeded rice-wheat systems, adoption of ZT with two irrigations in each crop improved topsoil physical properties and SOC content after four years with similar mean crop yields as with CT using four irrigations [112]. Conservation tillage improved soil aggregate stability and labile C pools in the surface layer, across different cropping systems both under rainfed and irrigated conditions in the Himalayas [57]. Introduction of a legume crop improved C retention in surface soils under conservation tillage even with only short-term adoption.

Table 22. Impacts of conservation tillage practices on carbon retention in the 0–15 cm layer and crop yield change in the Indian Himalayas

Rainfed Management Practices	Duration of Adoption (year)	Carbon Retention over Control/CT(Mg $ha^{-1}year^{-1}$)	Yield Change over	Yield Change (Mg $ha^{-1}year^{-1}$)
Zero tillage-irrigated	4	0.20	CT	−0.09 * (R) & −0.23 * (W)
Zero tillage-rainfed	4	0.61	CT	−0.44 (SEY)

SEY = Soybean equivalent yield, R = Rice and W = Wheat. * indicates not significant.

As mentioned earlier, CA has emerged as a new paradigm to achieve goals of sustainable agricultural production in South Asia [113]. Another technology is controlled traffic farming using permanent tram lines. For this system, all equipment on the farm needs a standardized track width. Soil between tram lines has better structure and is free of compaction, while the heavily compacted tramlines provide better trafficability and traction [114]. As a result, tillage cost is decreased and yield increase in the cropping area exceeds the loss of land due to tramlines. In the Indo-Gangetic Plain region, bed planting under CT and ZT generally saves irrigation water [115] and labor requirements without sacrificing crop productivity [116,117].

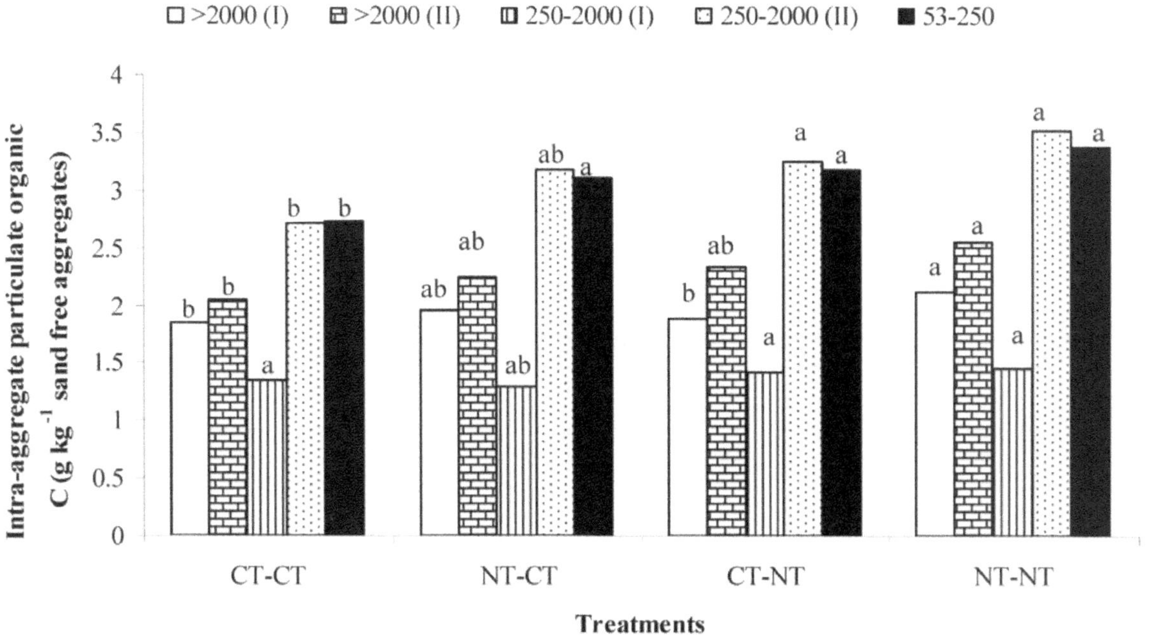

(a)

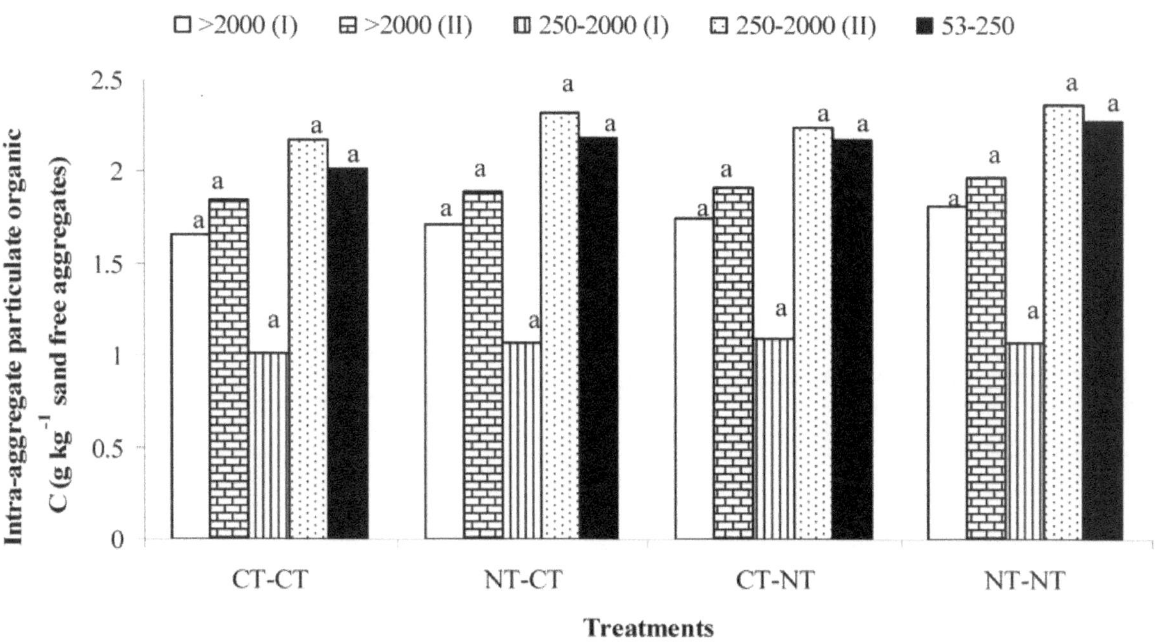

(b)

Figure 4. Intra-aggregate particulate organic matter-carbon (iPOM-C; g kg^{-1} of sand-free aggregates) in aggregate-size fractions at the (**a**) 0- to 5-cm and (**b**) 5- to 15-cm soil layers as affected by tillage practices after six years of rainfed cropping. Bars followed by a similar letter between treatments within an aggregate size class are not significantly different at $P < 0.05$ level of significance according to Tukey's HSD mean separation test. "(I)" and "(II)" in legend refer to coarse (250–2000 μm) and fine (53–250 μm) iPOM in the respective aggregate sizes (Source: Bhattacharyya *et al.* [57]).

In another study in the region, Das *et al.* [118] observed that plots under permanent broad bed with 20% cotton residue and 40% wheat residue retention had significantly higher economic profitability and crop productivity than farmers' practice under a CT cotton-wheat cropping system. In this study, Das *et al.* [118] concluded that 2-year mean seed cotton yield under ZT permanent broad-bed sowing with residue retention was about 24% and 51% greater compared with ZT narrow-bed sowing without residue retention (2.91 Mg ha^{-1}) and CT (2.59 Mg ha^{-1}), respectively (Table 23). Mean water productivity of the system in the permanent broad bed with residue retention (12.58 kg wheat grain ha^{-1}mm^{-1}) was 12%–48% greater compared with CT, narrow bed with and without residues, broad bed, and ZT plots (Table 24). Net return of the permanent broad bed plots with residue retention was 36% and 13% greater compared with CT and narrow bed plots, but was similar to other treatments. Some of the challenges that follow from continuous ZT practice are management of perennial weeds and strategies to combat yield reduction. Yields of ZT crops are often decreased by 5% to 10% on sandy loam soils of India compared with under CT in the initial years [119].

Table 23. Productivity (Mg ha^{-1}) of cotton, wheat and system productivity (Mg ha^{-1}) in terms of wheat equivalent yield (WEY) as affected by tillage, bed planting and residue management practices in the western Indo-Gangetic Plains (Data source: Das *et al.* [118])

Treatments *	2010–2011			2011–2012			2012–2013		
	Seed Cotton Yield	Wheat Grain Yield	System Productivity (WEY)	Seed Cotton Yield	Wheat Grain Yield	System Productivity (WEY)	Seed Cotton Yield	Wheat Grain Yield	System Productivity (WEY)
CT	2.44 c	4.85 a	10.30 b	2.73 c	4.29 b	11.16 c	2.70 c	4.46 b	12.25 b
PNB	2.71 bc	4.55 a	10.60 b	3.10 bc	4.37 b	12.17 bc	3.08 ab	4.83 ab	13.72 ab
PNB + R	2.96 b	4.61 a	11.23 ab	3.33 b	4.60 ab	12.97 b	3.38 a	4.98 a	14.74 a
PBB	3.13 ab	4.82 a	11.81 ab	3.42 ab	4.19 bc	12.80 b	3.11 ab	4.75 ab	13.72 ab
PBB + R	3.28 a	4.85 a	12.16 a	3.93 a	4.77 a	14.67 a	3.46 a	4.89 a	14.88 a
ZT + R	-	-	-	4.00 a	4.44 ab	14.50 a	3.21 ab	4.73 ab	13.99 ab
ZT	-	-	-	3.95 a	4.00 c	13.93 ab	3.02 bc	4.63 ab	13.35 ab

* Means followed by a similar lowercase letter within a column are not significantly different (at $P < 0.05$) according to Tukey's HSD Test. ZT = Zero tillage; CT = Conventional tillage; ZT + R = Zero tillage + residue retention; PNB = Permanent narrow bed; PBB = Permanent broad bed; PNB+R = Permanent narrow bed + residue retention; PBB + R = Permanent broad bed + residue retention.

Table 24. Impacts of tillage, bed planting and residue management practices on water productivity (kg wheat grain/ha.mm) under the cotton-wheat system (Source: Das *et al.*[118])

Treatments *	2011–2012		2012–2013	
	Total Water Applied in the System (mm)	System Water Productivity	Total Water Applied in the System (mm)	System Water Productivity
CT	1417	8.65 d	1331	8.38 d
PNB	1297	10.58 b	1208	10.07 c
PNB + R	1282	10.50 bc	1181	10.98 bc
PBB	1260	10.89 b	1160	11.03 bc
PBB + R	1222	12.18 a	1130	12.98 a
ZT + R	1312	10.66 b	1247	11.62 b
ZT	1387	9.62 c	1310	10.63 bc

Means followed by a similar lowercase letter within a column are not significantly different (at $P < 0.05$) according to Tukey's HSD Test. ZT = Zero tillage; CT = Conventional tillage; ZT + R = Zero tillage + residue retention; PNB = Permanent narrow bed; PBB = Permanent broad bed; PNB+R = Permanent narrow bed + residue retention; PBB + R = Permanent broad bed + residue retention.

Unlike conventional farming methods, CA minimizes soil disturbance and recycles crop residues. Soil bulk density may be decreased, soil aggregation may be improved, and SOC may increase to reverse land degradation with CA. Specific results from four years of wheat-based cropping system in the western Indo-Gangetic Plains indicate that ZT had higher C retention potential than CT in the 0–30 cm soil layer with 8.6% and 10.2% of the gross C input retained under CT and ZT, respectively (Figure 5; Das *et al.* [120]).

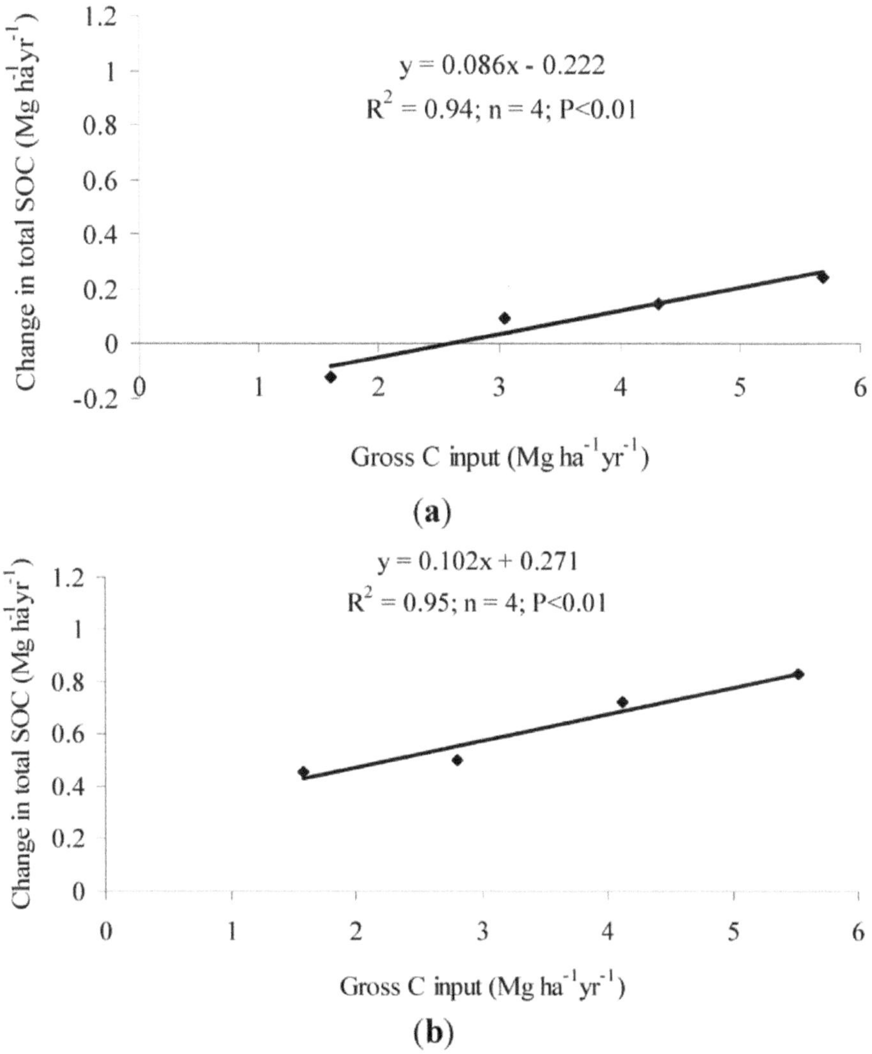

Figure 5. Total soil organic C (SOC) retention potential of residue management practices under (**a**) conventional tillage and (**b**) zero tillage under a wheat based cropping system in the Indo-Gangetic Plains (Source: Das *et al.* [120]).

In another study in the Indo-Gangetic Plains, topsoil under ZT with bed planting had greater concentration of macroaggregates (0.25–8 mm) and mean weight diameter with a concomitant lower silt + clay sized particles than under CT with bed planting and CT with flat planting after 4 years. Soil with both cotton/maize and wheat residue retention had greater macroaggregate concentration and mean weight diameter and similar bulk density than with residue removal (Figure 6). Soil aggregation is improved with larger aggregates and greater mean weight diameter [110].

Conservation agriculture has the potential to decrease sub-surface compaction and improve least limiting water range [121]. During the third year of a study at New Delhi, soil penetration resistance exceeded 2 MPa in the 15–30 cm soil layer beneath puddled and transplanted rice in rotation with wheat under CT, but under direct seeded rice with brown manuring and ZT penetration resistance values were <1.5 MPa throughout the 0–60 cm profile. Soil bulk density was lower under the ZT system than under the CT system in the 0–30 cm soil layer (Table 25). Retaining crop residues with permanent broad beds had significantly lower penetration resistance than with permanent narrow-beds with residue and other tillage and residuce management in a cotton-wheat system [121]. Retaining crop residue resulted in lower BD values in the 15–30 cm soil layer under the cotton-wheat system than removing them (Table 26). Retaining crop resicue also had ~12% higher least limiting water range than with CT (10.1%) in the 15–30 cm layer. In the 0–15 cm soil layer, retaining residues under ZT, permanent broad beds, and permanent narrow beds had 13%, 24% and 11% higher mean least limiting water range, respectively, than the same tillage systems without residue retention. This indicates that ZT without residue addition had deleterious impact on soil water availability and structural property and should be avoided. Overall, among the treatments, PBB + R and DSR + BM-ZTW were the best management practices for improved soil physical environment under cotton-wheat and rice-wheat systems, respectively, and therefore should be adopted.

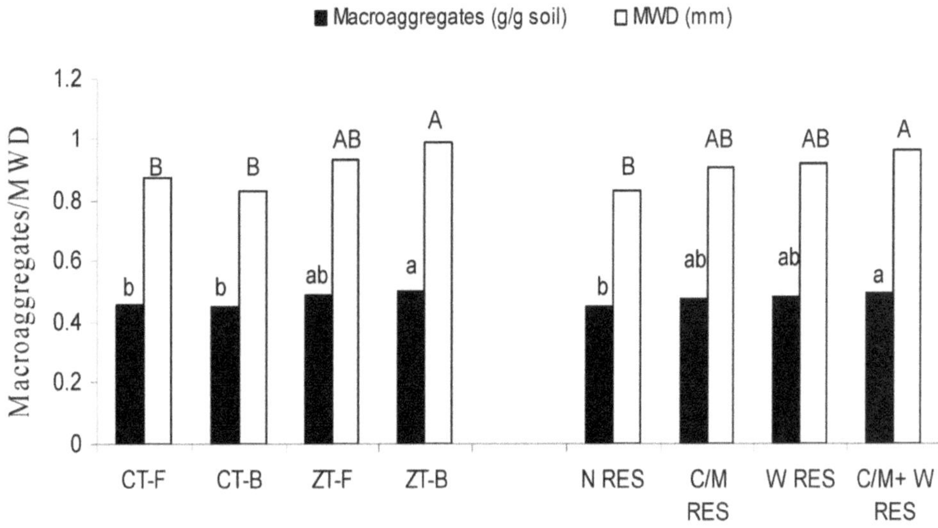

Figure 6. Impacts of conservation agriculture (CA) on soil aggregation in the 0–5 cm layer in the upper IGP (Source: Bhattacharyya *et al.* [110]).

Table 25. Soil bulk density of plots with different conservation agricultural practices in the rice–wheat system (Source: Mishra *et al.* [121])

Bulk Density (Mean of Two Sampling Events) during Rice (Mg m^{-3})			
Conservation Agricultural Practices	**0–15 cm**	**15–30 cm**	**30–45 cm**
PTR − CTW	1.45 b	1.70 a	1.72 a
DSR + BM − ZTW	1.47 ab	1.63 b	1.70 a
DSR − ZTW	1.50 a	1.64 ab	1.72 a

PTR − CTW = Puddled transplanted rice-conventionally tilled wheat; DSR − ZTW = Direct seeded rice-zero tilled wheat and DSR + BM − ZTW = Direct seeded rice + Brown manuring-zero tilled wheat; Means with similar lowercase letters within a soil depth and crop growth period are not significant at $P < 0.05$ according to Tukey's HSD test.

Adoption of CA, as a complete package, is one of the major strategies for increasing SOC stock. Although crop residue incorporation initially leads to immobilization of inorganic N, addition of 15–20 kg N ha^{-1} with straw incorporation eventually increases yield of rice and wheat. Incorporation/retention of rice residue in the soil returns essential organic C and N back to the field to favorably impact soil structural status. Surface residue placement had greater C retention than residue incorporation in a maize-wheat-greengram cropping system [120]. Zero tillage in particular can complicate manure application and may also contribute to nutrient stratification within the soil profile from repeated surface applications without mechanical incorporation. Conservation agriculture has tremendous potential to improve water use efficiency of crops and decrease water loss. If CA could be adopted on half of Haryana's rice-wheat area, the practice would decrease diesel use by 17.4 million liters per year. Using a conversion factor of 2.6 kg CO_2 per liter of diesel burned, this would represent a reduction of more than 25,000 tons each year in CO_2 emissions [116]. Because ZT takes immediate advantage of residual moisture from the previous rice crop, as well as cutting down on subsequent irrigation, water use is decreased by about 10 cm-hectares, or approximately 1 million liters $ha^{-1}year^{-1}$.

Table 26. Soil bulk density of plots with different conservation agricultural practices in the cotton-wheat system during crop growth period (Source: Mishra *et al.* [121])

Bulk Density (Mean of Four Sampling Events in Two Years) during Cotton (Mg m^{-3})			
Conservation Agricultural Practices	**0–15 cm**	**15–30 cm**	**30–45 cm**
CT	1.52 bc	1.65 ab	1.70 a
PNB	1.48 c	1.62 b	1.71 a
PBB	1.50 c	1.63 b	1.70 a
ZT	1.63 a	1.68 a	1.70 a
PNB + R	1.43 d	1.56 c	1.70 a
PBB + R	1.44 d	1.57 c	1.69 a
ZT + R	1.57 b	1.60 bc	1.70 a

Means with similar lowercase letters within a crop growth period are not significantly different at $P < 0.05$. CT = conventional tillage; PNB = Permanent narrow bed; PNB + R = Permanent narrow bed + residue retention; PBB = Permanent broad bed; PBB + R = Permanent broad bed + residue retention; ZT = Zero tillage; ZT + R = Zero tillage + residue retention

With greater aggregation due to fewer disturbances by tillage operations and addition of surface residues, total pore space in soil under CA increases. In a study from central India, CA generally improved water retention properties of soil through its effect on pore size distribution and soil structure. Volumetric soil water retention of the surface 0–15 cm soil was greater in ZT and decreased tillage systems than in CT (Table 27). Similarly at permanent wilting point, CA treatments retained more water than with CT. Difference in water storage between tillage treatments was less at permanent wilting point (2.5%) than at field capacity (4.2%). Conservation agriculture increased soil-water retention more at lower suctions due to increase in micro-pores and inter-aggregate pores caused by enhanced SOC content and higher activity of soil fauna e.g., earthworms and termites. At higher tensions close to permanent wilting point (1.5 MPa), nearly all pores were filled with air and surface area and thickness of water films on soil particle surfaces determined moisture retention. Following addition of organic matter, specific surface area of soils increased resulting in increased water holding capacity at higher tensions [122].

Table 27. Effect of different tillage systems on soil water retention of Vertisols (Source: Hati *et al.* [122])

Tillage	Soil Water Retention (%) (*v/v*)		
	Field Capacity (0.033 MPa)	Permanent Wilting Point (1.5 MPa)	Available Water Capacity
Conventional Tillage	33.5	22.6	10.9
Mouldboard tillage	35.4	24.7	10.7
Reduced Tillage	36.5	24.6	11.9
No Tillage	37.7 a	25.1	12.6
LSD ($P = 0.05$)	2.4	2.1	NS

Despite many benefits of CA practices as mentioned above, the adoption rate in India is very low. Farmers prefer to follow a partial adoption of CA practices, *i.e.*, transplanted rice in puddled soil in the *Kharif* season and CA (ZT with residue retention) for wheat in the *Rabi* season due to several factors, including (i) poor germination and low crop productivity under direct seeded rice, because puddling and waterlogged condition helps to reduce soil pH in alkaline soils and thus improve soil chemical health during the rice growing season; (ii) availability of rain and irrigation water for raising a good rice crop under puddled condition; and (iii) less care is needed for transplanted rice in puddled soil compared with direct seeding. Some farmers even grow CT maize/jowar in the *Kharif* season (for better weed control, aeration and reduction in surface compactness/crusting) and raise wheat under CA in the *Rabi* season. However, repeated puddling aggravates other problems like soil compaction, development of salt affected soils and decline in water table in the area (due to high evaporative demand in this climate). Looking at these facts and due to constant efforts by several institutions, some farmers of the district have started adopting full or complete CA (*i.e.*, direct-seeded rice followed by ZT wheat), but the duration of adoption is less than 3–4 years.

In drylands, Jat *et al.* [123] opined that the major constraints to the use of CA include insufficient amounts of residues due to water shortage and degraded nature of soil resource, competing uses of crop residues, resource poor smallholder farmers, and lack of in-depth research. Even then, CA holds considerable promise in the arid region, because it can control soil erosion by wind and water, reduce compaction and crusting. Due to limited production of biomass, competing uses of crop residues and shortage of firewood, farmers often find it hard to use crop residues to cover soil surface in dryland eco-systems, where only a single crop is grown in a year. With CA (soil cover with crop residues), it is sometimes possible to grow a second crop with residual soil moisture in the profile. It is, however, better to use the chopped biomass of semi-hard woody perennial plants instead of crop residues to cover the soil surface [37].

Intensive Cropping, Diversified Cropping and Integrated Farming Systems

There is already a greater emphasis on crop diversification due to growing concerns about the unsustainability of the rice–wheat system throughout the Indo-Gangetic Plains. The water requirement for rice is about 80% greater than for other crops. Growing non-rice crops in some areas and summer cropping with legumes such as green gram, cowpea (*Vigna unguiculata*L.) or dhaincha (*Sesbania sp.*) are essential for conserving resources and improving productivity. In Punjab, Haryana and Rajasthan, >95% of the area of rice, as well as a large portion of wheat, is under irrigation. Water use efficiency could be greatly increased with cover crops or growing of non-rice based cropping systems. Productivity of waterlogged soils in the eastern Indo-Gangetic Plains could be increased by practicing a raised-sunken bed system. Soybean can be grown on raised beds, while paddy rice can be grown in sunken beds during the rainy season. Productivity of rice fallows in eastern India could be increased by growing *rabi* legumes on raised beds and wheat in sunken beds.

In a study to develop sustainable agricultural intensification with CA in an alkaline soil in Karnal, Gathala *et al.* [124] compared four novel scenarios. Maize under ZT was as productive and almost as profitable as rice during the rainy (*kharif*) season, while using 90%–95% and 88%–91% less irrigation water than puddle transplanted rice and ZT rice, respectively (Table 28). Maize can therefore be an alternative to rice in areas with extreme labor and water scarcity. Avoiding puddling and use of ZT (for rice or maize) with full residue retention increased profitability and yield of the succeeding wheat crop each year (by 0.5 to 1.2 ton ha^{-1}) than farmers' practice. Inclusion of green gram in the cropping system resulted in greater system productivity and profitability than without green gram.

Table 28. Drivers of agricultural change, crop rotation, tillage, crop establishment method and residue management of the four scenarios as studied by Gathala *et al.*[124]

Scenario	Crop Rotation	Tillage	Crop Establishment	Residue Management
Farmers' practice; S1	Rice–Wheat	CT-CT *	Transplanted–Broadcasted	Removal
To deal with increasing food demand; S2	Rice–Wheat–Green gram	CT-ZT-ZT	Transplanted–ZT-ZT	Anchored–Removal–Incorporation
To deal with rising scarcity of labor, water and energy and degrading soil health (CA based); S3	Rice–Wheat–Green gram	ZT-ZT-ZT	Direct Drilling	Retention–Anchored–Retention
Futuristic intensified and diversified cropping system (CA based); S4	Maize–Wheat–Green gram	ZT-ZT-ZT	Direct Drilling / Planting	Retention–Anchored–Retention

* CA—Conservation agriculture, CT—Conventional tillage, ZT—Zero tillage.

In India, 65% of farming households are considered marginal in sustainability (<1 ha). These farms comprise nearly 400 million people and nearly 40% of them are vulnerable, marginalized and food insecure. Hence, integrated farming systems have emerged as a well-accepted, single window, and sound strategy for harmonizing simultaneously joint management of land, water, vegetation, livestock, and human resources. The goals of integrated farming systems are to meet soils' productive potential and reduce risks of environmental degradation. By including tree crops with a high quality of leaf litter and root binding ability, erodibility from rainfall/runoff can be reduced and physico-chemical conditions improve. Soil health can be managed and sustained through organic inputs.

Disaster (Tsunami) Management

The following management aspects are important in case of a tsunami: (i) traditional disaster detection systems should be integrated with current scientific techniques; (ii) early warning systems need to be installed in coastal regions; (iii) protection against tsunamis can be achieved through construction of sea walls, beach defenses, tree plantations, and making buffer zones like raised land masses and forests; (iv) awareness about tsunamis and their impact in coastal areas has to be created not only among the public but also among officials; and (v) enforcement of by-laws and 'Coastal Regulation Zone Norms' should be strictly implemented to minimize tsunami damage. A tsunami early warning system for the Indian Ocean was installed. The Indian Ocean Tsunami Warning System was agreed to in a United Nations conference.

CONCLUSIONS

Appropriate mitigation strategies of the nearly 147 Mha of existing degraded land in the sub-continent of India are of the utmost importance. With changing climate, land degradation is expected to only increase due to high intensity storms, extensive dry spells, and denudation of forest cover. Combating further land degradation and investing in soil conservation is a major task involving promotion of sustainable development and nature conservation. An integrated watershed approach should be given maximum attention to combat land degradation and environmental problems particularly in fragile areas. Sustainable agricultural intensification using innovative farming practices have tremendous potential of increasing productivity and conserving natural resources, particularly by sequestering SOC (both labile and recalcitrant) and improving soil quality. Conservation agriculture (CA) coupled with other technologies like micro-irrigation, fertigation, and management of problem soils using specific and necessary technologies hold great promise to increase productivity of crops and fruits and reverse soil degradation. Novel CA practices include: permanent broad bed with residue retention under maize/cotton/pigeon pea-wheat cropping systems and seasonal tillage alterations under rainfed and rice-based agro-ecosystems. These practices need to be evaluated in micro-environments of different agro-climatic regions with different farming practices for wider adaptability on a watershed basis. For sure, the non-edible (to animals) agricultural residues must not be burnt and should be used for mulching along with growing of cover crops, preferably legumes. Improved grazing practices, irrigation management, control on urban sprawl and control and management on mining are a few other solutions for preventing land degradation. Domestic and municipal wastes, sludges, pesticides, industrial wastes, *etc.* need to be used if possible to close nutrient cycles, but with caution to avoid the possibility of soil pollution. Future research should focus on enhancing nutrient and water use efficiencies and reduction in the pesticide use under CA.

For promotion of CA practices across diverse agro-ecologies, appropriate policy and institutional and technology support would be a prerequisite. Suitable economic incentives should be given to internalize land degradation wherever feasible. Many CA practices like rainwater harvesting through farm pond renovation or construction and its recycling are both capital and labor-intensive, which resource-poor farmers in rainfed areas may not be able to afford and hence need to be supported. Initial incentives, to procure appropriate machinery and to offset any economic loss due to residue retention or production loss, are also important to motivate irrigated farmers to follow CA. Involvement is needed of local communities at every stage in the implementation of resource conserving technologies, judicious irrigation water management, wasteland reclamation, watershed development, and afforestation. A well-defined integrated land use policy to include rural fuelwood and fodder grazing is urgently needed at the implementation level to guide sustainable management of land and forest with a scientific backing. Finally, another critical challenge is controlling fragmentation of land holdings. This could be achieved by providing security of land rights and land tenure and encouraging the efficient use of marginal lands.

ACKNOWLEDGMENTS

The authors are grateful to the anonymous reviewer and the Guest Editor for their valuable and editing comments.

AUTHOR CONTRIBUTIONS

Ranjan Bhattacharyya and Alan Joseph Franzluebbers conceived and designed the whole review work based on invitation from the Guest Editor. Ranjan Bhattacharyya, Birendra Nath Ghosh, Prasanta Kumar Mishra, Biswapati Mandal, ChSrinivasa Rao, Dibyendu Sarkar, Krishnendu Das, Kokkuvayil Sankaranarayanan Anil, Manickam Lalitha, Kuntal Mouli Hati wrote the paper. Alan Joseph Franzluebbers and Ranjan Bhattacharyya edited the manuscript.

REFERENCES

1. National Bureau of Soil Survey & Land Use Planning (NBSS&LUP). *Soil Map (1:1 Million Scale)*; NBSS&LUP: Nagpur, India, 2004.
2. Dhruvanarayan, V.V.N.; Ram, B. Estimation of soil erosion in India. *J. Irrig. Drain. Eng.* **1983**, *109*, 419–434.
3. NCA. *Report of the National Commission on Agriculture. National Commission of Agriculture*; Government of India: New Delhi, India, 1976; pp. 427–472.
4. MoA. *Indian Agriculture in Brief*, 17th ed.; Directorate of Economics and Statistics, Ministry of Agriculture, Department of Agriculture and Cooperation, Ministry of Agriculture and irrigation, Government of India: New Delhi, India, 1978.
5. Vohra, B.B. *A Policy for Land and Water*; Department of Environment, Government of India: New Delhi, India, 1980; Volume 18, pp. 64–70.
6. NWDB. *Ministry of Environment and Forests, National Wasteland Development Board Guidelines for Action*; Government of India: New Delhi, India, 1985.
7. Bhumbla, D.R.; Khare, A. *Estimate of Wastelands in India. Society for Promotion of Wastelands Development*; Allied: New Delhi, India, 1984; p. 18.
8. NRSA. *Waste Land Atlas of India*; Government of India Balanagar: Hyderabad, India, 2000.
9. MoA. *Indian Agriculture in Brief*, 20th ed.; Directorate of Economics and Statistics, Ministry of Agriculture, Department of Agriculture and Cooperation, Ministry of Agriculture, Goverment of India: New Delhi, India, 1985.
10. MoA. *Indian Agriculture in Brief*, 25th ed.; Directorate of Economics and Statistics, Ministry of Agriculture, Department of Agriculture and Cooperation, Goverment of India: New Delhi, India, 1994.
11. National Bureau of Soil Survey and Land Use Planning (NBSS&LUP). *Global Assessment of Soil Degradation (GLASOD) Guidelines*; NBSS&LUP: Nagpur, India, 1994.
12. National Bureau of Soil Survey and Land Use Planning (NBSS&LUP). *Annual Report 2005, Nagpur*; NBSS&LUP: Nagpur, India, 2005.
13. Joshi, P.K.; Agnihotri, A.K. An Assessment of the Adverse Effects of Canal Irrigation in India. *Ind. J. Agric. Ecol.* **1984**,*39*, 528–536.
14. Parikh, K.; Ghosh, U. *Natural Resource Accounting for Soil: Towards and Empirical Estimate of Costs of Soil Degradation for India'*; Discussion Paper No. 1995, 48; Indira Gandhi Institute of Development Research, IDEAS: Mumbai, India, 1995.
15. Joshi, P.K.; Wani, S.P.; Chopde, V.K.; Foster, J. Farmers Perception of Land Degradation: A Case Study. *Econ. Polit. Wkly.* **1996**, *31*, A89–A92.
16. Srinivasarao, C.H.; Venkateswarlu, B.; Lal, R.; Singh, A.K.; Sumanta, K. Sustainable management of soils of dry land ecosystems for enhancing agronomic productivity and sequestering carbon. *Adv. Agron.* **2013**, *121*, 253–329.
17. Reddy, V.R. Land degradation in India: Extent, costs and determinants. *Econ. Polit. Wkly.* **2003**, *38*, 4700–4713.
18. Sharda, V.N.; Dogra, P.; Prakash, C. Assessment of production losses due to water erosion in rainfed areas of India.*Indian J. Soil Water Conserv.* **2010**, *65*, 79–91.
19. NRSA. *National Remote Sensing Agency, IRS-Utilisation Programme: Soil Erosion Mapping*; Project Report National Remote Sensing Agency: Hyderabad, India, 1990.
20. ARPU. *Agro-Climatic Regional Planning: An Overview*; Agro-Climatic Regional Planning Unit: New Delhi, India, 1990.
21. Sehgal, J.; Abrol, I.P. *Soil Degradation in India: Status and Impact*; Oxford and IBH: New Delhi, India, 1994; p. 80.
22. Sahay, K.B. Problems of livestock population. Available online: http://www.tribuneindia.com/2000/20000411/edit.htm(accessed on 20 March 2015).
23. Sharma, K.D. Assessing the impact of overgrazing on soil erosion in arid regions at a range of spatial scales. Human impact on erosion and sedimentation. In Proceedings of the International Symposium of the Fifth Scientific Assembly of the International Association of Hydrological Sciences (IAHS), Rabat, Morocco, 23 April–3 May 1997; pp. 119–123.
24. MoEF. *National Forestry Action Programme of the Ministry of Environment and Forests*; Goverment of India: New Delhi, India, 1999; Volume 1, p. 79.
25. Sahu, H.B.; Dash, S. Land degradation due to Mining in India and its mitigation measures. In Proceedings of the Second International Conference on Environmental Science and Technology, Singapore, 26–28 February 2011.
26. Anon. *Dirty Metal, Mining Communities and Environment, Earthworks*; Oxfam America: Washington, WA, USA, 2006; p. 4.
27. Mythili, M. Intensive Agriculture and Its Impact on Land Degradation. 2013. Available online: http://coe.mse.ac.in/pdfs/coebreifs/Mythili.pdf (accessed on 13 October 2014).
28. FAO. *Land Degradation in South India; Its Severity, Causes and Effects on People*; World Soil Resources

Report; Food and Agriculture Organization: Rome, Italy, 1994; Volume 78, p. 100.

29. Barman, D.; Sangar, C.; Mandal, P.; Bhattacharjee, R.; Nandita, R. Land degradation: Its Control, Management and Environmental Benefits of Management in Reference to Agriculture and Aquaculture. *Environ. Ecol.* **2013**, *31*, 1095–1103.

30. Royal Commission on Agriculture in India Report; Agricole Publishing Academy: New Delhi, India, 1928; pp. 75–76.

31. Maheswarappa, H.P.; Nanjappa, H.V.; Hegde, M.R.; Biddappa, C.C. Nutrient content and uptake by galangal (*Kaempferia galanga* L.) as influenced by agronomic practices as intercrop in coconut (*Cocos nucifera* L.) garden. *J. Spices Arom. Crops* **2011**, *9*, 65–68.

32. Tandon, H.L.S. Assessment of Soil Nutrient Depletion. In Proceedings of the FADINAP Regional Seminar on Fertilization and the Environment, Chiangmai, Thailand, 7–11 September 1992.

33. Tandon, H.L.S. *Fertilizers in Indian Agriculture—From 20th to 21st Century*; FDCO: New Delhi, India, 2004; p. 240.

34. Prasad, R.N.; Biswas, P.P. Soil Resources of India. In *50 Years of Natural Resource Management*; Singh, G.B., Sharma, B.R., Eds.; Indian Council of Agricultural Research: New Delhi, India, 2000.

35. Hobbs, P.; Sayre, K.; Gupta, R. The role of conservation agriculture in sustainable agriculture. *Phil. Trans. R. Soc. B***2008**, *363*, 543–555. [PubMed]

36. MNRE. *Annual Report of the Ministry of New and Renewable Energy*; Government of India: New Delhi, India, 2009.

37. NAAS. *Management of Crop Residues in the Context of Conservation Agriculture*; Policy Paper No. 58; National Academy of Agricultural Sciences: New Delhi, India, 2012; p. 12.

38. Mandal, D.; Sharda, V.N.; Tripathi, K.P. Relative efficacy of two biophysical approaches to assess soil loss tolérance for Doon Valley soils of India. *J. Soil Water Conserv.* **2010**, *65*, 42–49.

39. Srinivasarao, C.H.; Venkateswarlu, B.; Lal, R. Long-term manuring and fertilizer effects on depletion of soil organic stocks under Pearl millet-cluster vean-castor rotation in Western India. *Land Degrad. Dev.* **2014**, *25*, 173–183.

40. CSWCR&TI Vision, 2030; Vision 2030 of the Central Soil and water Conservation Research and Training Institute, Allied publisher: Dehradun, India, 2011; pp. 1–46.

41. Mandal, K.G.; Hati, K.M.; Misra, A.K.; Bandyopadhyay, K.K.; Tripathy, A.K. Land surface modification and crop diversification for enhancing productivity of a Vertisol. *Int. J. Plant Prod.* **2013**, *7*, 455–472.

42. Singh, G.; Ram, B.; Narain, P.; Bhushan, L.S.; Abrol, I.P. Soil erosion rates in India. *Ind. J. Soil Conserv.* **1992**, *47*, 97–99.

43. Sharma, N.K.; Ghosh, B.N.; Khola, O.P.S.; Dubey, R.K. Residue and tillage management for soil moisture conservation in post maize harvesting period under rainfed conditions of north-west Himalayas. *Ind. J. Soil Conserv.* **2013**, *42*, 120–125.

44. Ghosh, B.N.; Sharma, N.K.; Dadhwal, K.S. Integrated nutrient management and intercropping/cropping system impact on yield, water productivity and net return in valley soils of north-west Himalayas. *Ind. J. Soil Conserv.* **2011**,*39*, 236–242.

45. Rao, J.V.; Khan, I.A. *Research Gaps in Intercropping Systems under Rainfed Conditions in India, an On Farm Survey*; CRIDA: Hyderabad, India, 2003.

46. Pathak, P.; Mishra, P.K.; Rao, K.V.; Wani, S.P.; Sudi, R. Best opt-options on soil and water conservation. In Best Bet Options for Integrated Watershed Management, Proceedings of the Comprehensive Assessment of Watershed Programs in India (ICRISAT. Pantancheru 502 324), Andhra Pradesh, India, 23–27 July 2009; Wani, S.P., Venkateshwarlu, B., Sahrawat, K.L., Rao, K.V., Ramakrishna, Y.S., Eds.; pp. 75–94.

47. Rao, K.P.C.; Steenhuis, T.S.; Cogle, A.L.; Srinivasan, S.T.; Yule, D.F.; Smith, G.D. Rainfall infiltration and runoff from an Alfisol in semi-arid tropical India. II. Tilled systems. *Soil Tillage Res.* **1998**, *48*, 61–69.

48. Palanisami, K., Suresh Kumar, D., Chandrasekharan, B., Eds.; *Watershed Management: Issues and Policies for 21st Century*; Associated Publishing Company: New Delhi, India, 2002; p. 341.

49. Joy, K.J.; Parnjpe, S.; Shah, A.; Badigar, S.; Lele, S. Scaling up of watershed development projects in India: learning from the first generation projects. In Proceedings of the Fourth IWMI-Tata Annual Partners Meet, International Water Management Institute, Anand, India, 24–26 February 2005; pp. 133–134.

50. Sharda, V.N. Land degradation and watershed management issues in Himalayan region: Status and Strategies. In*Workshop on "Mountain Agriculture in Himalayan Region: Status, Constraints and Potential"*; ICAR: Dehradun, India, 2011; pp. 1–22.

51. Adhikari, R.N.; Patil, A.; Raizada, D.; Ramajayam, M.; Prabhavathi, N.; Mondal, B.; Mishra, P.K. *Participatory Resource Conservation and Management in Semi-arid India—A Case Study from Netranahalli Watershed (Karnataka)*; Technical Bulletin; Central Soil & Water Conservation Research & Training Institute, Research Centre: Bellary, Karnataka, India, 2013; p. 65.

52. Natarajan, A.; Janakiraman, M.; Manoharan, S.; Kumar, A.K.S.; Vadivelu, S.; Sarkar, D. Assessment of land degradation and its impact on land resources of Sivagangai block, Tamil Nadu, India, 2010. In *Land Degradation, and Desertification: Assessment, Mitigation and Remediation*; Zdruli, P., Pagliai, M., Kapur, S., Faz Cano, A., Eds.; Springer Science: Berlin, Germany; pp. 235–252.

53. Kundu, S.; Bhattacharyya, R.; Ved-Prakash; Ghosh, B.N.; Gupta, H.S. Carbon sequestration and relationship between carbon addition and storage under rainfed soybean–wheat rotation in a sandy loam soil of the Indian Himalayas. *Soil Tillage Res.* **2007**, *92*, 87–95.

54. Bhattacharyya, R.; Prakash, V.; Pandey, S.C.; Kundu, S.; Srivastva, A.K.; Gupta, H.S. Effect of fertilization on carbon sequestration in soybean-wheat rotation under two contrasting soils and management practices in the Indian Himalayas. *Aust. J. Soil Res.* **2009**, *47*, 592–601.
55. Bhattacharyya, R.; Kundu, S.; Srivastva, A.K.; Gupta, H.S.; Prakash, V.; Bhatt, J.C. Long term fertilization effects on soil organic carbon pools in a sandy loam soil of the Indian Himalayas. *Plant Soil* **2011**, *341*, 109–124.
56. Sharda, V.N.; Sharma, N.K.; Mohan, S.C.; Khybry, M.L. Green manuring for conservation and production in western Himalayas: 2. Effect on moisture conservation, weed control and crop yields. *Indian J. Soil Conserv.* **1999**, *27*, 31–35.
57. Bhattacharyya, R.; Tuti, M.D.; Bisht, J.K.; Bhatt, J.C.; Gupta, H.S. Conservation tillage and fertilization impacts on soil aggregation and carbon pools in the Indian Himalayas under an irrigated rice-wheat rotation. *Soil Sci.* **2012**, *177*, 218–228.
58. Bhattacharyya, R.; Tuti, M.D.; Kundu, S.; Bisht, J.K.; Bhatt, J.C. Conservation tillage impacts on soil aggregation and carbon pools in a sandy clay loam soil of the Indian Himalayas. *Soil Sci. Soc. Am. J.* **2012**, *76*, 617–627.
59. Bhattacharyya, R.; Pandey, S.C.; Bisht, J.K.; Bhatt, J.C.; Gupta, H.S.; Titi, M.D.; Mahanta, D.; Mina, B.L.; Singh, R.D.; Chandra, S.; *et al.* Tillage and irrigation effects on soil aggregation and carbon pools in the Indian sub-Himalayas.*Agron. J.* **2013**, *105*, 101–112.
60. Mandal, B.; Majumder, B.; Bandopadhyay, P.K.; Hazra, G.C.; Gangopadhyay, A.; Samantaroy, R.N.; Misra, A.K.; Chowdhuri, J.; Saha, M.N.; Kundu, S. The potential of cropping systems and soil amendments for carbon sequestration in soils under long-term experiments in subtropical India. *Global Change Biol.* **2007**, *13*, 357–369.
61. Srinivasarao, C.H.; Ravindra Chary, G.; Venkateswarlu, B.; Vittal, K.P.R.; Prasad, J.V.N.S.; Singh, S.R.S.K.; Gajanan, G.N.; Sharma, R.A.; Deshpande, A.N.; Patel, J.J.; *et al. Carbon Sequestration Strategies under Rainfed Production Systems of India*; Central Research Institute for Dryland Agriculture, Hyderabad (ICAR): Hyderabad, India, 2009; p. 102.
62. CSWCR&TI. *Annual Report of the Central Soil Water Conservation Research and Training Institute*; CSWCR&TI: Dehradun, India, 2012.
63. Kannan, K.; Khola, O.P.S.; Selvi, V.; Singh, D.V.; Moharnraj, R. *Agronomical Management Practices for Higher Productivity, Resource Use Efficiency and Farm Income in Semi-Arid Region-A Case Study in Ayalur Watershed*; Technical Bulletin; Central Soil & Water Conservation Research & Training Institute, Research Centre, Udhagamandalam: Tamil Nadu, India, 2013; p. 20.
64. Sharma, P.D.; Sarkar, A.K. *Managing Acid Soils for Enchancing Productivity*; Technical Bulletin; NRM Division, KAB-II, Pusa Campus: New Delhi, India, 2005; p. 23.
65. Fageria, N.K.; Baligar, V.C. Ameliorating soil acidity of tropical Oxisols by liming for sustainable crop production.*Adv. Agron.* **2008**, *99*, 345–399.
66. Bhat, J.A.; Mandal, B.; Hazra, G.C. Basic slag as a liming material to ameliorate soil acidity in Alfisols of sub-tropical India. *Am.-Euras. J. Agric. Environ. Sci.* **2007**, *2*, 321–327.
67. Bhat, J.A.; Kundu, M.C.; Hazra, G.C.; Santra, G.H.; Mandal, B. Rehabilitating acid soils for increasing crop productivity through low-cost liming material. *Sci. Total Environ.* **2010**, *408*, 4346–4353. [PubMed]
68. Wilson, C.E.; Keisling, T.C., Jr.; Miller, D.M.; Dillon, C.R.; Pearce, A.D.; Frizzell, D.L.; Counce, P.A. Tillage influence on soluble salt movement in silt loam soils cropped to paddy rice. *Soil Sci. Soc. Am. J.* **2000**, *64*, 1771–1776.
69. Garg, V.K. Interaction of tree crops with a sodic soil environment: Potential for rehabilitation of degraded environments. *Land Degrad. Dev.* **1998**, *9*, 81–93.
70. Bhojvaid, P.P.; Timmer, V.R. Soil dynamics in an age sequence of *Prosopis juliflora* planted for sodic soil restoration in India. *Forest Ecol. Manage.* **1998**, *106*, 181–193.
71. Gupta, R.K.; Rao, D.L.N. Potential of wasteland for sequestering carbon by reforestation. *Curr. Sci.* **1994**, *66*, 378–380.
72. Gupta, R.K.; Abrol, I.P. *Salt-Affected Soils. Their Reclamation and Management for Crop Production*; Lal, R., Stewart, B.A., Eds.; Springer: New York, NY, USA, 1990; pp. 223–288.
73. Dhanushkodi, V.; Subrahmaniyan, K. Soil management to increase rice yield in salt affected coastal soil-a review. *Int. J. Res. Chem. Environ.* **2012**, *2*, 1–5.
74. Prapagar, K.; Indraratne, S.P.; Premanandharajah, P. Effect of soil amendments on reclamation of saline-sodic soil.*Trop. Agric. Res.* **2012**, *23*, 168–176.
75. Bandyopadhyay, A.K. Effect of Lime, Superphosphate, Powdered Oystershell, Rock phosphate and Submergence on Soil Properties and Crop Growth in Coastal Saline Acid Sulphate Soils of Sundarbans. In Proceedings of the International Symposium on Rice Production on Acid Soil of the Tropics, 26–30 June 1989.
76. Kumar, R.; Singh, R.D.; Sharma, K.D. Water resources of India. *Curr. Sci.* **2005**, *89*, 794–811.
77. Moukhtar, M.M.; El-Hakim, M.H.; Abdel-Mawgoud, A.S.A.; Abdel-Aal, A.I.N.; El-Shewikh, M.B.; Abdel-Khalik, M.I.I. Drainage and Role of Mole Drains for Heavy Clay Soils under Saline Water Table, Egypt. In Proceedings of the 9th ICID International Drainage Workshop, Utrecht, The Netherlands, 10–13 September 2003.
78. Abdel-Mawgoud, S.A.; El-Shewikh, M.B.; Abdel-Aal, A.I.N.; Abdel-Khalik, M.I.I. Open drainage and moling for desalinization of salty clay soils of Northeastern Egypt. In Proceedings of the 9th International Drainage Workshop, Utrecht, The Netherlands, 10–13 September 2003.

79. Mukhopadhyay, D.; Mani, P.K.; Sanyal, S.K. Effect of phosphorus, arsenic and farmyard manure on arsenic availability in some soils of West Bengal. *J. Indian Soc. Soil Sci.* **2002**, *50*, 56–61.

80. Ghosh, K.; Das, I.; Saha, S.; Banik, G.C.; Ghosh, S.; Maji, N.C.; Sanyal, S.K. Arsenic chemistry in groundwater in the Bengal Delta Plain: Implications in agricultural system. *J. Indian Chem. Soc.* **2004**, *81*, 1–10.

81. Elliot, P.; Grandner, J.; Allen, D.; Butcher, G. Completion Criteria for Alcoa of Australia Limited Bauxite Mine Rehabilitation. In Proceedings of 3rd International and 21st Annual Minerals Council of Australia Environmental Workshop, Newcastle, Australia, 14–18 October 1996.

82. Sharma, R.K.; Babu, K.S.; Chhokar, R.S.; Sharma, A.K. Effect of tillage on termites, weed incidence and productivity of spring wheat in rice-wheat system of North Western Indian plains. *Crop Prot.* **2004**, *23*, 1049–1054.

83. Sen, H.S.; Oosterbaan, R.J. *Research on Water Management and Control in the Sunderbans, India*; Annual Report; ILRI: Wageningen, The Netherlands, 1992.

84. Ambast, S.K.; Sen, H.S. Integrated water management strategies for coastal ecosystem. *J. Indian Soc. Coastal Agric. Res.* **2006**, *24*, 23–29.

85. Bandyopadhyay, K.K.; Ghosh, P.K.; Hati, K.M.; Misra, A.K. Efficient utilization of limited available water in wheat through proper irrigation scheduling and integrated nutrient management under different cropping systems in a Vertisol. *J. Indian Soc. Soil Sci.* **2009**, *57*, 121–128.

86. Biswas, A.K.; Mohanty, M.; Hati, K.M.; Misra, A.K. Soil organic carbon and aggregate stability effects of applied distillery effluents on a vertisol in India. *Soil Tillage Res.* **2009**, *104*, 241–246.

87. Grewal, S.S. Agroforestry systems for soil and water conservation in Shivaliks. In *Agroforestry in 2000 AD for the Semi-Arid and Arid Tropics*; NRC for Agroforestry: Janshi, India, 1993; pp. 82–85.

88. Ghosh, B.N. *Vegetative Barriers for Erosion Control in Western Himalayan Region*; CSWCR&TI: Dehradun, India, 2009; pp. 1–8.

89. Bhattacharyya, R.; Fullen, M.A.; Booth, C.A.; Kertesz, A.; Toth, A.; Kozma, K.Z.; Jakab, G.; Jankauskas, B.; Jankauskiene, G. Effeciveness of biological geotextiles on soil and water conservation in different agro-environments.*Land Degrad. Dev. Special Issue* **2011**, *22*, 495–504.

90. Bhattacharyya, R.; Zheng, Y.; Li, Y.; Tang, L.; Panomtarachichigul, M.; Peukrai, S.; Dao, C.T.; Tran, H.C.; Truong, T.T.; Jankauskas, B.; *et al.* Effects of biological geotextiles on aboveground biomass production in selected agro-ecosystems.*Field Crops Res.* **2012**, *126*, 23–36.

91. Fullen, M.A.; Subedi, M.; Booth, C.A.; Sarsby, R.W.; Davies, K.; Bhattacharyya, R.; Kugan, R.; Luckhurst, D.A.; Han, K.; Black, A.W.; *et al.* Utilizing biological geotextiles: Introduction to the BORASSUS Project and global perspectives.*Land Degrad. Dev. Spec. Issue* **2011**, *22*, 453–462.

92. Smets, T.; Poesen, J.; Bhattacharyya, R.; Fullen, M.A.; Subedi, M.; Booth, C.A.; Kertesz, A.; Toth, A.; Szalai, Z.; Jakab, G.; *et al.* Evaluation of biological geotextiles for reducing runoff and soil loss under various environmental conditions using laboratory and field plot data. *Land Degrad. Dev. Special Issue* **2011**, *22*, 480–494.

93. Bhattacharyya, R.; Fullen, M.A.; Booth, C.A. Using palm-mat geotextiles on an arable soil for water erosion control in the UK. *Earth Surf. Process. Landforms* **2011**, *36*, 933–945.

94. Bhattacharyya, R.; Fullen, M.A.; Booth, C.A.; Smets, T.; Poesen, J.; Black, A. Using palm mat geotextiles for soil conservation. I. Effects on soil properties. *Catena* **2011**, *84*, 99–107.

95. Bhattacharyya, R.; Fullen, M.A.; Davies, K.; Booth, C.A. Use of palm-mat geotextiles for rainsplash erosion control.*Geomorph.* **2010**, *119*, 52–61.

96. Bhattacharyya, R.; Fullen, M.A.; Davies, K.; Booth, C.A. Utilizing palm leaf geotextile mats to conserve loamy sand soil in the United Kingdom. *Agric. Ecosyst. Environ.* **2009**, *130*, 50–58.

97. Narain, P.; Singh, R.K.; Sindhwal, N.S.; Joshi, P. Agroforestry for soil and water conservation in the western Himalayan valley region of India. *Agrofor. Syst.* **1998**, *39*, 191–203.

98. Mahmud, M.K.; Chowdhury, N.H.; Md-Manjur. Mitigation of soil erosion with Jute geotextile aided by vegetation cover: Optimization of an integrated tactic for sustainable soil conservation system (SSCS). *Glob. J. Res. Eng. Civil Struct. Eng.* **2012**, *12*, 8–14.

99. Nair, P.K.R. *An Introduction to Agro Forestry*; Kluwer Academic Publishers: Dordrecht, The Netherland, 1993.

100. Ramesh, P.; Panwar, N.R.; Singh, A.B.; Ramana, S.; Yadav, S.K.; Shrivastava, R. Status of organic farming in India.*Curr. Sci.* **2010**, *98*, 1190–1194.

101. Nair, T. India to launch a brave new initiative to save the Critically Endangered Gharial. *SPECIES–Mag. Spec. Surv. Comm.* **2011**, *21*, 53.

102. Rosenstock, T.S.; Tully, K.L.; Arias Navarro, C.; Neufeldt, H.; Butterbach Bahl, K.; Verchot, L.V. Agroforestry with N2-fixing trees: Sustainable development's friend or foe? *Curr. Opin. Environ. Sustain.* **2014**, *6*, 15–21.

103. Singh, G. The Role of *Prosopis* in reclaiming high-pH soils and in meeting firewood and forage needs of small farmers. In *Prosopis: Semi-Arid Fuelwood and Forage: Tree Building Consensus for the Disenfranchised*; US National Academy of Science: Washington, DC, USA, 1996.

104. Mishra, A.; Sharmaa, S.D.; Pandeyb, R. Amelioration of degraded sodic soil by afforestation. *Arid Land Res. Manag.* **2004**, *18*, 13–23.

105. Kaur, B.; Gupta, S.R.; Singh, G. Soil carbon, microbial activity and nitrogen availability in agroforestry systems on moderately alkaline soils in Northern India. *Appl. Soil Ecol.* **2000**, *15*, 283–294.

106. Basavaraja, P.K.; Sharma, S.D.; Dhananjaya, B.N.; Badrinath, M.S. *Acacia nilotica*: A tree species for amelioration of sodic soils in Central dry zone of Karnataka, India. In Proceedings of the 19th World Congress of Soil Science, Soil Solutions for a Changing World, Brisbane, Australia, 1–6 August 2010; pp. 73–76.

107. Mishra, A.; Sharma, S.D.; Khan, G.H. Improvement in physical and chemical properties of sodic soil by 3, 6, and 9 year-old plantations of *Eucalyptus tereticornis* bio-rejuvenation of soil. *For. Ecol. Manag.* **2003**, *184*, 115–124.

108. Singh, A.; Kaur, J. Impact of conservation tillage on soil properties in rice wheat cropping system. *Agric. Sci. Res. J.* **2012**, *2*, 30–41.

109. Food and Agriculture Organization (FAO). *"Climate-Smart" Agriculture. Policies, Practices and Financing for Food Security, Adaptation and Mitigation*; FAO: Rome, Italy, 2010; p. 41.

110. Bhattacharyya, R.; Das, T.K.; Pramanik, P.; Ganeshan, V.; Saad, A.A.; Sharma, A.R. Impacts of conservation agriculture on soil aggregation and aggregate-associated N under an irrigated agroecosystem of the Indo-Gangetic Plains. *Nutr. Cycl. Agro-Ecosyst.* **2013**, *96*, 185–202.

111. VPKAS. *Annual Report of the Vivekananda Parvatiya Krishi Anusandhan Sansthan*; Vikas Publisher: Almora, India, 2011.

112. Bhattacharyya, R.; Kundu, S.; Pandey, S.; Singh, K.P.; Gupta, H.S. Tillage and irrigation effects on crop yields and soil properties under rice-wheat system of the Indian Himalayas. *Agric. Water Manag.* **2008**, *95*, 993–1002.

113. Jat, M.L.; Saharawat, Y.S.; Gupta, R. Conservation agriculture in cereal systems of South Asia: Nutrient management perspectives. *Karnataka J. Agric. Sci.* **2011**, *24*, 100–105.

114. RWC-CIMMYT. *Addressing Resource Conservation Issues in Rice-Wheat Systems of South Asia: A Resource Book*; Rice-wheat consortium for the Indo-Gangetic Plains–International Maize and Wheat Improvement Centre: New Delhi, India, 2003; p. 305.

115. Gathala, M.K.; Ladha, J.K.; Saharawat, H.S.; Kumar, V.; Sharma, P.K. Effect of tillage and crop establishment methods on physical properties of a medium-textured soil under a seven-year rice–wheat rotation. *Soil Sci. Soc. Am. J.* **2011**, *75*, 1851–1862.

116. Hobbs, P.R.; Gupta, R.K. Sustainable resource management in intensively cultivated irrigated rice-wheat cropping systems of Indo-Gangetic plains of south Asia: Strategies and options. In Proceedings of the International Conference on Managing Natural Resources for Sustainable Production in 21st Century, New Delhi, India, 14–18 February 2000; Singh, A.K., Ed.; Indian Society of Soil Science: New Delhi, India; pp. 584–592.

117. Ladha, J.K.; Kumar, V.; Alam, M.M.; Sharma, S.; Gathala, M.K.; Chandna, P.; Saharawat, Y.S.; Balasubramanian, V. Integrating crop and resource management technologies for enhanced productivity, profitability and sustainability of the rice–wheat system in South Asia. In *Integrated Crop and Resource Management in the Rice–Wheat System of South Asia*; Ladha, J.K., Singh, Y., Erenstein, O., Hardy, B., Eds.; IRRI: Los Banos, The Philippines, 2009; pp. 69–108.

118. Das, T.K.; Bhattacharyya, R.; Sudhishri, S.; Sharma, A.R.; Saharawat, Y.S.; Bandyopadhyay, K.K.; Sepat, S.; Bana, R.S.; Aggarwal, P.; Sharma, R.K.; *et al.* Conservation agriculture in an irrigated cotton–wheat system of the western Indo-Gangetic Plains: Crop and water productivity and economic profitability. *Field Crops Res.* **2014**, *158*, 24–33.

119. Jat, M.L.; Gathala, M.K.; Saharawat, Y.S.; Terawal, J.P.; Gupta, R.; Singh, Y. Double no-till and permanent raised beds in maize-wheat rotation of north-western Indo-Gangetic plains of India: Effects on crop yields, water productivity, profitability and soil physical properties. *Field Crops Res.* **2013**, *149*, 291–299.

120. Das, T.K.; Bhattacharyya, R.; Sharma, A.R.; Das, S.; Pathak, H. Impacts of conservation agriculture on total soil organic carbon retention potential under an irrigated agro-ecosystem of the western Indo-Gangetic Plains. *Eur. J. Agron.* **2013**,*51*, 34–42.

121. Mishra, A.K.; Aggarwal, P.; Bhattacharyya, R.; Das, T.K.; Sharna, A.R.; Singh, R. Least limiting water range for two conservation agriculture cropping systems in India. *Soil Tillage Res.* **2015**, *150*, 43–56.

122. Hati, K.M.; Chaudhary, R.S.; Mohanty, M.; Singh, R.K. Impact of conservation tillage on soil organic carbon content, its distribution in aggregate size fractions and physical attributes of Vertisols. In *IISS Contribution in Frontier Areas of Soil Research*; Kundu, S., Manna, M.C., Biswas, A.K., Chaudhary, R.S., Lakaria, B.L., Subba Rao, A., Eds.; Indian Institute of Soil Science: Bhopal, India, 2013; pp. 187–200.

123. Jat, R.A.; Wani, S.P.; Sahrawat, K.L. Conservation Agriculture in the Semi-Arid Tropics: Prospects and Problems. *Adv. Agron.* **2012**, *117*, 191–273.

124. Gathala, M.K.; Kumar, V.; Sharma, P.C.; Yashpal, S.; Saharawat, H.S.; Jat, M.S.; Kumar, A.; Jat, M.L.; Humphreys, E.; Sarma, D.K.; *et al.* Optimizing intensive cereal-based cropping systems addressing current and future drivers of agricultural change in the northwestern Indo-Gangetic plains of India. *Agric. Ecosyst. Environ.* **2013**, *177*, 85–97.

CHAPTER 13

THE ROLE OF SOIL TEST INFORMATION IN REDUCING GROUNDWATER POLLUTION

Ronald A. Fleming, Richard M. Adams and David E. Ervin

Keywords: Dynamic modeling, economic impacts, groundwater, hydrology, nitrate, nitrogen fertilizer, soil test, water policy

INTRODUCTION

Soil tests provide information on the amount of nitrogen and other nutrients available for crop consumption. This information is valuable to producers if used tos avoid the application of more nitrogen fertilizer than can be used by a crop. Eliminating excess fertilizer reduces crop production cost if fertilizer savings exceed the cost of soil testing. Reducing total fertilizer applications also may reduce the leaching of nutrients and improve groundwater quality. In agricultural regions with groundwater quality problems, producers may avoid environmental regulation by taking advantage of information that can lead to reductions in nutrient leaching and consequential improvements in groundwater quality.

Past studies have shown that soil test information can be valuable to producers. However, these studies tend to focus on cost savings, with little mention of the potential for improved groundwater quality. For example, Adams, Farris, and Menkhaus show that soil tests reduce nitrogen applications and improve per acre farm profit by $20 to $50 for sugar beets in Wyoming. A number of studies of Iowa, Nebraska, and Pennsylvania corn producers show that soil testing reduces average nitrogen fertilizer applications 15-41% while increasing per acre return (Babcock and Blackmer; Babcock, Carriquiry, and Stern; Fuglie and Bosch; Musser et al.). In an investigation of the Corn Belt, Lake States, and Northern Plains of the U.S., findings show a 25% reduction in nitrogen fertilizer reduced nitrate runoff and leaching while significantly increasing total net returns (Wu, Lakshminarayan, and Babcock).

The effect of soil testing on nitrogen applications is only an indicator of environmental performance (Musser et al.). Measuring actual environmental performance requires linking farm-level input decisions with ambient levels of an environmental contaminant. Because of the difficulty in linking nitrogen input decisions to groundwater nitrate concentration, to our knowledge no study has quantified changes in groundwater quality as a consequence of soil testing. While Babcock and Blackmer do not measure the value of changes in nitrate contamination of groundwater, they broach the possibility that soil testing may lower nitrogen applications sufficiently to decrease the likelihood of direct regulation.

Citation: Fleming, R., R. Adams, and D. Ervin. 1998. "The Role of Soil Test Information in Reducing Groundwater Pollution," Journal of Agr. and Resource Economics 23(1): 20-38. http://pdxscholar.library.pdx.edu/cgi/viewcontent.cgi?article=1030&context=econ_fac

The purpose of this analysis is to assess the potential for soil tests to improve ambient groundwater quality and producer profit. The efficacy of soil testing is assessed using a spatially distributed, dynamic simulation model which links changes in economic behavior as a result of soil testing with the physical processes that determine groundwater quality. By directly linking soil testing to ambient groundwater quality, we are able to examine the role of soil tests in reducing the demand for direct regulation.

STUDY REGION, GROUNDWATER CONCERNS, AND USE OF SOIL TESTS

The empirical focus is an irrigated agricultural region in Malheur County, Oregon. Located in eastern Oregon, this is a high desert environment; annual precipitation ranges from 5-16 inches, with an average of 10 inches. Irrigation is required for crop production, and surface (flood) irrition is the principal method of application. The specific region evaluated in this investigation lies within the land area bounded by the Snake and Malheur Rivers and encompasses 32 square miles, of which 17,860 acres (28 square miles) are farmed.

While many crops (including fruit, vegetables, and seed crops) are grown here, we focus on the five major crops in terms of acreage: soft white spring wheat, onions, potatoes, sugar beets, and hay (a composite of meadow hay and alfalfa). These five crops represent 72% of regional crop acreage and 54% of total crop sales in 1992 [Malheur County Extension Service (MCES)]. Onions, potatoes, and sugar beets have a large impact on the economy in terms ofjobs created by processing, handling, and field labor. Onions are the most valuable crop, accounting for 6% of regional acreage and 25% of crop sales in 1992.

Between August of 1988 and April of 1990, 199 wells were sampled; 32% had nitrate levels which exceeded the federal standard of 10 ppm for municipal water supplies. Because of these groundwater quality problems, the area was designated as a Groundwater Management Area (GMA) by the Oregon Department of Environmental Quality (ODEQ). This motivated several years of extensive data collection and analysis of geo-hydrologic conditions of the region.

An action plan was developed which relies on a voluntary approach to reduce nitrate concentrations through use of "best management practices." If nitrate levels at all monitoring wells do not reach 10 ppm by July 1, 2000, a regulatory approach with unstated mandatory action will be considered [Malheur County Groundwater Management Committee (MCGMC)].

Soil testing is identified in the plan as a method for reducing groundwater nitrate concentration. In response, the local Extension Service developed educational programs that encourage producers to test soils for nitrogen before applying fertilizer. These educational programs targeted onion and wheat producers. To date, 10% of the wheat fields and 80% of the onion fields are soil tested.

Soil tests cost $15 per sample for nitrogen and $35 per sample for a complete nutrient profile which includes nitrogen, phosphorous, potassium, soil organic material, and minor (or trace) nutrients. One sample is taken per field, where a sample consists of numerous random probes throughout the field.

A PROGRAMMING MODEL OF SOIL TESTING

While a conceptual model can provide insights into the potential impacts of soil testing, practical application of such a model is limited. For example, the impact of soil testing on groundwater quality is inferred, not specifically measured. Furthermore, the physical aspects of crop production and groundwater contamination are complex. Groundwater flow rates, the direction of groundwater flow, soils, crops grown, and crop production technology all vary across the region. Assessing the agronomic and environmental impacts of a policy that requires all fields to be tested for nitrogen requires a modeling approach capable of capturing these factors. A mathematical simulation or programming model that accounts for spatial linkages and variability in the physical parameters which determine groundwater quality is one means to assess the impact of soil testing on groundwater pollution.

Simulation or programming models which integrate geo-hydrology data with economic optimization models are commonly used to evaluate policies to control groundwater quality (Johnson, Adams, and Perry; Mapp et al.; Chowdhury and Lacewell). The integrated model developed here is a spatially distributed, dynamic simulation model and consists of three submodels: (a) economic, (b) soil water solute transport, and (c) groundwater solute transport. In the economic submodel, producers choose nitrogen fertilizer application rates to maximize profits subject to production constraints and assumptions concerning residual soil nitrogen. Results from the economic submodel become input parameters in the soil water solute transport submodel, which describes movement of water and nitrogen through the unsaturated or vadose zone of soil. This submodel also measures the level of residual soil nitrogen used in the economic submodel in the next period. Results from the soil water solute transport submodel become input parameters in the groundwater solute transport submodel, which tracks loading and movement of nitrates in the aquifer. With this last model, changes in optimal nitrogen input levels due to soil test information can be linked to changes in spatial groundwater nitrate concentration. This latter model component is an outgrowth of the extensive groundwater studies performed in the area because of its GMA status.

The Economic Submodel

Equations (1), (14), and (15) below are the primary, formal mathematical expressions of the model. Within this set of equations, equation (1) defines the economic submodel. Specifically, nitrogen input ($n_{c,s,t}$) and the crops to which this nitrogen is applied ($a_{c,s,t}$) are chosen to maximize profit (or net farm income) through time, subject to a level of residual soil nitrogen ($rsn_{c,s,t}$) that is measured or assumed and a series of production constraints. The production constraints are used to limit total nitrogen [equation (2)], fix maximum yield [equation (3)], restrict crop acreage [equation (4)], and determine crop rotation [equations (7)-(12)]. Onions, potatoes, and sugar beets are high-value crops, but with high yield and revenue variance. Hence, crop rotations arise as a result of crop diversification. Crop rotations also are used for weed, disease, and crop pest control, and as a form of nutrient management. Finally, because the model must account for crop rotations through time, both current-period and cross-period [equations (5) and (6)] crop acreage restrictions are necessary.

The economic submodel is specified in equations (1)-(12) below:

Maximize the objective function,

$$\underset{n_{c,s,t};a_{c,s,t};\lambda^{sw}_{c,s,t};\lambda^{gw}_{i,j,t};c^{sw}_{c,s,t};c^{gw}_{i,j,t}}{\text{Max } \Pi}$$

$$= \sum_{t=1}^{T}\left(\sum_{s=1}^{S}\sum_{c=1}^{C}\left[P_c Q(n_{c,s,t} + rsn_{c,s,t})_{c,s,t}\, qadj_c - np\cdot n_{c,s,t} - cfc_c\right]a_{c,s,t} - IC\right) \times \frac{1}{(1+r)^t}, \tag{1}$$

subject to the following production constraints:

$$n_{c,s,t} + rsn_{c,s,t} = ntot_{c,s,t} \le nmax_{c,s} \quad \forall\ c, s, \text{ and } t, \tag{2}$$

$$Q\left(ntot_{c,s,t}\right)_{c,s,t} \le cay_{c,s} \quad \forall\ c, s, \text{ and } t, \tag{3}$$

$$a_{c=w,s,t} + a_{c=o,s,t} + a_{c=p,s,t} + a_{c=s,s,t}\, a_{c=h,s,t} \le 40 \quad \forall\ s \text{ and } t, \tag{4}$$

$$a_{c=o,s,t-1} + a_{c=p,s,t-1} \le a_{c=w,s,t} + a_{c=s,s,t} + a_{c=h,s,t} \quad \forall\ s \text{ and } t, \tag{5}$$

$$a_{c=o,s,t} + a_{c=p,s,t} \le a_{c=w,s,t-1} + a_{c=s,s,t-1} + a_{c=h,s,t-1} \quad \forall\ s \text{ and } t, \tag{6}$$

$$a_{c=o,s,t} + a_{c=p,s,t} \le a_{c=w,s,t} + a_{c=s,s,t} \quad \forall\ s \text{ and } t, \tag{7}$$

$$a_{c=o,s,t} \le a_{c=p,s,t} \quad \forall\ s \text{ and } t, \tag{8}$$

$$a_{c=s,s,t} \le a_{c=w,s,t} \quad \forall\ s \text{ and } t, \tag{9}$$

$$0.25 a_{c=w,s=A,t} \le a_{c=h,s=A,t} \quad \forall\ t, \tag{10}$$

$$a_{c=w,s=C,t} + a_{c=s,s=C,t} \le a_{c=h,s=C,t} \quad \forall\ t, \tag{11}$$

and

$$0.25\left(a_{c=w,s=D,t} + a_{c=s,s=D,t}\right) \le a_{c=h,s=D,t} \quad \forall\ t. \tag{12}$$

The crop rotation constraints used here yield soil-specific crop mixes consistent with those observed prior to 1990 (when the area was listed as a Groundwater Management Area). More importantly, without these constraints, the model would not "grow" potatoes on onion/potato acreage, and wheat and hay would not be "grown" on sugar beet/wheat/hay acreage because these crops are less valuable. Also, nitrogen input is restricted to avoid an unbounded solution [equation (2)].

Common symbols across all equations include the c subscript, which denotes crop type (w = wheat, o = onions, p = potatoes, s = sugar beets, and h = hay); s, which represents the various production units differentiated by soil type (soil groups A, B, C, and D); and t, time. In equation (1), λ^{sw} and λ^{gw} are the co-state or shadow values for soil water and groundwater nitrates, c^{sw} and c^{gw} are the state or stock values for soil water and groundwater nitrates, P is crop price, Q is crop yield, qadj is an adjustment to crop yield resulting from storage losses, np is nitrogen fertilizer price, cfc represents crop production fixed costs, IC is information cost (cost of the soil test), and r is the discount rate. In equation (2), $ntot_{c,s,t}$ is the total amount of nitrogen (applied nitrogen $n_{c,s,t}$t, and residual soil nitrogen $rsn_{c,s,t}$) available for crop consumption. Total available nitrogen ($ntot_{c,s,t}$) cannot exceed $nmax_{c,s}$, the maximum amount of nitrogen needed to maximize crop yield. In equation (3), $cay_{c,s}$, is crop yield.

The agricultural sector consists of a large number of producers who use nitrogen fertilizer to produce the five major crops grown in the area. Soils in the study region are grouped into four general classifications, and each soil grouping produces a different mix of crops. Specifically, soil group A is planted mainly to wheat and hay (because of slope), soil group B is planted predominately to row crops (onions, potatoes, and sugar beets in rotation with wheat), soil group C is planted predominately to wheat, sugar beets, and hay (because of poor drainage), and soil group D is planted to wheat, potatoes, and hay. Differences in slope and soil type limit the productivity of onions, potatoes, and sugar beets on some soils. Hence, shifting crops to different soils as a strategy to reduce nitrogen losses is not practical. Finally, the soil zones are geographically contiguous and sloped such that groundwater from A flows through B to soil zones C and D, where it is discharged to the Snake (the east border of soil zones C and D) and the Malheur (the northern border of soil zone C) Rivers.

The basic building block in the groundwater solute transport submodel is a 40-acre "production unit" around which the economic submodel is constructed. Specifically, "representative" farms of 360 acres are composed of nine production units of similar soil type. All representative farms within a given soil zone are treated as identical (i.e., same-crop alternatives, management strategies, production technology, input use, returns, and costs), based on information from agricultural extension personnel. While each production unit within a soil zone is uniform with respect to production and soil water concentration (mentioned below), groundwater nitrate concentration beneath each can vary significantly.

Crop yield-nitrogen response (production) functions [$Q(ntot_{c,s,t})_{c,s,t}$ in equation (3)] are an important component of the economic submodel. Unfortunately, experimental data are not adequate to estimate statistical nitrogen-yield relationships for the study area, and well-established simulation models such as EPIC or CERES were not available for onions, potatoes and sugar beets at the time of this investigation.

There is considerable debate concerning what constitutes an appropriate fertilizer response function. Griffin, Montgomery, and Rister, in their review of different functional forms, conclude that none of the many selection criteria guarantee that the true relationship will be discovered, nor do any allow a totally objective choice to be made. Some researchers argue that the response is a smooth, concave function, while others contend that the correct function is the linear response and plateau (LRP) or von Liebig model (Berck and Helfand). Berck and Helfand, while noting that theoonal two functi forms are not mutually exclusive, conclude that aggregating across fields (as is done in this investigation) results in a response curve where yield is a smooth, concave function of nitrogen input.

Many functional forms, including the quadratic, are smooth, concave functions. An advantage of the quadratic function is that known endpoints can be used to mathematically derive the curve (Debertin; Fleming and Adams). However, some researchers object to the use of a smooth, concave function given that yields of some crops do not decrease with added nitrogen, as suggested by a concave production function (Babcock). Furthermore, uncertainty about growing conditions translates into uncertainty about the yield plateau and the amount of nitrogen at which this plateau is achieved. A smooth, concave function is ill suited to account for this uncertainty.

The nitrogen-yield functions used here are LRP, similar to those proposed by Babcock. Uncertainty in residual soil nitrogen [$rsn_{c,s,t}$ in equations (1), (2), and (13)] is incorporated in t the function. This uncertainty is not expected to impact the plateau (maximum) yield. Specifically, without a soil test, the producer must guess how much nitrogen is provided by soil, and then supplement with commercial nitrogen. In practice, producers monitor crop yields. If yields are low, then they infer that their estimate of rsn is too high; hence they lower their expectations and apply more nitrogen in the next year. Over time, producers come to "understand" their soils. However, for high-value crops like onions, the revenue loss associated with a small reduction in yield due to insufficient nitrogen induces growers to apply excess nitrogen.

Following Babcock, crop yield ($Q_{c,s,t}$) is a linear function of total available nitrogen ($ntot_{c,s,t}$) up to some maximum yield, after which yield is constant [equation (13)]. Here it is assumed that the soil-specific, long-run county average yield (cay) for any crop reflects the profit-maximizing yield. Given the "kinked" shape of the LRP production function, with nonlimiting factors, profit will be maximized at the kink. Hence, the maximum yield is also the profit-maximizing yield, which is assumed to be the county average yield in this model. Long-run county average yields were obtained from the Malheur County offices of the Farm Services Administration and the Natural Resource Conservation Service (table 1).

From the intercept to cays,, crop yield ($Q_{c,s,t}$) is a linear function of total available nitrogen ($ntot_{c,s,t}$). All the LRP functions pass through the origin. From the origin, then, increased total available nitrogen increases crop yield. When crop yield reaches the county average ($cay_{c,s}$ at the "kink"), total available nitrogen is equal to maximum required nitrogen ($nmax_{c,s}$). Further applications of nitrogen are possible, but yield will not change. Maximum required nitrogen is derived mathematically: $nmax_{c,s}$, is equal to $cay_{c,s}$, divided by $b_{c,s}$, the slope of the yield-nitrogen (production) function prior to the plateau [equation (13) and table 1].

Table 1. Crop Production Data and Model Parameters

Model Parameters / Crop	Soil Zone A	B	C	D
Average Yield (*cay*):				
▸ Wheat	109	125	106	111
▸ Onions	—	625	—	—
▸ Potatoes	—	405	—	380
▸ Sugar Beets	—	31	27	—
▸ Hay	5	6	4	5
Nitrogen Leached (*SoilNL*):[a]				
▸ Wheat	15	52	28	27
▸ Onions	—	143	—	—
▸ Potatoes	—	94	—	68
▸ Sugar Beets	—	51	28	—
Slope Coefficient (β):[b]				
▸ Wheat	0.482	0.437	0.465	0.466
▸ Onions	—	1.900	—	—
▸ Potatoes	—	1.535	—	1.640
▸ Sugar Beets	—	0.088	0.091	—
Maximum N (*n*max) to Produce *cay*:				
▸ Wheat	226.3	286.1	228.0	238.1
▸ Onions	—	329.0	—	—
▸ Potatoes	—	263.8	—	231.7
▸ Sugar Beets	—	353.9	296.1	—

	Region Crop Data: Wheat	Onions	Potatoes	Sug. Beets
High Yield (*YMax*)[c]	174	1,017	626	51
Crop N Uptake (*cnu*) at *YMax*[c]	277	314	251	425

Note: The computer package GAMS calculates values to the 13th digit. Hence, there can be rounding error for β using the values in this table.

[a] Derived from Fleming (p. 100).

[b] Percent nitrogen lost to surface wind and water erosion (*%SurfNL*) is 25% (Fleming, p. 57).

[c] Summarized from Malheur Experiment Station data reported in Fleming (pp. 191–203).

The slope term ($b_{c,s}$) is derived from experimental and simulation data (Fleming) (table 1). Specifically, Pc is the experimentally measured maximum crop yield (YMax) divided by the quantity of nitrogen needed to obtain that yield [equation (13)]. Unfortunately, crop nitrogen uptake (cnu), not the quantity of nitrogen needed, is measured. However, mass balance (from soil physics) requires that the quantity needed equals crop uptake (cnu) plus that lost to surface wind and water erosion (%SurfNL) and deep percolation (SoilNL). Quantities for leached nitrogen (from deep percolation) were obtained from simulation data using the crops and soils in the study region (Fleming) (table 1). Simulation data were used because soil mineralization 2 and leached nitrogen were measured as a residual in the experimental data.

Equation (13) specifies the crop yield-nitrogen response (production) functions of the economic submodel:

$$Q_{c,s,t} = \beta_{c,s} ntot_{c,s,t} = \beta_{c,s}\left(n_{c,s,t} + rsn_{c,s,t}\right) \quad \text{for } 0 \le ntot_{c,s,t} \le n\max_{c,s},$$

$$= Q\max_{c,s} = cay_{c,s} \quad \text{for } ntot_{c,s,t} > n\max_{c,s},$$

$$n\max_{c,s} = \frac{cay_{c,s}}{\beta_{c,s}},$$

$$\beta_{c,s} = \frac{YMax_c}{cnu_c\left(1 + \dfrac{\%SurfNL}{100}\right) + SoilNL_{c,s}}. \tag{13}$$

While the slope of the production function in equation (13) is from physical data, the plateau of this function is for profit-maximizing yield ($cay_{c,s}$). This "adjustment" is necessary because the profit-maximizing solution is an interior corner-point. These production functions represent the regional response, not the frontier function, of an individual producer. Crop production technology is fixed through time (although technology, e.g., the type of irrigation system, does vary across soils in the model). The producer has flexibility with respect to nitrogen input and crop choice.

Regional crop enterprise budgets from the Oregon State University Extension Service were used to determine per acre crop production fixed costs [cfc in equation (1)]. Field and farm sizes are consistent with those assumed in the crop enterprise budgets, since enterprise fixed costs (which include, among other things, a machinery complement) are calculated for a specific field size. These budgets are based on best management practices; hence they may overstate fixed costs because area average yields are used. Constant returns to scale are assumed for each crop. The cost of testing soil for nitrogen [IC in equation (1)] is also fixed ($15 per field). Depending on field size, the per acre cost of soil testing ranges from $0.47 to $1.50 annually. For wheat and onions, where soil tests are not the norm, IC is initially set to zero.

The Soil Water and Groundwater Solute Transport Submodels

The soil water solute transport submodel determines leached soil water and nitrogen, given nitrate applications ($n_{c,s,t}$) the number of acres of each crop grown ($a_{c,s,t}$) from the economic submodel, and irrigation information. This submodel is based on the Nitrogen Leaching Simulator (NLEACH), version 3.0 simulation model developed in the Department of Bioresource Engineering, Oregon State University, to investigate the dynamics of nitrogen in irrigated soils (English). NLEACH is a one-dimensional, mass balance model that simulates the movement and transformation of nitrate-nitrogen in an irrigated crop system. This model is based on commonly accepted principles of nitrogen dynamics without adjustments for unique features of soil or the crop (Mills; Rao, Davidson, and Hammond).

Pulses of soil nitrogen are tracked by the soil water solute transport submodel to a depth of 48 inches. For potatoes and sugar beets, the historic mean crop yield (the plateau or profit-maximizing yield) reflects the long-time use of soil tests. For potatoes, a soil test is assumed to measure the nitrogen in the top foot of soil. Specifically, the simulated content of nitrogen in soil no more than 12 inches deep at the start of the growing season is summed and assigned to the variable residual soil nitrogen [$rsn_{c,s,t}$, in equations (1), (2), and (13)]. The process is similar for sugar beets, but nitrogen in the top two feet of soil is measured. Historically, onion and wheat fields were not soil tested. For these crops, the producer makes an informed guess concerning the value of $rsn_{c,s,t}$. Note that $rsn_{c,s,t}$ is exogenous to the optimization problem in the economic submodel regardless of whether it is estimated by the soil water submodel or guessed.

Equations (14) and (15) comprise the groundwater solute transport submodel:

$$c^{lgw}_{i,j,t,t_c} = c^{gw}_{i,j,t,t_c} + \left(\frac{wl_{i,j,t}}{deep \times n_a \times 12}\right)\left(\frac{dt_c}{365}c^{sw}_{i,j,t} - c^{gw}_{i,j,t,t_c}\right); \quad (4)$$

$$c^{gw}_{i,j,t,t_c+dt_c} = c^{lgw}_{i,j,t,t_c} - \frac{dt_c VI_{i,j}}{n_a di}\left(c^{lgw}_{i,j,t,t_c} - c^{lgw}_{i+di,j,t,t_c}\right)$$

$$- \frac{dt_c VJ_{i,j}}{n_a dj}\left(c^{lgw}_{i,j,t,t_c} - c^{lgw}_{i,j-dj,t,t_c}\right). \quad (15)$$

Leached nitrogen and soil water from NLEACH (in the pulses that reach 48 inches in depth) are converted to a concentration in parts per million (ppm or milligrams per liter). In equation (14), c^{lgw} is the concentration of nitrate in groundwater after loading, c^{gw} is the concentration of nitrate in groundwater before loading, c^{sw} is soil water nitrate concentration, wl is the depth of soil water leached (in inches), deep is the depth of the aquifer (in feet), and n_a is porosity or actual pore space containing groundwater. Groundwater nitrate concentration, hence nitrate loading, is calculated every five (dt_c) days.

The geo-hydrology model is based upon the work of Walker. Groundwater flow is modeled utilizing a mesh of grid points superimposed over a map of the study region. Each grid point is centered in a 40-acre production unit. Also, each point is characterized by a set of physical and production data (for the production unit) that is used to calculate groundwater nitrate concentration.

Nitrates in groundwater [equation (14)] are transported throughout the study region. It is assumed that (a) nitrogen is not chemically bound by soil, (b) nitrogen does not chemically decay through time, and (c) nitrogen (through a chemical process called dispersion) does not move upstream against the flow of groundwater. These assumptions give rise to equation (15), the reduced discrete direfully form of the explicit finite difference expression [for groundwater velocities in the i and j directions (VI_{ij} and VJ_{ij}) ^{3} 0]. Equation (15) is used to solve for groundwater nitrate concentration at each production unit.

PROCEDURES

The integrated simulation model is constructed around the 40-acre production unit because this area was deemed sufficiently small to measure changes in groundwater nitrate concentration every five "days." With equation (15), groundwater nitrate concentration is estimated (or predicted) at every production unit in the study region. Because the model is dynamic, a rule was needed to determine when to stop. Specifically, the simulation ends when the annual change in groundwater nitrate concentration across all cells is less than 0.3 ppm; this is the steady state .Furthermore, the locations of wells in the region used to monitor groundwater nitrate levels are known. From these data, specific production units are identified as monitoring wells in the simulation and used to predict groundwater nitrate concentrations (high, average, and low values for a soil zone or the region) in tables 2-4.

Estimating the impact of soil testing on producer profit and groundwater quality is accomplished in two steps. First, base case results are established assuming pre-1990 conditions, where only sugar beet and potato fields are soil tested. Residual soil nitrogen (RSN) for wheat and onions is fixed, respectively, at 100 pounds and 30 pounds per acre. The base simulation starts with no nitrogen in the soil profile and a pristine aquifer. From this pristine state, regional economic and geophysical data are compiled, and simulated agricultural production begins. In time, steady-state soil water and groundwater nitrate concentrations are reached and base case results measured.

Next, starting with base case levels, "test case" results are estimated. Here all modeled fields, including onion and wheat fields, are soil tested. The expected consequence of soil testing all crop fields is that new (lower) steady-state soil water and groundwater nitrate concentrations are achieved. Base case and test case results are then compared to determine the impact of additional soil testing on producer profit and groundwater nitrate concentration.

The results of the base case depend on the assumption that soil provides 100 pounds and 30 pounds of nitrogen per acre to wheat and onions, respectively. To test the sensitivity of results to changes in assumed soil nitrogen, three additional scenarios were tested. In the first, an extreme case is tested where it is assumed that soil does not provide residual nitrogen to wheat and onions. In the second scenario, RSN available to wheat and onions is decreased 20 pounds per acre from base levels (wheat 80 pounds/ acre and onions 10 pounds/acre). In the third scenario, available RSN is the same for wheat and onions (60 pounds/acre). For each scenario, test case (all fields soil tested) results are estimated. All model runs were compiled using the General Algebraic Modeling System (GAMS) release 2.25 solver.

RESULTS

Base Case Results

The results for the base case are reported in table 2. In general, the model performed adequately in terms of replicating key hydrologic endpoints. Groundwater movement is from soil zone A, through B, to soil zones C and D; the rate of groundwater flow slows dramatically beneath soil zone B. These spatial heterogeneities in the rate and direction of groundwater flow account for the reported different in groundwater nitrate concentration across the region. Excluding soil zone C, predicted mean concentrations are nearly equal to the observed mean. In soil zone C, the observed mean is 8 ppm greater than predicted. Predicted high concentrations tend to be less than observed levels (by 7 ppm on average). The model accounts for much of the spatial variation in nitrate concentration in soil zones A and D, and less so in soil zones B and C.

Table 2 reports both the high and mean predicted groundwater nitrate concentration across simulated observation well sites for the region and individual soil zones. The high concentration is reported because a mean concentration implies that some individuals still may be consuming potentially harmful levels of nitrates even if groundwater quality meets the target level of 10 ppm (Lee, Howitt, and Marino). Highest predicted groundwater nitrate concentrations occur beneath soil zones B and D (both exceed 20 ppm) due to slow groundwater flow rates and the production of onions and potatoes. Soil zone C also has high groundwater concentrations (15 ppm), but water quality in soil zone A is good. For the region, both the high and mean concentrations exceed the standard set by the Oregon Department of Environmental Quality (7 ppm). None of the production units in soil zone A had an underlying groundwater nitrate concentration in violation of the federal standard (10 ppm; see table 2). However, 74%, 12%, and 87% of the production units in soil zones B, C, and D, respectively, were in violation, resulting in an overall regional compliance rate of 45%.

With respect to base case crop production, the model estimates average per acre return to be \$637 for onions, \$266 for potatoes, \$178 for sugar beets, \$40 for wheat, and \$27 for hay. These returns are consistent with published data. Acreage in wheat and onions ranges from 25-80% across the soil zones, with 47.5% of regional acreage in these crops (table 2). Producers apply 147 pounds of nitrogen to wheat crops and 299 pounds to onion crops. Average nitrogen application rates (across crops) within a soil zone range from 77-223 pounds per acre, with a mean of 155 pounds. These estimates are consistent with current application rates of 284, 215, 205, and 136 pounds per acre, respectively, for onions, potatoes, sugar beets, and wheat.

In table 2, results for average per acre return (across the crops) and yield are reported as "modeled" and "tested RSN." The modeled return and yield is what the producer will receive if the assumed levels of residual soil nitrogen (here 100 pounds and 30 pounds for wheat and onions, respectively) are correct. The tested RSN return and yield are what the producer actually receives based on predicted levels of residual soil nitrogen. If modeled (anticipated) soil nitrogen levels equal tested levels, then average per acre returns are the same (as is yield).

Table 2. Base Case Results: Crop Acreage, Farm Income, Crop Yield, Nitrogen Use, and Groundwater Quality

Description	Soil Zone				Region (Total or Mean)
	A	B	C	D	
Acreage[a]	3,620	8,520	4,400	1,320	17,860
% of Acreage in:					
▸ Wheat	80	25	25	40	37
▸ Onions	—	25	—	—	12
Per Acre Return ($):					
▸ As Modeled	55	311	41	110	178
▸ Tested RSN	-1	302	35	72	158
Wheat Yield (bu./acre):					
▸ As Modeled	109	125	106	111	114
▸ Tested RSN	87	113	98	81	97
Onion Yield (cwt/acre):					
▸ As Modeled	—	625	—	—	625
▸ Tested RSN	—	625	—	—	625
N Required (lbs./acre):					
▸ Average	181	308	131	188	230
▸ Wheat	226	286	228	238	247
▸ Onions	—	329	—	—	329
N Applied (lbs./acre):					
▸ Average	102	223	77	104	155
▸ Wheat	126	186	128	138	147
▸ Onions	—	299	—	—	299
Residual N (lbs./acre):[b]					
▸ Average	44	84	52	60	66
▸ Wheat	55	74	82	37	64
▸ Onions	—	53	—	—	53
Soil Nitrogen Leached (ppm)	6.2	21.1	2.8	20.6	13.5
Groundwater Quality:					
▸ High N Concentration (ppm)					
- Observed	8.1	31.0	32.9	14.4	32.9
- Predicted	3.4	20.5	15.0	20.6	20.6
▸ Mean N Concentration (ppm)					
- Observed	2.7	12.8	12.2	14.2	11.0
- Predicted	2.4	13.7	4.3	14.9	8.7
▸ % Nodes above 10 ppm	0	74	12	87	45

Note: Potatoes and sugar beets, not wheat and onions, are soil tested, and it is assumed that wheat and onions utilize 100 pounds and 30 pounds, respectively, of residual soil nitrogen (RSN).

[a] Total acreage by crop: wheat = 6,654 acres, onions = 2,130 acres, potatoes = 2,658 acres, sugar beets = 3,230 acres, and hay = 3,188 acres.

[b] Includes sugar beets, where residual soil nitrogen (RSN) is measured to 24 inches.

Across the soil zones, tested per acre return is less than modeled (table 2). For the region, tested return is $158 per acre, $20 less than modeled (expected), due to an average yield loss in wheat of 17 bushels per acre. Specifically, soil was anticipated to provide 100 pounds of nitrogen, but actually provided much less (26-63 pounds less), resulting in reduced wheat yields. Onion yields were not affected; more residual nitrogen than anticipated was provided by soil. However, excess nitrogen was applied to onions, resulting in a lower per acre return. Finally, per acre return is greatest in soil zone B because of the high-valued crops grown here. This, and the fact that there is more acreage in soil zone B, inflates regional per acre return. At $158, tested per acre regional return is $86 to $159 larger than tested returns in soil zones A, C, and D.

Impact of Mandatory Soil Testing

With soil testing, modeled and tested RSN returns are identical, as are yields (table 3). The use of soil tests on wheat and onion fields decreased modeled regional per acre return $3, but increased tested per acre return $17 (compare tables 2 and 3). With soil testing, per acre return across the soil zones is $2 to $7 less than modeled return in the base case. This reduction is due to the cost of soil testing and the fact that insufficient nitrogen is applied in the base model. However, soil testing increased per acre returns $5 to $49 relative to tested RSN returns. This increase is due to improved nutrient management that improves yield. Specifically, wheat yields are increased 17 bushels per acre from tested (actual) yields in the base case. Furthermore, the cost savings associated with these yield gains exceed the cost of the fertilizer test ($15 per field).

One surprising result is that soil testing increases applied nitrogen (compare tables 2 and 3). While nitrogen applications to onions are reduced 25 pounds per acre, applications to wheat are increased 35 pounds, resulting in a region wide average increase of nine pounds per acre. With soil testing, the economically efficient level of nitrogen is achieved, which is higher than the base case. The increased expenditures on nitrogen are offset by the revenue gains associated with increased yields.

Although nitrogen use is increased relative to the base scenario, groundwater quality is slightly improved. With higher yields, more nitrogen is taken up (consumed) by the crop. Specifically, better nutrient management allows the crop to produce more (and deeper) roots which take up more nitrogen. Here, more efficient nitrogen use offsets the increase in applied nitrogen and the slight increase in residual soil nitrogen and reduces the amount of nitrogen being leached to groundwater, i.e., leached nitrate is reduced 1-6 ppm across the soil zones for an average reduction of 3 ppm (compare tables 2 and 3). This reduction in leached nitrate reduces the high and mean predicted levels of groundwater nitrate concentration by approximately 2.1 ppm. Furthermore, the percentage of fields with groundwater in violation of the federal drinking water standard is reduced from 45% to 36%.

Sensitivity Analysis

A set of sensitivity scenarios was estimated to measure the degree to which model results change under alternative levels of residual soil nitrogen. Table 4 reports the results of the three sensitivity model runs. For the base case and sensitivity scenarios, separate test case scenarios (where all crops are soil tested) were estimated. The test case results are not sensitive to changes in assumed residual soil nitrogen levels. As shown in column 4, table 4, the test case results are identical.

Table 3. Soil Testing Results: Crop Acreage, Farm Income, Crop Yield, Nitrogen Use, and Groundwater Quality

	Soil Zone				Region (Total or Mean)
Description	A	B	C	D	
Acreage[a]	3,620	8,520	4,400	1,320	17,860
% of Acreage in:					
▸ Wheat	80	25	25	40	37
▸ Onions	—	25	—	—	12
Per Acre Return ($)	48	309	40	104	175
Wheat Yield (bu./acre)	109	125	106	111	114
Onion Yield (cwt/acre)	—	625	—	—	625
N Required (lbs./acre):					
▸ Average	181	308	131	188	230
▸ Wheat	226	286	228	238	247
▸ Onions	—	329	—	—	329
N Applied (lbs./acre):					
▸ Average	131	226	79	129	164
▸ Wheat	163	217	145	222	182
▸ Onions	—	274	—	—	274
Residual N (lbs./acre):[b]					
▸ Average	51	82	52	60	67
▸ Wheat	64	69	83	16	65
▸ Onions	—	55	—	—	55
Soil Nitrogen Leached (ppm)	0.0	19.2	1.6	14.8	10.6
Predicted Groundwater Quality:					
▸ High N Concentration (ppm)	0.0	18.5	13.3	14.8	18.5
▸ Mean N Concentration (ppm)	0.0	11.1	3.2	10.4	6.6
▸ % Nodes above 10 ppm	0	63	6	54	36

[a] Total acreage by crop: wheat = 6,654 acres, onions = 2,130 acres, potatoes = 2,658 acres, sugar beets = 3,230 acres, and hay = 3,188 acres.
[b] Includes sugar beets, where residual soil nitrogen (RSN) is measured to 24 inches.

In all cases, model returns were less than or equal to tested (or actual) returns (table 4). An unreported supplementary analysis shows that if all producers in the region accurately predicted residual soil nitrogen levels without the aid of a soil test, average return is $175.50 per acre. With soil testing, average return is $174.94 per acre; hence the average cost of requiring all producers in the region to soil test is $0.56 per acre. However, this cost is spread across potato and sugar beet acreage that is already soil tested. The average cost of producing wheat and onions in the study region is increased by $1.13 per acre when these crops are required to be soil tested.

Assuming no residual nitrogen (scenario 1) also resulted in modeled and tested returns that were equal (table 4). This is the "conservative" or risk-minimizing strategy because a producer ignores soil nitrogen and applies sufficient nitrogen to guarantee optimal yield. In this case, nitrogen is over applied and returns are reduced $5 per acre. But this loss is less than the $6 to $17 actual loss reported in the other scenarios. In fact, scenario 1 behavior is observed in many of the high-valued cropping regions of Oregon and Washington. Such behavior will remain common as long as nitrogen is relatively inexpensive (Johnson, Adams, and Perry).

Table 4. Sensitivity Analysis: Crop Acreage, Farm Income, Crop Yield, Nitrogen Use, and Groundwater Quality for the Region

	Scenario [a]			
Description	1	2	3	4
Per Acre Return ($):				
▸ As Modeled	170	178	175	175
▸ Tested RSN	170	158	169	175
Wheat Yield (bu./acre):				
▸ As Modeled	114	114	114	114
▸ Tested RSN	114	97	113	114
Onion Yield (cwt/acre):				
▸ As Modeled	625	625	625	625
▸ Tested RSN	625	625	616	625
N Required (lbs./acre):				
▸ Average	230	230	230	230
▸ Wheat	247	247	247	247
▸ Onions	320	329	329	329
N Applied (lbs./acre):				
▸ Average	191	155	193	164
▸ Wheat	247	147	187	182
▸ Onions	329	299	269	274
Residual N (lbs./acre): [b]				
▸ Average	76	66	67	67
▸ Wheat	81	64	67	65
▸ Onions	46	53	55	55
Soil Nitrogen Leached (ppm)	16.2	13.5	11.4	10.6
Predicted Groundwater Quality:				
▸ High N Concentration (ppm	27.4	20.6	19.2	18.5
▸ Mean N Concentration (ppm)	10.7	8.7	7.1	6.6
▸ % Nodes above 10 ppm	49	45	37	36

Note: There are 17,860 total crop acres in the region: wheat = 6,654 acres (37%), onions = 2,130 acres (12%), potatoes = 2,658 acres (15%), sugar beets = 3,230 acres (18%), and hay = 3,188 acres (18%).

[a] Scenario Descriptions:

1 = The extreme case where wheat and onions are not soil tested and it is assumed that wheat and onions do not utilize residual soil nitrogen.

2 = The base case where wheat and onions are not soil tested and it is assumed that wheat and onions utilize 100 pounds and 30 pounds of residual soil nitrogen, respectively.

3 = A sensitivity scenario where wheat and onions are not soil tested and it is assumed that wheat and onions utilize 60 pounds of residual soil nitrogen. At the regional level, these results are nearly identical to assuming that wheat and onions utilize 80 pounds and 10 pounds of residual soil nitrogen, respectively.

4 = The test case where all crops are soil tested.

[b] Includes sugar beets, where residual soil nitrogen (RSN) is measured to 24 inches.

Changes in the assumed level of residual soil nitrogen (RSN > 0) have some impacts on production and profits, but little effect on groundwater quality (compare columns 2 and 3, table 4). Groundwater quality results were most sensitive to the case where residual soil nitrogen is ignored (scenario 1). Here, substantially more nitrogen is applied to wheat and onions, and average residual soil levels increase. Potatoes and sugar beets benefit from this increase (this is why the average application is less than that in column 3), but increased consumption of nitrogen by these crops is not sufficient to prevent groundwater damage, as the high predicted concentration is increased at least 7 ppm.

Summary of Results

A key finding of this analysis is that requiring all fields in the study region to be soil tested improves groundwater quality. However, soil tests are not sufficient by themselves to meet Oregon's quality target of 7 ppm for nitrate based on highest predicted concentration (not mean concentration) (Lee, Howitt, and Marino). This is true even though the predicted high groundwater nitrate concentrations in the base case are substantially less than observed (table 2). Soil testing thus has limited potential as a policy tool in that such tests are not likely to reduce the demand for direct regulation.

In addition to improving groundwater quality, soil testing has the added benefit of increasing modeled producer profit. While this win-win situation is desirable, this result does raise the question of why wheat and onion fields are not currently soil tested if doing so is profitable. The roles of education and risk attitudes, which could play a major role in a producer's decision to utilize soil test information, are not included in this analysis. Requiring that all fields be soil tested does not mean that a producer has to utilize this information. In that case, farm profits would be less since the farmer would pay the cost of testing. Furthermore, the potential of soil testing as a policy to improve groundwater quality is weakened if test information is not utilized. These issues are addressed below.

IMPLICATIONS AND CONCLUSIONS

The primary purpose of this investigation was to assess the potential benefits of soil testing on ambient groundwater quality. Simulation results indicate that soil testing can improve groundwater quality beneath the study region by 2-6 ppm (2.5 ppm on average). While other studies have shown soil testing to improve farm income, and perhaps soil water quality (which serves as a proxy for groundwater quality), this study directly links changes in nitrogen input as a result of soil testing to changes in groundwater quality. The analysis thus addresses an important policy question of whether soil testing can reduce groundwater nitrate concentration sufficiently to eliminate the need for direct regulation of this form of agricultural nonpoint pollution.

The answer to this question, at least in our study region, is no. While model results indicate that soil testing all fields reduces high predicted groundwater nitrate levels within a soil zone and across the region, the reduction is not sufficient to meet the Oregon state groundwater quality target (7 ppm). Further, predicted levels for the high groundwater nitrate concentration are significantly less than observed levels in the base case. With soil testing, the regional mean groundwater nitrate concentration is in compliance, but specific zones are in violation. Generally, these results are not sensitive to the chosen level of residual soil nitrogen. The exception is the case where the producer ignores residual soil nitrogen (scenario 1, table 4). Under this scenario, groundwater quality is significantly reduced.

The second purpose of this analysis was to assess the impact of soil testing on farm income. As found in previous research, soil nitrogen testing improved per acre return. However, unlike previous research, soil testing also increased nitrogen applications. The improvement in per acre return is the result of increased revenue from increased yields (particularly in wheat), which exceed the cost of the additional fertilizer and the soil test. However, the improvement in per acre returns shown here is more modest than in earlier studies ($5 to $49, for an average of $17-rather than $20 to $50 per acre found in previous work). Note, however, that the crops, soils, and climate of the study region are very different from those of other studies which focus primarily on corn in the Corn Belt region.

This analysis assumes all fields will be soil tested when, in fact, soil testing in the region is voluntary. While identified as a best management practice to improve groundwater quality, in practice, it is left to the producer to soil test fields and, if tested, to utilize that information. Hence, actual improvements in groundwater quality are expected to be less (and the demand for regulation greater) because not all producers will test their soils. In fact, model results show that soil testing only improves regional average per acre return $5 over a strategy of ignoring residual soil nitrogen levels. Onions are a high-value crop, and yield is sensitive to climatic factors and nutrient availability. In this case, a producer would have to demonstrate a great deal of confidence in soil test results. Soil testing will have a greater impact on net returns to less valuable crops where there is less yield risk, but it is still unreasonable to expect all wheat fields to be tested.

Soil testing, while not solving the groundwater quality problem in the region, is a potential win-win alternative; it can improve groundwater quality and increase farm profits. Extension education, technical assistance, and cost sharing are all policy tools that promote voluntary adaption. There is evidence, however, that educational programs need to be accompanied with a regulatory program to ensure improvements in groundwater quality (Bosch, Cook, and Fuglie). While regulation may be necessary to meet the desired quality goal for the study region, alternative technologies and production practices need to be evaluated (particularly changes in irrigation application rates and methods). In fact, changes in irrigation technology may do more for improving groundwater quality than changes in nitrogen application rate or technology (Larson, Helfand, and House).

The integrated assessment framework developed in this investigation is a comprehensive representation of a complex, site-specific problem, namely groundwater nitrate contamination. Without a representation of nitrate transport, such as used here, policy makers cannot evaluate the effect of technology change (for example, the use of a soil test) on groundwater quality. The results of this study apply to hydrological regions with similar soil characteristics and shallow aquifers that discharge to streams or rivers. Further extensions of this type of analysis include an accounting for an increase in the plateau yield as a result of increased soil testing (due to precision farming) and the addition of technology substitution. Taking advantage of soil test information may require adaption of new nitrogen application and irrigation technologies, which could improve returns to farmers and further reduce groundwater pollution.

REFERENCES

1. Adams, R. M., P. J. Farris, and D. J. Menkhaus. "Response Functions and the Value of Soil Test Information: The Case of Sugar Beets." N. Cent. J. Agr. Econ. 5(July 1983):77-82.
2. Babcock, B. A. "The Effects of Uncertainty on Optimal Nitrogen Applications." Rev. Agr. Econ. 14(July 1992):271-80.
3. Babcock, B. A., and A. M. Blackmer. "The Value of Reducing Temporal Input Nonuniformities." J. Agr. and Resour. Econ. 17(1992):335-47.
4. Babcock, B. A., A. L. Carriquiry, and H. S. Stern. "Evaluation of Soil Test Information in Agricultural Decision Making." Appl. Statis. 45(1996):447-61.
5. Berck, P., and G. Helfand. "Reconciling the von Liebig and Differentiable Crop Production Functions." Amer. J. Agr. Econ. 72(November 1990):985-96.
6. Bosch, D. J., Z. L. Cook, and K. 0. Fuglie. "Voluntary versus Mandatory Agricultural Policies to Protect Water Quality: Adoption of Nitrogen Testing in Nebraska." Rev. Agr. Econ. 17(1995):13-24.
7. Chowdhury, M. E., and R. D. Lacewell. "Implications of Alternative Policies on Nitrate Contamination of Groundwater." J. Agr. and Resour. Econ. 21(1996):82-95.
8. Debertin, D. L. "Developing Realistic Agricultural Production Functions for Use in Undergraduate Classes." S. J. Agr. Econ. 17(December 1985):207-14.

9. English, M. Professor, Department of Bioresource Engineering, Oregon State University, Corvallis. Personal communication, August 1997.

10. Fleming, R. A. "The Economics of Agricultural Groundwater Quality: Effects of Spatial and Temporal Variability on Policy Design." Unpub. Ph.D. diss., Dept. of Agr. and Resour. Econ., Oregon State University, Corvallis, 1996.

11. Fleming, R. A., and R. M. Adams. "The Importance of Site-Specific Information in the Design of Policies to Control Pollution." J. Environ. Econ. and Manage. 33(1997):347-58.

12. Fuglie, K. 0., and D. J. Bosch. "Economic and Environmental Implications of Soil Nitrogen Testing: A Switching-Regression Analysis." Amer. J. Agr. Econ. 77(November 1995):891-900.

13. Griffin, R. C., J. M. Montgomery, and M. E. Rister. "Selecting Functional Form in Production Function Analysis." West. J. Agr. Econ. 12(1987):216-27.

14. Johnson, S. L., R. M. Adams, and G. M. Perry. "The On-farm Costs of Reducing Groundwater Pollution." Amer. J. Agr. Econ. 73(1991):1064-73.

15. Larson, D. M., G. E. Helfand, and B. W. House. "Second-Best Tax Policies to Reduce Nonpoint Source Pollution." Amer. J. Agr. Econ. 78(1996):1108-17.

16. Lee, D. J., R. E. Howitt, and M. A. Marino. "A Stochastic Model of River Water Quality: Application to Salinity in the Colorado River." Water Resour. Res. 29(1993):3917-23.

17. Malheur County Extension Service (MCES). "Malheur County Agriculture, 1992." Oregon State University Extension Service, Malheur County Office, Ontario OR, 1992.

18. Malheur County Groundwater Management Committee (MCGMC). Northern Malheur County Groundwater Management Action Plan. Pub. No. GW\WH4760.5, State of Oregon Department of Environmental Quality, Portland, December 1991.

19. Mapp, H. P., D. J. Bernardo, G. J. Sabbagh, S. Geleta, and K. B. Watkins. "Economic and Environmental Impacts of Limiting Nitrogen Use to Protect Water Quality: A Stochastic Regional Analysis." Amer. J. Agr. Econ. 76(1994):889-903.

20. Mills, W. B. Water Quality Assessment: A Screening Procedure for Toxic and Conventional Pollutants in Surface and Ground Water, Parts 1 and 2. Pub. Nos. EPA/660/6-85/002a and EPA/660/6-85/002b. Environmental Research Laboratory, Office of Research and Development, U.S. Environmental Protection Agency, Athens GA, 1985.

21. Musser, W. N., J. S. Shortle, K. Kreahling, B. Roach, W. Huang, D. B. Beegle, and R. H. Fox. "An Economic Analysis of the Pre-Sidedress Nitrogen Test for Pennsylvania Corn Production." Rev. Agr. Econ. 17(1995):25-35.

22. Oregon State University Extension Service. "Enterprise Budgets for Oregon." Compiled by Dept. ofAgr. and Resour. Econ., Oregon State University, Corvallis. Various years and crops, 1980-92.

23. Perry, G. P., R. A. Fleming, and B. Conway. "Survey of Farming Practices in Oregon's Treasure Valley." Staff Pap., Dept. of Agr. and Resour. Econ., Oregon State University, Corvallis, 1992.

24. Rao, P. S. C., J. M. Davidson, and L. C. Hammond. "Estimation of Nonreactive and Reactive Solute Front Locations in Soils." In Residual Management by Land Disposal: Proceedings of the Hazardous Waste Research Symposium, Tucson AZ, 2-4 February 1976, ed., W. H. Fuller. Environmental Protection Agency, Office of Research and Development, Municipal Environmental Research Laboratory, Cincinnati OH, 1976.

25. Walker, D. S. "Groundwater Flow Model for the Lower Malheur Basin Near Ontario, Oregon." Unpub. master's thesis, Dept. of Civil Engineering, Oregon State University, Corvallis, 1989.

26. Wu, J., P. G. Lakshminarayan, and B. A. Babcock. "Impacts of Agricultural Practices and Policies on Potential Nitrate Water Pollution in the Midwest and Northern Plains of the United States." Work. Pap. No. 96-WP-148, Center for Agriculture and Rural Development (CARD), Dept. of Econ., Iowa State University, Ames, February 1996.

CHAPTER 14 AN INNOVATIVE AGRO-FORESTRY SUPPLY CHAIN FOR RESIDUAL BIOMASS: PHYSICOCHEMICAL CHARACTERISATION OF BIOCHAR FROM OLIVE AND HAZELNUT PELLETS

Ilaria Zambon, Fabrizio Colosimo, Danilo Monarca, Massimo Cecchini, Francesco Gallucci, Andrea Rosario Proto, Richard Lord and Andrea Colantoni

Keywords: Biochar; biomass; soil fertilizer; olive; hazelnut

INTRODUCTION

Biochar is a carbon-rich material produced by thermal decomposition of biomass under oxygen-limited conditions [1]. According to the International Biochar Initiative (IBI), biochar is primarily used for soil applications for both agricultural and environmental gains [2]. The IBI definition differentiates biochar from charcoal, whose use is as a fuel for heat, as an absorbent material, or as a reducing agent in metallurgical processes [1,3]. Thermo-chemical processes include (i) slow pyrolysis (conventional carbonization); (ii) fast pyrolysis; (iii) flash carbonization; and (iv) gasification [4,5,6,7]. During the last two decades, pyrolysis process received more attention from the scientific community, since it is an efficient method for converting biomass into bio-fuel [5,8]. The pyrolysis process and its parameters, such as final temperature, pressure, heating rate, and residence time, greatly influence biochar quality [5]. The advantage of slow pyrolysis is to retain up to 50% of the carbon (C) feedstock in stable biochar [9], which makes it suitable as soil fertilizer. High-temperature pyrolysis (>550 °C) produces biochar with high aromatic content and, therefore, recalcitrant to decomposition [10]. Biochars produced through low-temperature processes (<550 °C) typically have a less-condensed C structure and are expected to give a better contribution to soil fertility [11]. The nature of the biomass feedstock also influences the properties of the produced biochar [3,12]. The relation between biochar properties and its potential to improve agricultural soils is a nascent focus area and the appropriate pyrolysis conditions are still unclear [13]. Numerous recent studies focused on methodologies for the chemical characterization of biochars [13,14,15], other studies investigated the intrinsic potential of biochar as a soil amendment [16,17], although further efforts are required to obtain biochar with suitable properties [3]. One of the attractive characteristics of biochar as a soil amendment is its porous structure, which improves water retention and increases soil surface area [2]. Moreover, the concentration of biochar into soil has been related to an improved nutrient use efficiency, either through nutrients contained in biochar or through physicochemical processes that allow a better uptake of soil-inherent or fertilizer-derived nutrients [2]. The application of biochar increases physical and chemical

Citation: Zambon, I.; Colosimo, F.; Monarca, D.; Cecchini, M.; Gallucci, F.; Proto, A.R.; Lord, R.; Colantoni, A. An Innovative Agro-Forestry Supply Chain for Residual Biomass: Physicochemical Characterisation of Biochar from Olive and Hazelnut Pellets. Energies 2016, 9, 526. doi:10.3390/en9070526.

qualities of soils, resulting in greater productivity of the agro-ecosystem [18]. Biochar, due to its biological and chemical stability, can also act as a C sink. The recalcitrance of biochar to microbial degradation enables the long-term sequestration of C in soil [2,19].

Biochar application in agriculture, positively affects the water holding capacity; this property derives from the distribution and the degree of cohesion of the pores in biochar, which depends on the particles size and aggregation, as well as the organic matter (OM) content. The effect of biochar on water holding capacity is dependent on both the high internal surface area of biochar and the capability to aggregate soil particles with OM, minerals, and microorganisms. The increase in soil porosity also allows a better percolation of excess water towards the deeper layers of the soil, therefore increasing ventilation.

This work aims to determine the opportunity to create an innovative supply chain from agricultural waste biomass, especially regarding olive (*Olea europaea*) and hazelnut (*Corylus avellana*) in order to evaluate the production of biochar from their pruning residues. Biomass residues in Mediterranean areas come mainly from agricultural and agro-industrial activities, as well as forest by-products. Only a few woody residues are used to produce fertilizers and as renewable energy resources [20]. In contrast, typical management strategies in the agricultural industry do not provide any valorisation of these biomasses, which are burnt in the field to prevent proliferation of plant diseases [20]. However, this landfill choice affects the soil structure since OM in woody biomass residues must be completely decomposed before used as fertilizer.

MATERIALS AND METHODS

Biomass from Olive and Hazelnut Prunings

In the area of Viterbo, pruning residues from olive and hazelnut are rarely utilized as a source of energy in burning stoves or boilers; they are, instead, burnt in situ, therefore reducing the formation of soil organic carbon. During summer, besides pruning residues, suckers are removed before the harvest, representing another significant loss of biomass. Approximately 15 m^3 of biomass samples from both olive and hazelnut have been collected in farms of the Viterbo province. Recent studies [21,22] have investigated the possibility of enhancing olive and hazelnut residue waste management as a means to produce soil fertilizers and energy, therefore reducing the environmental impact of such residual organic wastes. Biomass from pruning crop operations (Figure 1a,b) represents an attractive resource that could be exploited for (i) fuel production (combustion and/or gasification) and (ii) biochar production (pyrolysis) that can be used as soil fertilizer.

Figure 1. Pruning residues from (**a**) olive and (**b**) hazelnut after crop operations; and (**c**) bio-shredding.

A pelletization procedure was developed and applied on bio-shredding obtained from olive and hazelnut residues (Figure 1c). Pruning residues were collected on site and immediately transferred to the laboratory for sifting and exsiccation (Figure 2) until a water content of 15% was achieved. Final water content as low as 15% is necessary for further refining of the product and pellet production. The humidity concentration in the prunings is very notable, because we can improve the technical process for pellet production by biomass. In Italy there are not many companies and total supply chains that work the prunings for pellet production and for use of these residual agriculture sub-products (Figure 3).

Figure 2. Schematic of the pelletization process showing (**a**) pellet mill; (**b**) olive and hazelnut pellet; and (**c**) packaging.

Figure 3. Biochar production from pellets showing (**a**) the Elsa Research carbonization system; (**b**) a schematic representation of the conversion process; and (**c**) the final product (biochar).

Pyrolysis Process

Pyrolysis of biomass is commonly considered as a thermochemical conversion process [7,23]. Pyrolysis is carried out under partial (or complete) absence of oxygen and relies on capturing the off-gases from thermal decomposition of the organic materials [19]. The physicochemical characteristics of biochar are determined by the type of feedstock and by the temperature of pyrolysis. For example, higher salt and ash contents are expected in wheat straw than in wood-derived biochar [24], and C content and N content are greater in pine chips than in poultry litter-derived biochar [25]. A higher pyrolysis temperature results in greater surface area, lower biochar recovery, higher ash content, elevated pH, minimal total surface charge [26], and lower cation exchange capacity [24]. Removal of volatile compounds at higher pyrolysis temperatures also cause biochars to have higher C content and lower hydrogen (H) and O content [26]. Pyrolysis of agro-forestry residues is typically carried out with temperatures between 400 and 800 °C. With these conditions, the feedstock is converted to liquid products (so-called tar or pyrolysis oil) and/or gas (syngas), which can be used as fuels or raw materials for subsequent chemical transformation. The residual solid carbonaceous material obtained (biochar) could be further refined to products, such as activated carbon.

Biochar Production form Olive and Hazelnut Pellets

The carbonization system Elsa Research (Blucomb Ltd., based in Udine, Italy) was used to produce the biochar from olive and hazelnut pellets; biomass conversion was achieved by pyrolytic micro-gasification (Figure 3). The Elsa Research carbonization system works with natural ventilation and does not require being powered by batteries or electricity. A chimney is typically used to increase the air draft for fuels that have difficulties igniting.

Physicochemical characterization of the biochar obtained from the Elsa Research carbonization system was performed at the European Biochar Institute, which released the EBC based on the quality of the biochar.

RESULTS

From Biomass to Biochar: Conversion Rates Analyses

Auto-thermal conversion of biomass was carried out under natural ventilation. Quantitative analyses of pyrolysed biomass and produced biochar, as well as the conversion rates, are reported for 10 and four sessions of pyrolysis, respectively, for olive and hazelnut pellets (Table 1). A statistical comparison between olive and hazelnut performances during pyrolysis is reported in Table 2, showing the total conversion rates, mean, and standard deviation (SD) of the results obtained in the experiments.

Table 1. Conversion rates of biomass obtained from each pyrolysis session

Olive				Hazelnut			
Session	Biomass (kg)	Biochar (kg)	Conversion Rate	Session	Biomass (kg)	Biochar (kg)	Conversion Rate
1	38.35	8.11	0.209	1	37.69	8.11	0.215
2	39.07	8.21	0.210	2	36.25	7.96	0.220
3	38.88	8.19	0.211	3	37.03	8.09	0.218
4	38.96	8.16	0.209	4	37.11	8.09	0.218
5	34.09	7.10	0.208				

6	39.02	8.23	0.211				
7	38.89	8.19	0.211				
8	38.93	8.19	0.210				
9	38.97	8.20	0.210				
10	38.81	8.13	0.209				

Table 2. Comparisons and statistical values of conversion rates of olive and hazelnut

	Olive		**Hazelnut**	
	Biochar (kg)	**Conversion Rate**	**Biochar (kg)**	**Conversion Rate**
Total	384.47	80.71	148.08	32.25
Mean		0.210		0.218
SD		0.00088		0.00188

Further analyses were carried out to investigate the calorific power of the two biochars produced. Composition, structure, heat value of the gas, tar liquid, and semi-char solid products depend on the pyrolysis temperature [7]. Quantity and quality of resulting outputs from biomass pyrolysis are related to the chemical composition of the operating temperature and the feedstock [7,27]. The calorific values calculated were compared with those provided by the producers in order to make energy considerations on the process. The results obtained are consistent with other pyrolysis processes. The latter led to the volatilization of a fraction of biomass with a calorific value ranging between 75% and 85% of the starting biomass. The calorific value is measured in terms of the high calorific value [28]. Table 3 distinguishes two types of calorific value (usually expressed in MJ/kg): (i) the higher calorific value that it is the amount of heat produced by a complete combustion of a mass unit of a sample, at constant volume, in an atmosphere rich of oxygen at standard conditions (25 °C, 101.3 kPa); and (ii) the lower calorific value (PCI) that does not include the heat of the condensation of water [28].

Table 3. Analysis of the calorific power of pyrolysis reaction for the two biochars produced in this study

Olive Wood	**Units**	**Pellet**	**Biochar**
Higher calorific value	MJ/kg	19.47	31.71
Lower calorific value	MJ/kg	16.17	30.48
Calorific value from pyrolysis	MJ/kg	12.37	
Percentage of calorific value from pyrolysis	%	0.76	
Hazelnut Wood	Units	Pellet	Biochar
Higher calorific value	MJ/kg	19.02	26.62
Lower calorific value	MJ/kg	16.71	25.66
Calorific value from pyrolysis	MJ/kg	14.21	
Percentage of calorific value from pyrolysis	%	0.85	

Pyrolysis does not produce energy from heat; rather, it leads to the production of gas from biomass. In general, pyrolysis involves the heating of biomass to temperatures greater than 400 °C in the absence of oxygen [29]. At these temperatures, biomass thermally decomposes releasing a vapor phase and biochar (solid phase). On cooling the pyrolysis vapor, polar and high-molecular-weight compounds condense out as bio-oil (liquid phase) while low-molecular-weight volatile compounds remain in the gas phase (syngas) [6]. The physics and chemistry of pyrolysis process results are extremely complex, and are dependent depending on both the rector conditions and the nature of the biomass [29]. The combustion of gas in the Elsa Research system occurs in "close-coupled combustion" (micro-gasification). Biochars produced by Blucomb Ltd. (Udine, Italy) (spin-off) for the European project were analyzed by Eurofins laboratories, accredited for the certification of the EBC. International biochar experts developed the EBC in order to consider it in the European context as a voluntary industrial standard [30]. The EBC guarantees a sustainable biochar production, with a low-risk use in agronomic systems. Biochar produced in accordance with the standards of the EBC fulfils all of the requirements of sustainable production and environmental impact by certifying (i) sustainable production and provision of biomass feedstock; (ii) energy efficient, low emission pyrolysis technique; (iii) low contaminant level in the biochar; and (iv) low hazard use and application of the biochar. These standards are in compliance with current environmental European regulations [31].

Elemental Analysis

The chemical composition of biochar is determined also by the source of biomass employed. Biochar produced from wood, for example, is denser and has higher C content (~80%) [32]. These properties reflect the chemical complexity of lignin, which makes it more resistant to thermal degradation. The elemental composition, plotted as H/C vs. O/C ratios (Figure 4), is often used to describe maturity, decomposition rate, and combustion behavior of fossil chars and coal [33,34]. When applied to biochar, the H/C and O/C ratios can be suitable indicators of the degree of carbonization. High ratios typically point to primary plant macromolecules, such as simple carbohydrates and cellulose [35]. An H/C ratio of ≤0.2 indicates C of plant origin with elevated carbonization [36].

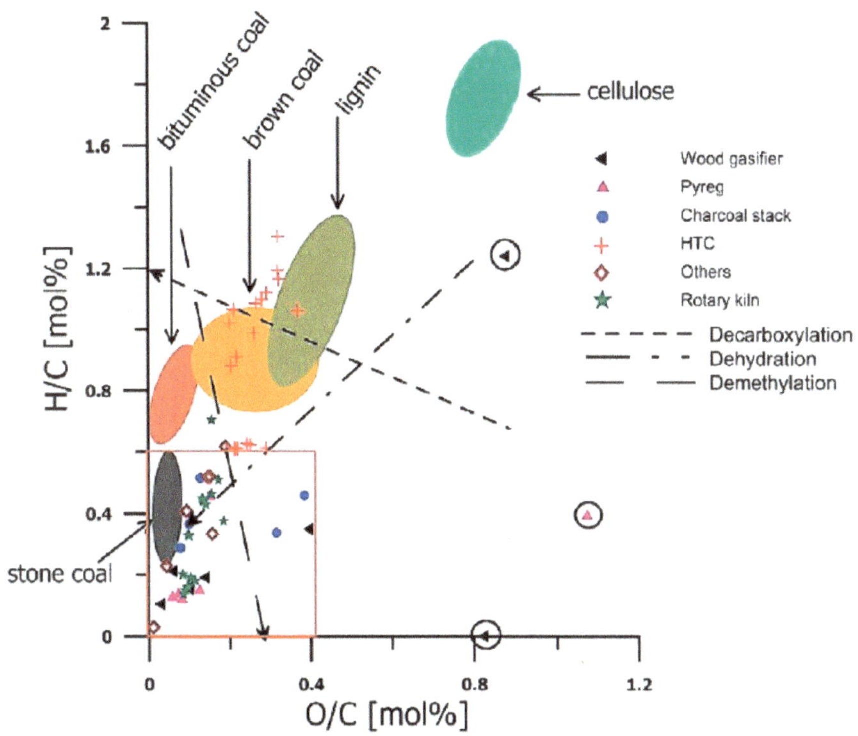

Figure 4. Example of Van Krevelen diagram of biochars obtained through different pyrolisis processes. The red square shows the optimum elemental ratio values of H/C and O/C for biochar production.

The O/C ratio is an indicator of the presence of polar functional groups, which influence the stability of biochar by preventing a dense, graphite-like structure of the material [37]. Therefore, the O/C ratio is useful to assess hydrophilicity and hydrophobicity of the charred material. The ratios of H, O, and C can also be used to differentiate between materials obtained by different processes. In the view of C sequestration and for material with complex aromatic structure and low presence of functional groups, optimum ratios of H/C and O/C are approximately ≤0.6 and ≤0.4, respectively [38]. Nitrogen in biochar is an important nutrient, its concentration is related to the concentration in the starting material, with values between 1.8 and 56 $g \cdot kg^{-1}$, although N in biochar is in a form often not readily bioavailable [39]. The C/N ratio, an indicator of the bioavailability of an organic compound, is highly variable and ranges between 7 and 500 [38].

The results of elemental analyses of the two biochars investigated in this study are reported in Table 4. Both biochars are characterized by values well below the limits established by the EBC, in particular the olive and hazelnut biochars have high values of C and low H/C and O/C ratios. A low H/C ratio indicates that the produced biochars are also recalcitrant to microbial degradation. These results indicate that our production process yield high-quality biochars with a level of carbonization that makes it suitable for C sequestration, as confirmed by the H/C ratios.

Table 4. Elemental analyses from EBC (Method DIN 51732)

Elements	Units	Hazelnut Biochar	Olive Biochar	EBC Biochar Base	EBC Biochar Premium
H (Hydrogen)	% w/w	1.21	1.58	-	-
C (Carbon, total)	% w/w	78.1	90.1	>50	>50
N (Nitrogen, total)	% w/w	0.64	0.42	-	-
O (Oxygen)	% w/w	1.2	1.7	-	-
Carbonate as CO_2	% w/w	2.62	1.17	-	-
Carbonate (organic)		75.5	89.8		

H/C ratio (molar)		0.18	0.21	<0.6	<0.6
O/C rate (molar)		0.012	0.014	<0.4	<0.4
Sulphur (total)	% w/w	**0.07**	**<0.03**		

Nutrients and Trace Elements

Biomass residues containing high concentrations of minerals, such as those obtained from herbaceous plants produce biochars with high ash content [32], maintain in the biochar matrix most of the nutrients present in the starting material (Table 5). These types of biochar have a lower total carbon (TC) content and cohesion than those obtained from wood-pruning biomass. The low C content, together with elevated concentrations of nutrients, makes biochars from herbaceous material more readily available for microorganisms [2]. The concentration of phosphorus (P) and potassium (K) in the biochar is related to the initial content in the feedstock. The content of P and K are typically between 2.7 and 480 $g \cdot kg^{-1}$ and 10 to 58 $g \cdot kg^{-1}$, respectively [39].

Table 5. Determination from microwave digestion (method: DIN 22022-1)

Elements	Units	Methods	Hazelnut Bio-char	Olive Bio-char	EBC Biochar Base	EBC Biochar Pre-mium
P (Phosphorus)	mg/kg	ISO 11885	590	330	-	-
Mg (Magnesium)	mg/kg	ISO 11885	2900	1400	-	-
Ca (Calcium)	mg/kg	ISO 11885	38,000	11,000	-	-
K (Potassium)	mg/kg	ISO 11885	5500	3500	-	-
Na (Sodium)	mg/kg	ISO 11885	2100	260	-	-
Fe (Iron)	mg/kg	ISO 11885	6500	1500	-	-
Si (Silicon)	mg/kg	ISO 11885	25,000	9700	-	-
S (Sulphur)	mg/kg	ISO 11885	910	200	-	-
Pb (Lead)	mg/kg	ISO 17294-2	66	20	<150	<120
Cd (Cadmium)	mg/kg	ISO 17294-2	<0.2	<0.2	<1.5	<1
Cu (Copper)	mg/kg	ISO 17294-2	100	6	<100	<100
Ni (Nickel)	mg/kg	ISO 17294-2	9	8	<50	<30
Hg (Mercury)	mg/kg	DIN EN 1483	<0.07	<0.07	<1	<1
Zn (Zinc)	mg/kg	ISO 17294-2	340	84	<400	<400
Cr (Chromium total)	mg/kg	ISO 17294-2	22	15	<90	<80
B (Boron)	mg/kg	ISO 17294-2	32	10	-	-
Mn (Manganese)	mg/kg	ISO 17294-2	350	380	-	-

EBC biochar base and premium report the limits required by the EBC protocol of certification. The total ash content ranged between 6.2% and 18.8% (w/w) for biochar from pellets of olive and hazelnut wood. The nutrient content is much greater in hazelnut biochar than olive, which was evident especially for Mg, Ca, Fe, S, Cu, and Zn. Biochar from hazelnut pellets could bring a greater contribution of nutrients to the soil and, therefore, be less resistant to microbial decomposition. Heavy metal content in both biochars was well below the EBC limits. Only Cu in the hazelnut biochar was close to the maximum value established by the EBC.

PAHs (Polycyclic Aromatic Hydrocarbons) Composition

PAHs are ubiquitous in the environment, being by-products of the incomplete combustion of organic material [40]. The chemical structure of PAHs makes them highly resistant to biodegradation and oxidation [41]. The presence of PAHs in pyrolytic reactions above 700 °C is well established [42], although they can be produced in pyrolysis reactions of less than 700 °C at low concentration [43]. It is, therefore, critical to ensure PAH concentrations remain below the limits established by the EBC. The 16 priority US EPA PAHs are typically used to assess the total PAH content; the limits established by the EBC are of <12 and <4 mg/kg for biochar standard and premium, respectively. The PAH composition of the two biochars analysed in this study (Table 6), shows that both biochars are well below the EBC limits, with values ranging from <0.1 to 1.1 mg/Kg.

Table 6. PAHs determination from toluene extract

Elements	Units	Methods	Limits		Hazelnut Biochar	Olive Biochar
			GW 1 *	GW 2 *		
Naphthalene	mg/kg	DIN EN 15527	-	-	0.9	1.1
Acenaphthylene	mg/kg	DIN EN 15527	-	-	<0.1	<0.1
Acenaphthene	mg/kg	DIN EN 15527	-	-	<0.1	<0.1
Fluorene	mg/kg	DIN EN 15527	-	-	<0.1	<0.1
Phenanthrene	mg/kg	DIN EN 15527	-	-	0.3	0.3
Anthracene	mg/kg	DIN EN 15527	-	-	<0.1	<0.1
Fluoranthene	mg/kg	DIN EN 15527	-	-	0.1	0.1
Pyrene	mg/kg	DIN EN 15527	-	-	0.1	0.1
Benz(a)anthracene	mg/kg	DIN EN 15527	-	-	<0.1	<0.1
Chrysene	mg/kg	DIN EN 15527	-	-	<0.1	<0.1
Benzo(b)fluoranthene	mg/kg	DIN EN 15527	-	-	<0.1	<0.1
Benzo(k)fluoranthene	mg/kg	DIN EN 15527	-	-	<0.1	<0.1
Benzo(a)pyrene	mg/kg	DIN EN 15527	-	-	<0.1	<0.1
Indeno(1,2,3-cd)pyrene	mg/kg	DIN EN 15527	-	-	<0.1	<0.1
Dibenz(a,h)anthracene	mg/kg	DIN EN 15527	-	-	<0.1	<0.1
Benzo(g,h,i)perylene	mg/kg	DIN EN 15527	-	-	<0.1	<0.1
SUM PAHs (EPA)	mg/kg	calculated	<12	<4	1.20	1.60

* (GW 1 = quality level basic related dry bases; GW 2 = quality level premium related dry bases).

Total PAH content of the two biochars are 1.2 and 1.6 mg/kg for olive and hazelnut, respectively. Therefore, both biochars can be considered suitable for soil applications, since both are well below the EBC threshold limit of 4 mg/kg for biochar premium.

pH, Electrical Conductivity (EC), and Density

In general, the pH of biochar is relatively homogeneous and varies from neutral to basic pH. Feedstock of various origins produce biochar with an average pH between 6.2 and 9.6 [39]. Lower pH is typically found for biochars obtained from green pruning feedstock and organic waste, while the highest values are to be attributed to poultry litter biochar. The Table 7reports the elements values, according to their pH, electrical conductivity, salt content and density.

Table 7. Elements value (pH, electrical conductivity, and density)

Elements	Units	Hazelnut Biochar	Olive Biochar
pH values ($CaCl_2$)	-	9.9	8.4
Electrical conductivity	µS/cm	332	217
Salt content	g/kg	0.655	1.18
Salt content cal. with bulk density	g/L	0.287	0.527

The two biochar have a pH of 8.4 and 9.9 for olive and hazelnut, respectively. The EBC indicates a maximum limit of 10; therefore, biochar produced from these types of wood residues is slightly below the limit established by the certification. The EC is of particular importance when adding biochar to soils with high EC and salinity. The two biochars had an EC of 217 and 332 mS/cm, respectively, for olive and hazelnut (as shown in Table 7). Both values are very low and do not represent a real risk for the addition to soil even under conditions of high EC. In general, biochar has a lower density than soil, with an average of 0.4 $g \cdot cm^{-3}$ compared to a soil of medium texture, with average of 1.3 $g \cdot cm^{-3}$. When adding biochar to soils with little ventilation, this property can help to reduce the density by mitigating issues related to the compaction of soil. The olive and hazelnut biochars produced in this study have a density of 0.45 and 0.44 $g \cdot cm^{-3}$, respectively.

CONCLUSIONS

The two biochars analyzed in this study show excellent physicochemical properties, which makes them suitable for agricultural applications. Both biochars can be certified as Biochar Premium according to the regulations of the EBC; this allows a potential commercialization of the biochars, with higher prices than Biochar Base, typically less expensive, but with a higher content of PAHs. The benefits of using Biochar Premium as soil fertilizer includes improved productivity, increased water holding capacity of the soil (e.g., [44,45,46]), and a better retention of nutrients and agrochemicals in soils, all of which should offset initial investment and provide added profits per application. Biochar fuel commands a high-value application, offering numerous benefits, and an authentic alternative to develop the biomass utilization efficiency [4,47]. The added value of biochar is also linked to other issues, such as those involving agricultural and environmental sustainability. As claimed by many studies [1,16], biochar as a soil amendment can improve soil health and increase agricultural productivity with further environmental benefits related to global warming mitigation [16,48,49,50,51,52]. Based on our results, we intend to define an agro-forestry chain to use the residual waste biomass for the production of high quality biochar for agronomic and commercial purposes. We are proceeding to evaluate the properties of biochar for soil improvement.

ACKNOWLEDGMENTS

This work was supported during my visiting professor at Strathclyde University.

AUTHOR CONTRIBUTIONS

All authors contributed equally to the work done.

REFERENCES

1. Lehmann, J.; Stephen, J. *Biochar for Environmental Management—Science and Technology*; Routledge: Abingdon, UK, 2009.
2. Sohi, S.; Lopez-Capel, E.; Krull, E.; Bol, R. Biochar, climate change and soil: A review to guide future research. *Civ. Eng.* **2009**, 6618–6664.
3. Manyà, J.J. Pyrolysis for biochar purposes: A review to establish current knowledge gaps and research needs. *Environ. Sci. Technol.* **2012**, *46*, 7939–7954.
4. Anupam, K.; Sharma, A.K.; Lal, P.S.; Dutta, S.; Maity, S. Preparation, characterization and optimization for upgrading Leucaena leucocephala bark to biochar fuel with high energy yielding. *Energy* **2016**, *106*, 743–756.
5. Jahirul, M.I.; Rasul, M.G.; Chowdhury, A.A.; Ashwat, N. Biofuels production through biomass pyrolysis—A technological review. *Energies* **2012**, *5*, 4952–5001.
6. Hoogwijk, M.; Faaij, A.P.C.; van den Broek, R.; Berndes, G.; Gielen, D.; Turkenburg, W. Exploration of the ranges of the global potential of biomass for energy. *Biomass Bioenergy* **2003**, *25*, 119–133.
7. Panwar, N.L.; Kothari, R.; Tyagi, V.V. Thermo chemical conversion of biomass – Eco friendly energy routes. *Renew. Sustain. Energy Rev.* **2012**, *16*, 1801–1816.
8. Demirbas, A. Partly chemical analysis of liquid fraction of flash pyrolysis products from biomass in the presence of sodium carbonate. *Energy Convers. Manag.* **2002**, *43*, 1801–1809.
9. Gaunt, J.L.; Lehmann, J. Energy balance and emissions associated with biochar sequestration and pyrolysis bioenergy production. *Environ. Sci. Technol.* **2008**, *42*, 4152–4158.
10. Singh, B.P.; Cowie, A.L. A novel approach, using ^{13}C natural abundance, for measuring decomposition of biochars in soil. In Proceedings of the Carbon and Nutrients Management in Agriculture, Palmerston North, New Zealand, 13–14 February 2008; Currie, L.D., Yates, L., Eds.;
11. Steinbeiss, S.; Gleixner, G.; Antonietti, M. Effect of biochar amendment on soil carbon balance and soil microbial activity. *Soil Biol. Biochem.* **2009**, *41*, 1301–1310.
12. Di Blasi, C.; Branca, C.; Lombardi, V.; Ciappa, P.; di Giacomo, C.; Chimica, I. Effects of particle size and density on the packed-bed pyrolysis of wood. *Energy Fuels* **2013**, *27*, 6781–6791.
13. Brewer, C.; Schmidt-Roht, K.; Satrio, J.; Brown, R. Characterization of Biochar from fast pyrolysis and gasification systems. *Environ. Prog.* **2009**, *28*, 386–396.
14. Sánchez, M.E.; Lindao, E.; Margaleff, D.; Martínez, O.; Morán, A. Pyrolysis of agricultural residues from rape and sunflowers: Production and characterization of bio-fuels and biochar soil management. *J. Anal. Appl. Pyrolysis* **2009**,*85*, 142–144.
15. Keiluweit, M.; Nico, P.S.; Johnson, M.G.; Kleber, M. Dynamic molecular structure of plant biomass-derived black carbon (biochar). *Environ. Sci. Technol.* **2010**, *44*, 1247–1253.
16. Agegnehu, G.; Bass, A.M.; Nelson, P.N.; Muirhead, B.; Wright, G.; Bird, M.I. Biochar and biochar-compost as soil amendments: Effects on peanut yield, soil properties and greenhouse gas emissions in tropical North Queensland, Australia. *Agric. Ecosyst. Environ.* **2015**, *21*, 372–385.
17. Beesleya, L.; Moreno-Jiménez, E.; Gomez-Eyles, J.L.; Harris, E.; Robinsond, B.; Sizmure, T. A review of biochars' potential role in the remediation, revegetation and restoration of contaminated soils. *Environ. Pollut.* **2011**, *159*, 3269–3282.
18. Laird, D.A.; Fleming, P.; Davis, D.D.; Horton, R.; Wang, B.; Karlen, D.L. Impact of biochar amendments on the quality of a typical Midwestern agricultural soil. *Geoderma* **2010**, *158*, 443–449.
19. Lehmann, J. A handful of carbon. *Nature* **2007**, *447*, 143–144.
20. Di Giacomo, G.; Taglieri, L. Renewable energy benefits with conversion of woody residues to pellets. *Energy* **2009**, *34*, 724–731.
21. Colantoni, A.; Longo, L.; Evic, N.; Gallucci, F.; Delfanti, L. Use of Hazelnut's pruning to produce biochar by gasifier small scale plant. *Int. J. Renew. Energy Res.* **2015**, *5*, 873–878.
22. Di Giacinto, S.; Longo, L.; Menghini, G.; Delfanti, L.; Egidi, G.; de Benedictis, L.; Riccioni, S.; Salvati, L. Model for estimating pruned biomass obtained from Corylus avellana L. *Appl. Math. Sci.* **2014**, *8*, 6555–6564.
23. White, J.E.; Catall, W.J.; Legendre, B.L. Biomass pyrolysis kinetics: A comparative critical review with relevant agricultural residue case studies. *Anal. Appl. Pyrol.* **2011**.
24. Kloss, S.; Zehetner, F.; Dellantonio, A.; Hamid, R.; Ottner, F.; Liedtke, V.; Schwanninger, M.; Gerzabek, M.H.; Soja, G. Characterization of slow pyrolysis

biochars: Effects of feedstocks and pyrolysis temperature on biochar properties. *J. Environ. Qual.* **2012**, *41*, 990–1000.

25. Gaskin, J.W.; Steiner, C.; Harris, K.; Das, K.C.; Bibens, B. Effect of low-temperature pyrolysis conditions on biochar for agricultural use. *Trans. Asabe* **2008**, *51*, 2061–2069.
26. Novak, J.; Lima, I.; Xing, B. Characterization of designer biochar produced at different temperatures and their effects on a loamy sand. *Ann. Environ. Sci.* **2009**, *3*, 195–206.
27. Ceylan, R.; B-son Bredenberg, J. Hydrogenolysis and hydrocracking of the car-bonoxygen bond. 2. Thermal cleavage of the carbon–oxygen bond in guaiacol. *Fuel* **1982**, *61*, 377–382.
28. Jothibasu, K.; Mathivanan, G. Enhancing combustion efficiency using nano nickel oxide catalyst in biomass gasifier.*Int. J. Core Eng. Manag. (IJCEM)* **2015**, *2*, 31–41.
29. Laird, D.A. Review of the pyrolysis platform for coproducing bio-oil and biochar. *Biofuels Bioprod. Bioref.* **2008**, *3*, 547–562.
30. European Biochar Certificate (EBC). *Guidelines for a Sustainable Production of Biochar*; European Biochar Foundation: Arbaz, Switzerland, 2012; pp. 1–23.
31. European Parliament, Council of the European Union. Regulation (EC) No. 1907/2006 of the European Parliament and of the Council of 18 December 2006 concerning the Registration, Evaluation, Authorisation and Restriction of Chemicals (REACH), establishing a European Chemicals Agency, amending Directive 1999/45/EC and repealing Council Regulation (EEC) No 793/93 and Commission Regulation (EC) No 1488/94 as well as Council Directive 76/769/EEC and Commission Directives 91/155/EEC, 93/67/EEC, 93/105/EC and 2000/21/EC. Available online: http://eur-lex.europa.eu/legal-content/EN/TXT/?uri=CELEX%3A02006R1907-20140410 (accessed on 10 April 2014).
32. Demirbas, A. Effects of temperature and particle size on bio-char yield from pyrolysis of agricultural residues. *J. Anal. Appl. Pyrolysis* **2004**, *72*, 243–248.
33. Hammes, K.; Smernik, R.J.; Skjemstad, J.O.; Herzog, A.; Vogt, U.F.; Schmidt, M.W.I. Synthesis and characterisation of laboratory-charred grass straw (Oryza sativa) and chestnut wood (Castanea sativa) as reference materials for black carbon quantification. *Org. Geochem.* **2006**, *37*, 1629–1633.
34. Krevelen, V. *Coal Science*; Elsevier: Amsterdam, The Netherlands, 1957.
35. Chun, Y.; Sheng, G.; Chiou, G.T.; Xing, B. Compositions and sorptive properties of crop residue-derived chars. *Environ. Sci. Technol.* **2004**, *38*, 4649–4655.
36. Kuhlbusch, T.A.J.; Crutzen, P.J. Toward a global estimate of black carbon in residues of vegetation fires representing a sink of atmospheric CO_2 and a source of O_2. *Glob. Biogeochem. Cycles* **1995**, *9*, 491–501.
37. Laine, J.; Yunes, S. Effect of the preparation method on the pore size distribution of activated carbon from coconut shell. *Carbon N. Y.* **1992**, *30*, 601–604.
38. Schimmelpfennig, S.; Glaser, B. One step forward toward characterization: Some important material properties to distinguish biochars. *J. Environ. Qual.* **2012**, *41*, 1001–1013.
39. Chan, K.Y.; Xu, Z. Biochar: Nutrient properties and their enhancement. *Biochar Environ. Manag. Sci. Technol.* **2012**, 67–84.
40. Keith, L.H.; Telliard, W.A. Priority pollutants I-a perspective view. *Environ. Sci. Technol.* **1979**, *13*, 416–423.
41. Preston, C.M.; Schmidt, M.W.I. Black (pyrogenic) carbon: A synthesis of current knowledge and uncertainties with special consideration of boreal regions. *Biogeosciences* **2006**, *3*, 397–420.
42. Ledesma, E.B.; Marsh, N.D.; Sandrowitz, A.K.; Wornat, M.J. Global kinetic rate parameters for the formation of polycyclic aromatic hydrocarbons from the pyrolysis of catechol, a model compound representative of solid fuel moieties. *Energy Fuels* **2002**, *16*, 1331–1336.
43. Garcia-Perez, M. *The Formation of Polyaromatic Hydrocarbons and Dioxins during Pyrolysis: A Review of the Literature with Descriptions of Biomass Composition, Fast Pyrolysis Technologies and Thermochemical Reactions*; Washington State University: Pullman, WA, USA, 2008; pp. 1–58.
44. Fellet, G.; Marchiol, L.; delle Vedove, G.; Peressotti, A. Application of biochar on mine tailings: Effects and perspectives for land reclamation. *Chemosphere* **2011**, *83*, 1262–1267.
45. Karhu, K.; Mattila, T.; Bergstrom, I.; Regina, K. Biochar addition to agricultural soil increased CH4 uptake and water holding capacity—Results from a short-term pilot field study. *Agric. Ecosyst. Environ.* **2011**, *140*, 309–313.
46. Streubel, J.D.; Collins, H.P.; Garcia-Perez, M.; Tarara, J.; Granatstein, D.; Kruger, C.E. Influence of contrasting biochar types on five soils at increasing rates of application. *Soil Sci. Soc. Am. J.* **2011**, *75*, 1402–1413.
47. Liu, Z.; Quek, A.; Hoekman, S.K.; Balasubramanian, R. Production of solid biochar fuel from waste biomass by hydrothermal carbonization. *Fuel* **2013**, *103*, 943–949.
48. Reza, M.T.; Andert, J.; Wirth, B.; Busch, D.; Pielert, J.; Lynam, J.G.; Mumme, J. Hydrothermal carbonization of biomass for energy and crop production. *Appl. Bioenergy* **2014**, *1*, 11–29.
49. Hill, J.; Nelson, E.; Tilman, D.; Polasky, S.; Tiffany, D. Environmental, economic, and energetic costs and benefits of biodiesel and ethanol biofuels. *Proc. Natl. Acad. Sci. USA* **2006**, *103*, 11206–11210.
50. Chandra, R.; Takeuchi, H.; Hasegawa, T. Hydrothermal pretreatment of rice straw biomass: A potential and promising method for enhanced methane production. *Appl. Energy* **2012**, *94*, 129–140.
51. Kim, D.; Yoshikawa, K.; Park, K.Y. Characteristics of biochar obtained by hydrothermal carbonization of cellulose for renewable energy. *Energies* **2015**, *8*, 14040–14048.
52. Civitarese, V.; Spinelli, R.; Barontini, M.; Gallucci, F.; Santangelo, E.; Acampora, A.; Scarfone, A.; del Giudice, A.; Pari, L. Open-air drying of cut and windrowed short-rotation poplar stems. *Bioenergy Res.* **2015**, *8*, 1614–1620.

CHAPTER 15 BIOCHAR AS A SAND-BASED ROOTZONE

Shane R. Brockhoff and Nick E. Christians

INTRODUCTION

Bicohar may be valuable as an amendment in sand-based turfgrass rootzones. Currently, peat moss is the most common organic amendment mixed with sand when sand systems are constructed. Peat moss increases water retention and nutrient holding capacity of the sand; however, peat moss is prone to decomposition over a relatively short period of time. Biochar is very stable in the soil profile and may prove to be a viable organic amendment for sand-based turfgrass. In this study, fast pyrolysis switchgrass biochar was used. The objectives of this research were to

- quantify soil water retention capabilities,
- determine water infiltration rates, and
- measure rooting depths of creeping bentgrass in sand and biochar soils.

MATERIALS AND METHODS

Soil water retention. This experiment was conducted in the Soil Physics Laboratory using pressure head methodology. Metal cylinders were filled with biochar and sand mixes, saturated, and placed in pressure chambers. Water was collected at each pressure level to calculate soil water retention. Six biochar and sand treatments were evaluated ranging from 0 to 25% biochar increasing by 5% increments. Four replications per treatment were evaluated at matric potentials -0.010, -0.025, -0.049, -0.098, -0.196, and -0.327 bars (n=4). Two replications per treatment were evaluated at -0.5, -1.0, -5.0 and -15.0 bars (n=2).

Saturated hydraulic conductivity. This experiment was conducted in the Soil Physics Laboratory using Marriott bottle methodology. The same cylinders were re-saturated from the soil water retention study. Water was allowed to flow through the cylinders maintaining a constant ponded water level. The volume of outflow was measured over a given time frame.

Rooting depth. This experiment was conducted in a controlled environment greenhouse in the Horticulture greenhouse complex. Four replications of the six biochar and sand treatments, ranging from 0 to 25% biochar increasing by 5% increments, were evaluated. T-1 creeping bentgrass was established via seed in 60 cm long root tubes containing the treatment mixes with a 20 cm layer of pea gravel at the bottom. After 110 days, the root tubes were sliced open and roots measured.

Citation: Brockhoff, Shane R. and Christians, Nick E, Biochar as a Sand-based Rootzone, (2009). Iowa State Research Farm Progress Reports. Paper 318. http://lib.dr.iastate.edu/farms_reports/318.

RESULTS AND DISCUSSION

Saturated hydraulic conductivity. Biochar decreases saturated hydraulic conductivity in sand-based rootzones (Table 1).

Table 1. Saturated hydraulic conductivity and creeping bentgrass rooting depth evaluated in six biochar treatments. Rooting depth was evaluated using root tubes in which T-1 creeping bentgrass was grown in a greenhouse for 110 days. Saturated hydraulic conductivity values represent means of four replications[x,y]

Biochar Percentage biochar (v/v)	Saturated hydraulic conductivity cm hr^{-1}	Rooting depth cm
0	85a[z]	46a
5	56b	45a
10	53b	43a
15	29c	34b
20	16d	30b
25	7d	25b

[x]Means within columns followed by the same letter are not different according to Fisher's Protected $LSD_{0.05}$.
[y]Rooting depth values represent means of twelve replications.

Rooting depth. At levels greater than 10% (v/v) biochar in the sand, rooting depth is decreased (Table 1).

Soil water retention. Biochar increases soil water retention in sand. Biochar increases plant available water in sand-based rootzones (Figure 1).

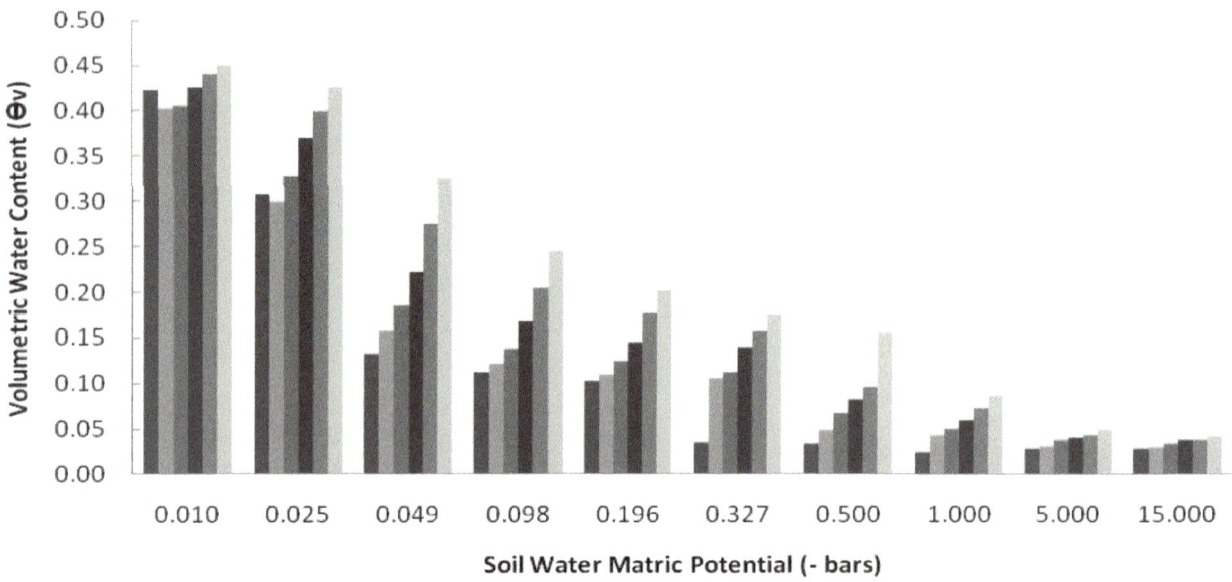

Figure 1. Volumetric water content in relation to the matric potential. At each matric potential level, six treatments were evaluated. Treatments from left to right at each matric potential level: 0% biochar to 25% biochar increasing by 5% biochar increments (v/v). Four replications per treatment were evaluated at matric potentials -0.010, -0.025, -0.049, -0.098, -0.196, and -0.327 bars (n=4). Two replications per treatment were evaluated at -0.5, -1.0, -5.0, and -15.0 bars (n=2)

CHAPTER 16

WATER INFILTRATION AND TIME TO RECHARGE THE PROFILE OF SOILS

Afranio Almir Righes, Galileo Adeli Buriol, Valduino Estefanel

Keywords: Macro Porosity; Rainfall; Drought; Soil Water Storage

INTRODUCTION

In Rio Grande do Sul (RS) State in Brazil, during the colder months of the year, all of the territory has excess water, and in the warmer months, especially in December, January, and February, droughts occur.

These conditions are more frequent in the southern area of the state [1]. Normal medium rainfall data indicate that in the southern part of the state, the probability of potential evapotranspiration is higher than the total monthly rainfall in the summer months, approximately 70%, but the northern area showed no water deficit [2]. However, despite these results, the deficiencies of soil water for crops, even in soils more than 8 m deep and with high water storage capacity, appear to have increased, especially in the northern part of the state. In this part of the state, at Passo Fundo, the soil type is oxisol [3,4]. The soil macroporosity layer of 0.025 m to 0.15 m depth has only 10% of volume in macropores [5].

The mean total annual rainfalls for the north, center, and south of the state are approximately 2000, 1700, and 1200 mm, respectively [6,7]. The normal values of average annual potential evapotranspiration range from approximately 600 mm in Sierra Northeast to 1000 mm in the Lower Valley of Uruguay [8]. Considering these values, the areas probably would not have water deficits if all, or most, of the volume of water precipitated was stored in the soil and the plant root system could extract it [9]. Thus, other contributing factors are present, so that within periods of the same number of days without precipitation, soil water deficits occur with more intensity.

The intensive soil is mobilized by mechanical equipment, such as plows and harrows, that occurred in the 1970s accelerated the soil structure degradation and increased soil compaction. The intensive soil mobilization changed its original structure by fractionation of the aggregates into smaller units, reducing the percentage of organic matter and the percentage of macropores and increasing the soil bulk density and the volume of micropores. As a consequence, there is a reduction in the soil water infiltration rate, resulting in increased runoff [10-12]. A similar situation has also occurred in areas used by livestock with a high number of animals per area. Being trampled by the animals has compacted on the soil surface, with hydrological results similar to those of heavily exploited agricultural areas. Intensive traffic of agricultural machines

Citation: Righes, A. , Buriol, G. and Estefanel, V. (2013), Water infiltration and time to recharge the profile of soils, Agricultural Sciences, 4, 12-18. doi: 10.4236/as.2013.411A002.

and animal trampling in areas with a crop-livestock system have increased soil compaction. In this system of management, the use of heavy machinery must be prevented and animals must be removed from the common crop-livestock area when the soil water content is over the range of friability [13].

The biggest problems of water erosion, siltation of rivers, and soil loss in the RS occurred in the 1970s. By that time, different mechanical techniques were used to control soil erosion, such as terraces, sub-soiling, and scarifications. These techniques were not sufficient to control soil erosion satisfactorily. Finally, the no-tillage system, ideal for tropical regions, was introduced. However, when implemented on degraded soils with impediment layers (compacted layers with a low rate of water infiltration) located below the mobilized layer in the conventional tillage system, the no-tillage system reduced water infiltration into the soil.

The effectiveness of no-tillage in controlling soil erosion led the farmers in the northern region of the RS to almost entirely removing the existing terraces in their crops and fields, because of the increased operating capacity of agricultural machinery and even sowing in the slope direction. However, the farmers have not used crop rotation in the no-tillage system. The "foot plow" was not eliminated and remains in use today. Thus, the no tillage system has effectively controlled soil loss, but the water losses by runoff are even higher than in the conventional tillage system. The losses of nutrients and organic matter in runoff sediment in the no-tillage system are higher than those in the same soil with conventional tillage [14]. Low values of infiltration dramatically reduce the water storage in the soil and aquifer recharge, increasing runoff and flooding. In the north regions of the RS, in the oxisols mapping units Passo Fundo and Santo Angelo, most wells with depths between 10 and 15 m are dry during drought periods, and even after heavy rainfall in the summer time, the water levels do not return to normal. This shortage is evident that the macropores are not filled with water from rainfall and, as a consequence, there is little free gravity water to feed the wells, rivers, and strands. Another finding of this effect can be seen in road cuts in deep soils, that, during heavy rainy periods, after only a few hours, there is no more water to be gravity drained from the soil profile.

This study was conducted in three soil-mapping units in three regions of the RS, with the following objectives:

1. determine the variation of annual and seasonal precipitation over time,
2. determine which seasons of the year effective rainfall can refill the soil water profile, and
3. estimate the time required to fill the pore space of the soil profile with water available from the effective rainfall with different scenarios of water infiltration into the soil.

METHODS

The study was conducted in the north of the RSon the Passo Fundosoil-mapping unit, classified as oxisol;in the center of the RS on the São Pedro soil-mapping unit, classified as red-yellow podzolic loam texture; and in the south of the RS onthe Vacacaí soil-mapping unit, classified as planosolo hydromorphic eutrophic, used with flood-irrigated rice [3,4].

The study of temporal rainfall was done using monthly and annual totals data from three weather stations: Passo Fundo (latitude 28°15'39"S, longitude 52°24'33"W, and altitude 678 m), Santa Maria (latitude 29°41'25"S, longitude 53°48'42"W, and altitude 138 m), and Santa Vitória do Palmar (latitude 33°31'14"S, longitude 53°21'47"W, and altitude 6 m), located, respectively, in the north, center, and south of the RS. Data were obtained from the Eighth District of Meteorology (8th DISME) of the National Institute of Meteorology (INMET), in Porto Alegre. The observation period of rainfall for the three sites was 1914 to 2004. From the monthly totals, annual and seasonal monthly averages were obtained, taking into consideration the summer months of December, January, and February; the autumn months of March, April, and May; the winter months of June, July, and August; and the spring months of September, October, and November.

At the three weather stations, the rainfall was found to be of type iso-hygro, the rain is distributed similarly over the 12 months of the year, and the Spearman correlation coefficient showed that the rainfall is homogeneous [15]. This coefficient shows that the rainfall is homogeneous for the three stations, not that the system is iso-hygro. The iso-hygro quality was checked visually. In a preliminary exploratory analysis, as suggested by the World Meteorological Organization [16], scatter diagrams were plotted, relating the calendar year to the original data and also to the centered moving average, with a period of 9 years. To verify the existence of abrupt changes in each weather station, the historical data series was divided into two periods and the Mann-Whitney test applied, comparing the averages of these periods. The dividing point was held year to year from 1918 to 1998, involving 81 test applications. The year in which the test showed a lower minimum level of significance MLS was considered the date of abrupt change. A year in which no test had $p \leq 0.05$ was considered a year of no abrupt changes.

The Spearman correlation coefficient and the MannWhitney nonparametric test were used because rainfall data do not meet all the requirements of parametric equivalent tests, particularly the assumption of independence of errors [17].

The potential evapotranspiration (PET) was calculated by the model of Thornthwaite and Mather [18]. For this, the average monthly temperatures for the period 1931- 1960 were used [7]. The effective rainfall to refill the soil profile was obtained by the difference between the rainfall computed for the period and average PET, the normal value for the same period.

To estimate the time needed to fill the soil profiles with water from the effective rainfall, we used precipitation data and mean values of the soil physical parameters for water and dates from its mapping-unit descriptions. These values were obtained from the reconnaissance soil survey of the RS [4] and experiments performed in the areas under study [5,19]. The total porosity (Tp), in% of volume was calculated by Equation (1):

$$Tp = 100 - \left[\frac{(100 \times Sd)}{Pd} \right], \tag{1}$$

where Sd is the soil bulk density, and Pd is the particle density.

The maximum storage of water for each mapping unit, Equation (2), was determined by considering the depth of the soil profile:

$$V = \frac{(Fc - Wp)}{100} Sd \cdot d \cdot S, \quad (2)$$

where V is the maximum storage of water ($m^3 \cdot ha^{-1}$), Fc is field capacity (% mass), Wp is wilting point (% mass), Sd is soil bulk density, S is the area (m^2), and d is soil depth (m).

To estimate the time required to refill the pore space of the soil profile it was assumed that 50% of the maximum soil water storage capacity (field capacity less wilting point) was already filled with water. Three scenarios of water infiltration into the soil were simulated using 1/2, 2/3, and 3/4 of the normal average effective precipitation. The volume needed to fill the macropores (V_{macro} = pore space between field capacity and % of soil water saturation) was calculated, using Equation (3):

$$V_{macro} = \frac{(Tp - Fc)}{100} \times d \times S, \quad (3)$$

where V_{macro} is the volume of water needed ($m^3 \cdot ha^{-1}$) to fill the soil macropores, Tp is the total porosity (% volume), Fc is field capacity (% vol.), d is profile depth (m), and S is area (m^2). If the area S is 10,000 m^2, to find the values in depth of water h (mm) the value V_{macro} must be divided by 10.

RESULTS AND DISCUSSION

Figure 1 shows the total annual rainfall scatter points for the counties of Passo Fundo, Santa Maria, and Santa Vitóriado Palmarduring the period of 1914-2004 and average normal values for PET. It shows that for Santa Maria and Passo Fundo, only in 1917 was the average PET higher than the total annual rainfall. For Santa Vitóriado Palmar PET was higher than average rainfall in 4 years. The results in Figure 1 indicate that the volume of water from rainfall is not the limiting factor for the lack of water in the soil.

When the existence of sudden changes in precipitation curves for two observation periods were studied, a change in the volume of precipitation was found in or near the year 1981.

After 1981, the averages were higher for the three weather stations: Passo Fundo, 241.5 mm; Santa Maria, 219.7 mm; and Santa Vitóriado Palmar, 129.1 mm (Table 1). From these results, it is evident that in recent years no reduction in the volume of rainfall occurred in the three regions of the RS, although the impacts of drought periods on water shortages have been higher.

The drought effects that occur today are extremely more impactful on crop yields and on base flow of the rivers than formerly occurred because in that time more water was probably stored in the soil profile. To analyze this impact, Table 2 presents the average values of physical parameters of the soil and water in the three soil mapping units used in this work, which form the basis to simulate future scenarios on the time required for the soil profile to recover the saturated flow to sustain the streams and rivers.

In Table 2, the high value for total depth of water needed to saturate the soil profile for the Passo Fun- dosoil-mapping unit is because of the 8-meter depth and high percentage of water storage (10%). The data show that if the water infiltration rate is not the limiting factor, this soil can store the total amount of one year's rainfall at once. However, that is not possible because of low infiltration. The estimation of the time required to fill the pore space of the soil profile with water available from the effective rainfall with different scenarios of water infiltration into the soil can be found in Table 3. When PET values were higher than the effective rain fall, it was assumed that there was no water storage in the soil. The recharge of ground water throughout the year occurs during the winter, followed by spring and autumn, with reversing spring through autumn in the Passo Fundo region because of higher values of normal annual rainfall.

Table 1. Average total annual rainfall for the three weather stations in the RS during the periods of 1914-1981 and 1982- 2004

Weather Stations	1914-1981	1982-2004	Difference
		(mm)	
Santa Maria	1672.2	1891.9	219.7
Passo Fundo	1697.1	1938.6	241.5
Santa Vitória do Palmar	1204.4	1333.5	129.1
Average	1524.8	1721.3	196.7

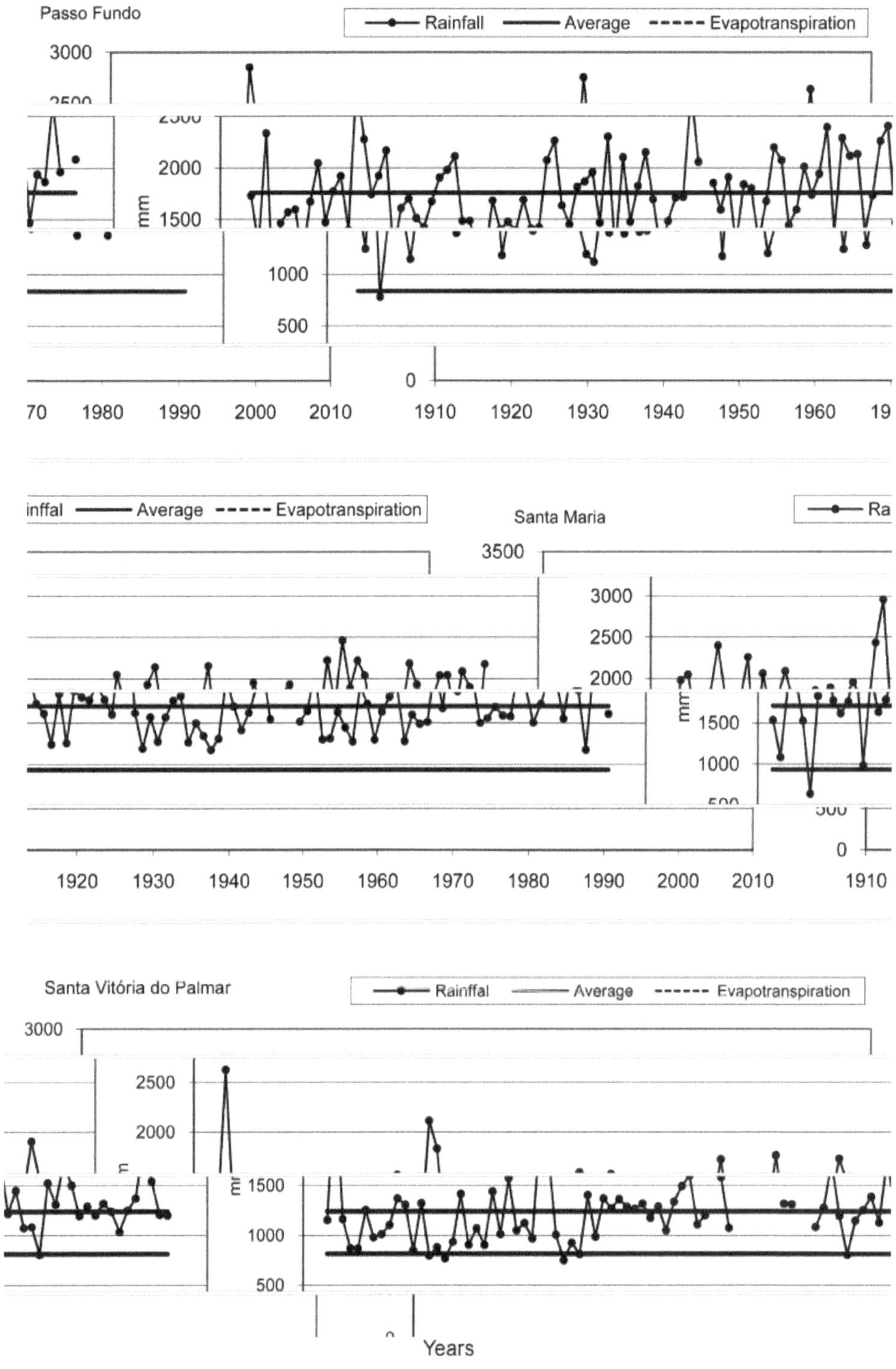

Figure 1. Scatter plots of the total annual rainfall and average annual PET from the meteorological stations of Passo Fundo, Santa Maria, and Santa Vitóriado Palmar.

The number of years required to saturate the soil profile can be seen in Figure 2. In the Passo Fundo soil mapping unit, considering a scenario in which 1/2 of the effective rainfall infiltrates the soil, it would take 16.6 years to fill the pores with water. This observation probably explains why the current droughts have a greater impact on the availability of water in rivers and on slopes compared with the droughts that occurred before the 1940s. The soil macropores under current conditions are practically empty.

In periods of drought, reducing recharge of macropores has a direct impact on water availability in rivers and reservoirs. Limited recharging incapacitates these environments, making them unable to resist the climatic oscillations that were evaluated by analysis of the time series of rainfall. The storage capacity depends on soil water infiltration, pore space, and the depth of the soil profile. Thus, the stability of the flow in rivers in summertime depends more on storage in deep soils (oxisols) than in lowland soils (planosols).

Table 2. Average values of soil physics and hydric parameters of soil-mapping units of Passo Fundo oxisols, Vacacaí planosolo hydromorphic eutrophic, and São Pedro podzólico hapludox

	County region		
Physical parameters[1]	Passo Fundo	Santa Vitória do Palmar	Santa Maria
Soil-mapping unit	Passo Fundo	Vacacaí	São Pedro
Depth of soil profile (m)	8.00	1.00	3.00
Field capacity (% mass)	24.00	15.00	19.00
Wilting point (% mass)	14.00	7.00	8.00
Soil bulk density	1.24	1.52	1.41
Particle density	2.69	2.61	2.63
Total porosity (% vol.)	53.90	41.76	46.39
Microporosity (% vol.)	29.76	22.80	26.79
Macroporosity at field capacity (% vol.)	24.14	18.96	19.60
Maximum storage in the soil profile (mm)	992.00	121.60	465.30
Depth of water (50%); maximum storage (mm)	496.00	60.80	232.65
Depth of water to fulfill the soil macroporosity (mm)	1931.47	189.62	587.93
Total depth of water to saturate the soil profile (mm)	**2427.47**	**250.42**	**820.58**

[1]Source: The values of soil physical parameters used are the average results from published research papers [5,8,19].

Table 3. Water storage in the soil, normal and effective rainfall, PET, and net storage of water in the soil for the seasons, with three scenarios and estimation of infiltration recharge of the pore space for the Passo Fundo (oxisols), São Pedro (podzolic hapludox) and Vacacaí (eutrophic planosolo hydromorphic) mapping units

Simulated Scenarios	Mapping Unit/Parameters	Seasons of the Year				Total Annual
		Spring	Summer	Autumn	Winter	
	Passo Fundo Soil			**(mm)**		
	Total depth of water to saturate the soil profile (mm)*	-	-	-	-	2427.5
	Normal rainfall (mm)	489.2	439.9	392.8	433.6	1755.5
	PET (mm)	213.4	325.3	191.6	107.0	837.3
1/2 of the precipitation infiltrates	1/2 Rainfall (mm)	244.6	219.9	196.4	216.8	
	Net storage in the soil (mm)	31.2	0.0	4.8	109.8	145.8
	Time to saturate the soil profile (years)					**16.6** years
2/3 of the precipitation infiltrates	2/3 Rainfall (mm)	326.1	293.3	261.9	289.1	1,170.4
	Net storage in the soil (mm)	112.7	0.0	70.3	182.1	335.0
	Time to saturate the soil profile (years)					**6.6** years
3/4 of the precipitation infiltrates	3/4 Rainfall (mm)	366.9	329.9	294.6	325.2	
	Net storage in the soil (mm)	153.5	4.7	103.0	218.2	479.4
	Time to saturate the soil profile (years)					**5.1** years
	São Pedro Soil					
	Total depth of water to saturate the soil profile (mm)*	-	-	-	-	820.6
	Normal rainfall (mm)	428.8	416.1	430.3	426.8	1702.0

	PET (mm)	221.9	385.0	216.3	111.0	934.2
1/2 of the precipitation infiltrates	1/2 Rainfall (mm)	214.4	208.1	215.2	213.4	
	Net storage in the soil (mm)	0.0	0.0	0.0	102.4	102.4
	Time to saturate the soil profile (years)					**8.0** years
2/3 of the precipitation infiltrates	2/3 Rainfall (mm)	285.9	277.4	286.9	284.5	
	Net storage in the soil (mm)	64.0	0.0	70.6	173.5	308.1
	Time to saturate the soil profile (years)					**2.7** years
3/4 of the precipitation infiltrates	3/4 Rainfall (mm)	321.6	312.1	322.8	320.1	
	Net storage in the soil (mm)	99.7	0.0	106.4	209.1	415.2
	Time to saturate the soil profile (years)					**2.0** years

	Vacacaí Soil					
	Total depth of water to saturate the soil profile (mm)*	-	-	-	-	250.4
	Normal rainfall (mm)	280.6	290.3	326.4	324.0	1221.3
	PET (mm)	190.0	325.0	204.0	94.0	813.0
1/2 of the precipitation infiltrates	1/2 Rainfall (mm)	140.3	145.1	163.2	162.0	
	Net storage in the soil (mm)	0.0	0.0	0.0	68.0	68.0
	Time to saturate the soil profile (years)					**3.7** years
2/3 of the precipitation infiltrates	2/3 Rainfall (mm)	187.1	193.5	217.6	216.0	
	Net storage in the soil (mm)	0.0	0.0	13.6	122.0	135.6
	Time to saturate the soil profile (years)					**1.8** years
3/4 of the precipitation infiltrates	3/4 Rainfall (mm)	210.5	217.7	244.8	243.0	
	Net storage in the soil (mm)	20.5	0.0	40.8	149.0	210.3
	Time to saturate the soil profile (years)					**1.2** years

*Data taken from **Table 2**.

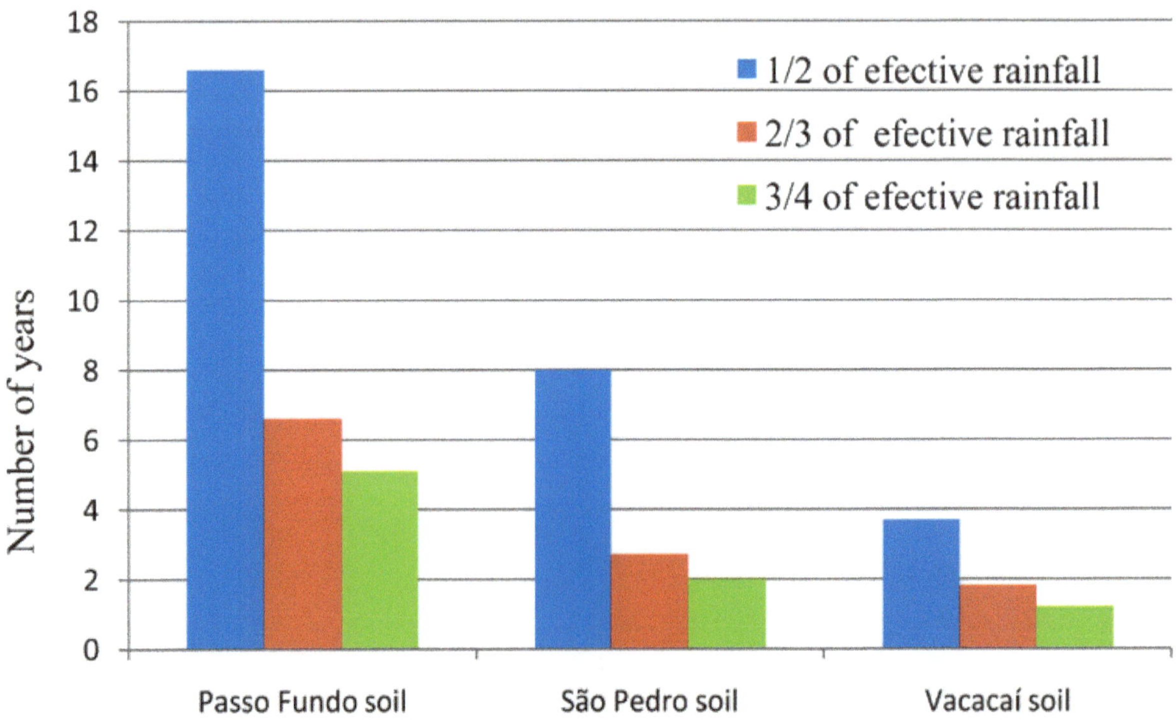

Figure 2. Number of years with normal rainfall necessary to saturate the soil profile of the Passo Fundo, São Pedro, and Vacacaí mapping units for three scenarios of infiltration, 1/2, 2/3, and 3/4 of the effective rainfall.

CONCLUSIONS

1. In the north, center, and south regions of Rio Grande do Sul State, based on the data of normal rainfall in Passo Fundo, Santa Maria, and Santa Vitória do Palmar between 1914 and 2004, there is no reduction in the annual volume of rainfall.
2. During the summer, the three simulated conditions showed that there is not enough recharge of soil macropores. The recovery of water storage in the soil profile must occur mainly in winter, followed by the spring and fall seasons.
3. In the simulation scenario, infiltrating 1/2, 2/3, and 3/4 of effective rainfall shows that the time needed to refill the porous spaces is 16.6, 6.6, and 5.1 years in Passo Fundo soil, is 8.0, 2.7, and 2.0 years in São Pedro soil, is 3.7, 1.8, and 1.2 years in Vacacaí soil, respectively.

REFERENCES

1. Buriol, G.A., Estefanel, V., Saccol, A.V., Heldwein, A.B. and Schneider, F. (1980) Possibility occurrence of available soil water in the state of Rio Grande do Sul. Journal of Rural Sciences Center, 10, 1-141.
2. Avila, A.M.H. (1994) Rainfall regime in the state of Rio Grande do Sul based on long-term series. Dissertation, UFRGS, Faculty of Agronomy, UFRGS Porto Alegre, 75 p.
3. Brazil Ministry of Agriculture (1973) Reconnaissance soil survey of the state of Rio Grande do Sul Recife. Technical Bulletin, 30, 431.
4. Embrapa (1999) Brazilian system of soil classification. Production Information Service, Brasilia, 412 p.
5. Righes, A.A., Nishijima, T., Denardin, J.E., Kochhann, R. A. and Garcia, S.M. (2002) Vertical mulching and runoff inno-tillage system. XIV Brazilian Meeting of Management and Conservation of Soil and Water, Cuiabá.
6. Machado, F.P. (1950) Contribution to the study of the climate of Rio Grande do Sul. Brazilian Institute of Geography and Statistics, Rio de Janeiro, 91 p.
7. Agronomical Research Institute (1989) Agroclimatic Atlas of the State of Rio Grande do Sul Porto Alegre: Secretariat of Agriculture and Supply, 1, 102 p.
8. Ministry of Agriculture (1972) Water Balance of Brazil. National Department of Meteorology, 94 p.
9. Righes, A.A. (2002b) Water: Sustainability, use and availability for irrigation. Science and Environment, 21, 90- 102.
10. Bertol, I., Beutler, J.F., Milk, D. and Batistela, O. (2001) Physical properties of a humic cambisol affected by the type of soil management. Scientia Agricola, 58, 555-560. http://dx.doi.org/10.1590/S0103-90162001000300018
11. Denardin, J.E. (1992) Soil: Formation anddegradation. In: Marcantonio, G., Ed., Soilsand irrigation, UFRGS/ FEDERACITE, Porto Alegre, 13-28.
12. Spera, S.T, dos Santos, H.P., Fontaneli, R.S., Kochhann, R.A., Denardin, J.E. and Spera, M.R.N. (2002) Soil compaction in southern Brazilian oxisols under no-till system. In: International Conference on Land Degradation and Meeting of The IUSS Sub-commission C-Soil and Water Conservation, Rio de Janeiro, 1-3.
13. Albuquerque, J.A., Sangoi, L. and Ender, M. (2001) Effects of crop/livestock integration in soil physical properties and characteristics of corn. Brazilian Journal of Soil Science, 25, 717-723.
14. Denardin J.E., Kochhann, R.A. and Righes, A.A. (2005) Vertical mulching: Technical management of runoff in tillage. Tillage Magazine, 14, 37-39.
15. Pashiardis, S. (2009) Trends of precipitation in Cyprus rainfall analysis for agricultural planning. www.fao.org/ rd/climagrimed/pdf/ws01_08.pdf
16. WMO (2000) Detecting trends and other changes in hydrological data. World Meteorological Organization (WMO/TD No. 1013), Geneva, 157 p.
17. de Campos, H. (1983) Non-parametric experimental statistical. 4th Edition, ESALQ, Piracicaba, 349 p.
18. Thornthwaithe, C.W. and Mather, J.R. (1955) Water balance climatology. Publications in Climatology, 8, 1-104.
19. Peraza, J.E.S. (2003) Water retention and pedofunctions for soils of Rio Grande do Sul. Masters Thesis, Graduate Program in Agricultural Engineering, Federal University of Santa Maria, Santa Maria, 117 p.

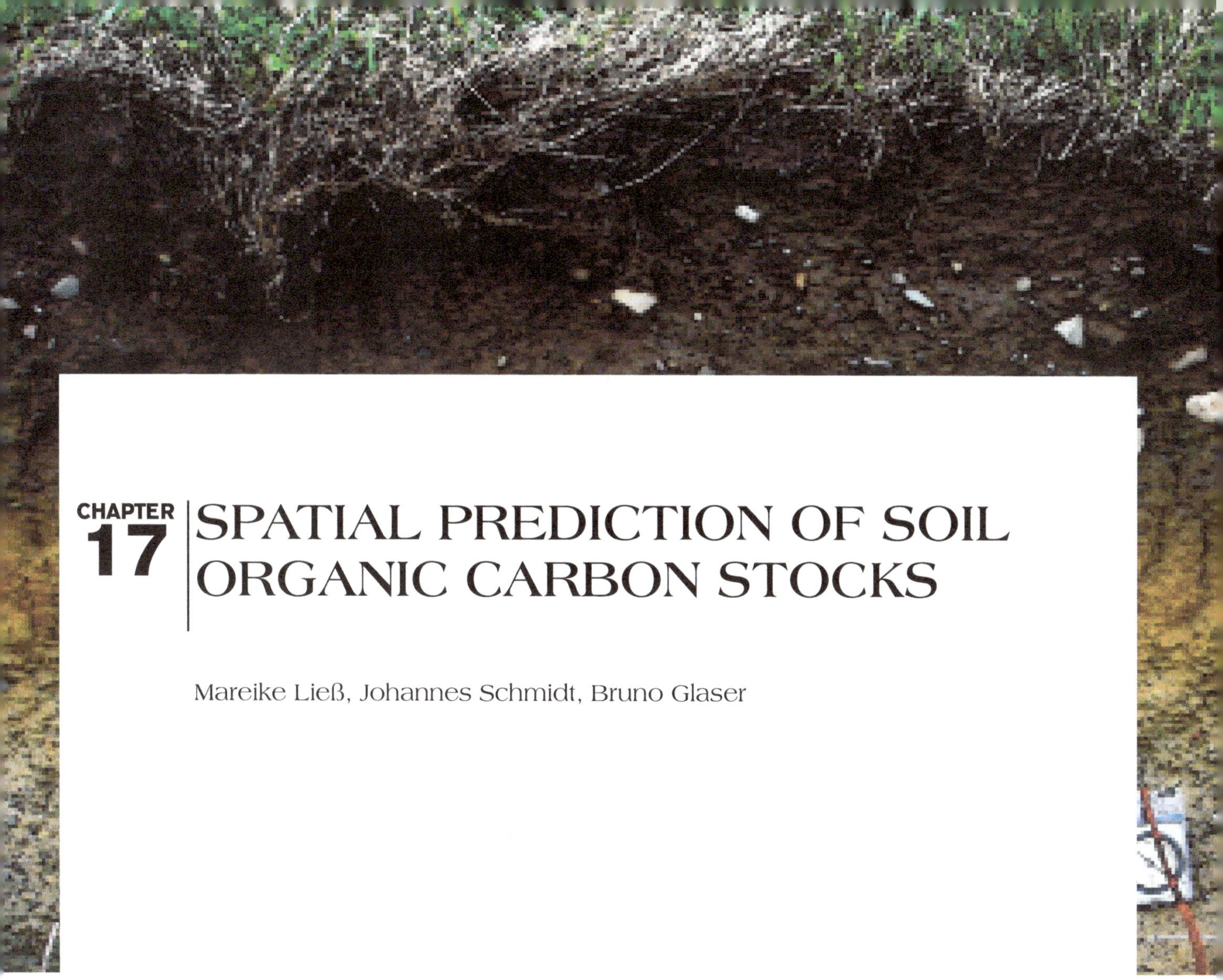

CHAPTER 17 SPATIAL PREDICTION OF SOIL ORGANIC CARBON STOCKS

Mareike Ließ, Johannes Schmidt, Bruno Glaser

INTRODUCTION

Tropical forests play a key role in the global carbon cycle storing a total of 471 Pg carbon [1,2]. The soils' carbon storage potential is generally even greater than that of the vegetation [3]. Don et al. [4] report, that 36 to 60% of the tropical ecosystem's carbon is stored in soil. But, land use change from primary forest to other land uses leads to a decrease in soil organic carbon (SOC) stocks [4,5]. Ecuador in particular has the highest annual deforestation rate in South America [6]. Tapia-Armijos et al. [7] report a reduction of the area covered by natural vegetation by 46% (Southern Ecuadorian provinces). Local farmers make "extensive use of fire" to convert primary forest into farming land and pastures [8]. According to Bahr et al. [9] 9 to 13 Mg SOC per hectare are lost due to land use changes from forest to crop land and pastures. Finally, spatial estimates of SOC are increasingly important to acknowledge the soils' carbon storage potential in the context of climate change. However, it is particluarly the tropical mountain areas with their thick organic layers which are highly complex and difficult to access [10]. SOC stock data of tropical mountain forest soils are scarce, SOC stock data of the organic layer hardly exist.

Regression based digital soil mapping (DSM) provides a means of regionalizing soil data from a limited amount of samples to a landscape level by making use of the factors of soil formation [11] as predictors. Spatial continuous predictors representing topography and vegetation are obtained from digital elevation models (DEMs) and satellite images. However, for many soil properties, spatial regression modelling may not produce a robust model. According to Ryan et al. [12], low r^2 values may result from one or more of the following causes:

1. poor relation to the available environmental predictor variables,
2. extreme local variation due to unknown or random effects, or
3. the collected data spans a very small interval in the total range of the response variable.

While the latter can be mostly avoided by a good sampling design which follows a good representation of the predictor space [13], the former two causes provide real challenges.

DEMs are often used at their original raster resolution with a 3x3 window size for the calculation of the derived predictors. However, a number of studies suggest that predictor-response relationships are strongly landscape and scale dependent [12]. Cavazzi et al.[14] investigated the interacting effect between

Citation: Ließ M, Schmidt J, Glaser B (2016), Spatial Prediction of Soil Organic Carbon Stocks, PLoS ONE 11(4): e0153673. doi: 10.1371/journal.pone.0153673

window and raster cell size and found cell size to be significant in all considered areas whereas the interaction between window and cell size was significant in morphological rough areas. Finally, soil-forming factors (predictors) vary and respond at different scales [15]. Maynard and Johnson [16] found a strong scale-dependency for total carbon having the best model performance at coarse neighborhood extents (150 to 300 m); DEM resolution affected soil-terrain correlations to a much lesser degree. Samuel-Rosa et al. [17] have shown that investigating the impact of scale in predictors is more important when the predictor-response relationships are weak. Finally, multi-scale as well as feature selection approaches according to [15] deserve more research to obtain a better prediction accuracy.

The term feature selection refers to the process of removing irrelevant predictors from the predictor set to enhance a model's performance and generalization capability (e.g. [18]). It requires an exhaustive search of all possible subsets of predictors in order to decide which subset performs best. With a large set of predictors, this procedure is simply not applicable [19]. Therefore, due to practicability reasons the selection of a subset which is just good enough if not optimal, might have to be sufficient [20]. Predictor selection procedures can be described by two main categories [21]: (1) filter methods and (2) wrapper methods. Filter methods make a predictor assessment based on general characteristics of the dataset (e.g. predictor–response correlation), independently from the particular machine learning algorithm and hence ignore the predictors's effectiveness within the particular model. Wrapper methods, evaluate predictor performance by running the particular machine learning algorithm on the dataset [22]. Accordingly, most filter methods evaluate each predictor individually and ignore possibly important predictor interactions, whereas wrapper methods result in an increase in computation time [23]. It is often argued that particularly recursive partitioning methods do not require predictor selection as they are at least theoretically resistant to irrelevant predictors [23,24]. However, Witten and Frank [18] show that decision trees are affected by non-informative predictors and Kuhn and Johnson [23] show that ANN and SVM are affected to an even larger extent.

A number of machine learning algorithms are commonly used in DSM, such as e.g. tree-based methods, artificial neural networks (ANN), multivariate adaptive regression splines (MARS) and support vector machines (SVM). Each of them has its strengths and pitfalls and in dependence of the particular application and soil-landscape there is no single algorithm which serves all. Grimm et al. [25] and Guo et al. [26] use random forest to map SOC. Stepwise regression was applied by Gessler et al. [27], Gasparini et al. [28] and Zhang et al. [29]. Martin et al. [30,31] applied boosted regression trees. Pastick et al. [32] and Bou Kheir et al. [33] used decision tree models. ANN were applied by Dai et al. [34]. However, comparisons of various machine learning approaches to spatially predict SOC in tropical mountain areas are scarce. Indeed we only found one: Were et al. [35] compare SVM to ANN and random forest to map SOC stocks in an afromontane landscape with SVM showing the best performance.

Typically the machine learning algorithms have one or several tuning parameters, and the estimation of these tuning parameters should be based on an estimate of the prediction error [24]. The number of layers and neurons is a crucial decision to be made when constructing ANNs. A network with too few neurons cannot differentiate between complex data patterns, whereas too many neurons would lead to overfitting [36]. In support vector regression the model parameters C and ε as well as the kernel parameters must be tuned to obtain sensible results [37,38]. According to Hastie et al. [24] random forests do remarkably well with little tuning. In our experience [10,39] tuning random forest and boosted regression tree models can improve prediction results, but tuning needs to be tested carefully for it can also cause overfitting.

Particularly in the complex soil-landscape settings of tropical mountain areas, the choice of methodology for spatial prediction is of high importance to improve the expected poor model results. Accordingly within this study four aspects will be considered to obtain the best possible model:

- Considering different spatial settings/scales for the predictors
- Apply predictor selection strategies to reduce noise and enhance the model's performance
- Try different machine learning algorithms to capture the complex predictor-response relation. Among the vast choice of algorithms, we selected algorithms that follow different adaptation strategies
- Apply model tuning to optimize the model's performance

MATERIAL AND METHODS

Research Area

The research permit for the respective area was granted by the Ecuadorian Ministry of the Environment. The research area is situated on the eastern escarpment of the southern Ecuadorean Andes between the provincial capitals Loja and Zamora (Fig 1). It forms part of the Natural Reserve Podocarpus—El Condor and is mainly covered by tropical montane forest vegetation changing into shrub and grassland vegetation above the tree line of ca. 2800 m a.s.l.. The area is influenced by an altitudinal decrease in temperature and increase in rainfall. This corresponds to a mean annual air temperature ranging from 19.4°C to 9.4°C and a mean annual total rainfall from 2050 mm to 4400 mm [40,41]. Geologically the area is part of a metamorphic belt of palaeozoic age. Litherland et al. [42] describe the underlying bedrock as part of the Chiguinda Unit of the litho-tectonic section Loja Terrane, consisting of pelite, schist, phyllite, meta-siltstone, sandstone and quartzite. The area is known for its immense organic layers leading to its description as sloping mire soil-landscape [10]. Its formation is partly related to the dominance of soil hydromorphic properties in the topsoil caused by the silty soil texture, heavy rainfall and shallow slope parallel subsurface flow [43,44]. The occurrence of frequent landslides adds to the complexity of this remote tropical mountain landscape.

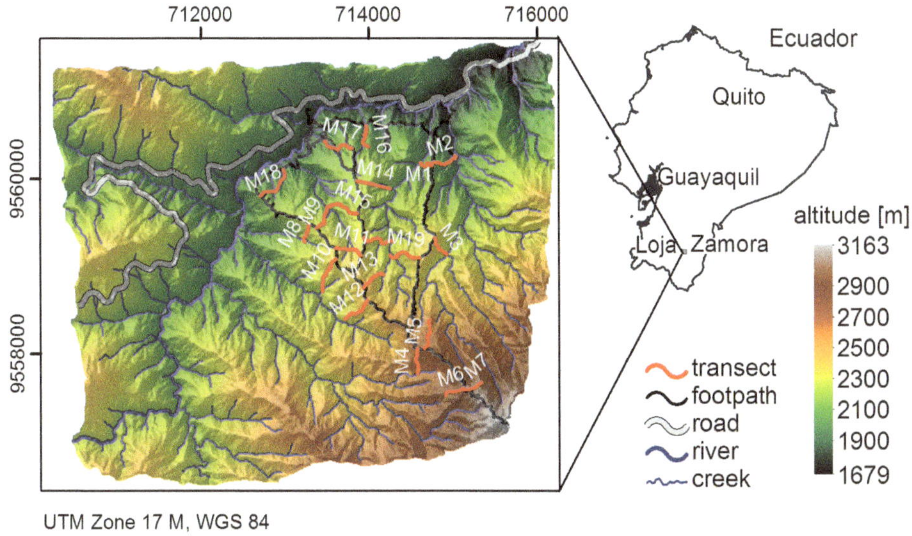

Figure 1. Research area with sampling transects (adapted from [45]).

Dataset

Carbon Data–Sampling and Analysis

The organic layer of the research area was sampled along transects as indicated in Fig 1. The transects were laid to cover the complete hillslope from the ridge to the valley bottom. They were positioned to cover terrain units formed by an overlay of two altitudinal, three slope and two aspect classes, while inclination had to be considered to permit accessibility of the very steep terrain [45]. Each transect was then sampled randomly at three positions to cover the upper, middle and foot slope. Sampling was conducted by using a 20 by 20 cm metal frame of 10 cm height. Samples were oven-dried at 45°C until mass consistency. After the removal of fresh roots, organic carbon contents were determined by a VARIO MAX elemental analyzer at the soil laboratory of the University of Halle.

Fig 2 gives an overview of the SOC stocks of the organic layer [46] used for this study. Interestingly, the highest stocks are found at the upper slope position of transects M1 and M19, both positioned at rather low altitudes, while transects M4 to M7 which are positioned at high altitudes have very low carbon stocks.

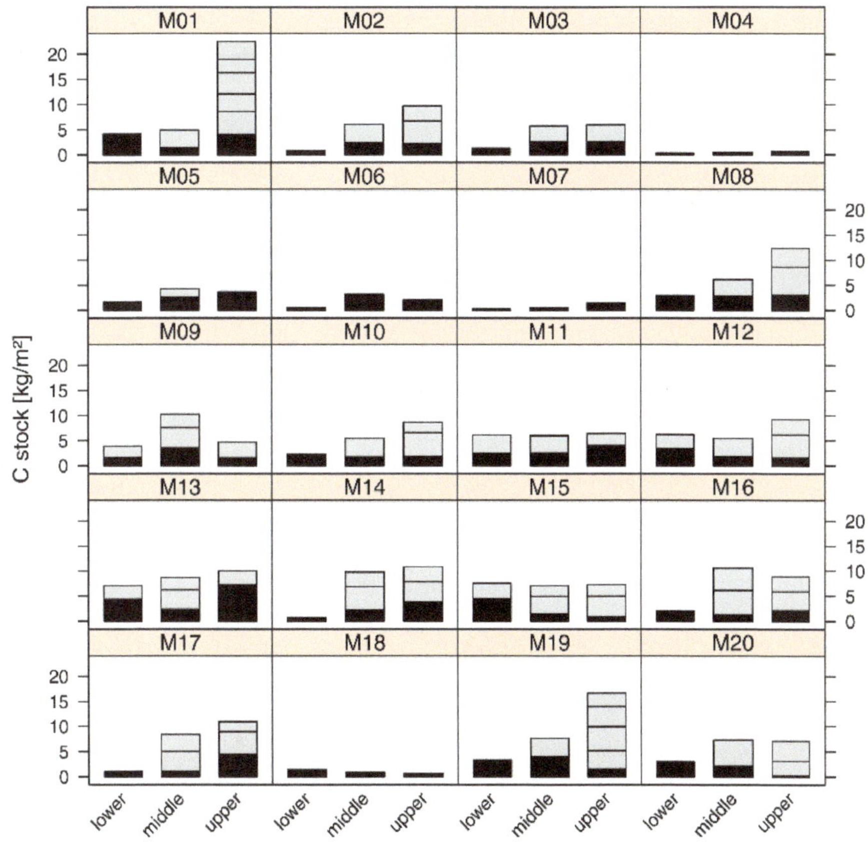

Figure 2. Carbon stocks of the organic layer in each of the sampled profiles at the upper,middle and lower part of the transects M1 to M20. Each part of the stacked bar plots refers to the carbon stock of a layer of 10 cm with exception of the lowermost layer, here indicated by black color.

Predictors

Predictors used for model development include parameters derived from a satellite image and a DEM. The Landsat 8 OLI/TIRS image from 2014, February 18th, was provided as image courtesy of the U.S. Geological Survey (USGS). The provided image bands were transformed to Top of the Atmosphere Reflectance (TOA) and At-Satellite Brightness Temperature (SatTEMP) using the image's metadata and information provided by the USGS [47]. The normalized difference vegetation index (NDVI), and the normalized difference moisture index (NDMI) were calculated according to Jackson et al. [48]. The perpendicular vegetation index (PVI) and the transformed soil adjusted vegetation index (TSAVI) are using the soil line concept introduced by Richardson and Wiegand [49]. The soil line concept demonstrates the observed linear relationship between Red and NIR reflectance of bare soil. The parameters β_1 and β_0 in Eqs 1 and 2 refer to the slope and intercept of this relation and were automatically determined according to Fox et al. [50] as $\beta_1 = 0.71$ and $\beta_0 = 0.11$. X in Eq 2 is a constant usually assumed to be 0.08 [50] [51] [52].

$$PVI = \frac{1}{\sqrt{\beta_1^2 + 1}}(NIR - \beta_1 R - \beta_0) \tag{1}$$

$$TSAVI = \frac{\beta_1(NIR - \beta_1 R - \beta_0)}{\beta_1 NIR + R - \beta_1\beta_0 + X(1 + \beta_1^2)} \tag{2}$$

All predictors calculated from the mentioned landsat image are shown in Table 1. Landsat bands 1 to 11 refer to the predictor ID 1 to 10, the calculated vegetation indices refer to ID 11 to 14.

Table 1. Predictors derived from the landsat image

ID	Predictor	Input	Calculation
1	Aerosol	band 1	1.2901E-02*x-64.50640
2	Blue	band 2	1.3211E-02*x-66.05534
3	Green	band 3	1.2174E-02*x-60.86943
4	Red	band 4	1.0266E-02*x-51.32853
5	NIR	band 5	6.2821E-03*x-31.41050
6	SWIR1	band 6	1.5623E-03*x-7.81151
7	SWIR2	band 7	5.2658E-04*x-2.63290
8	Panchromatic	band 8	1.1618E-02*x-58.08977
9	TIRS1	band 10	3.3420E-04*x+0.10000
10	TIRS2	band 11	3.3420E-04*x+0.10000
11	NDVI	band 4, band 5	
12	PVI	band 4, band 5	
13	TSAVI	band 4, band 5	
14	NDMI	band 5, band 6	

doi:10.1371/journal.pone.0153673.t001

Geomorphological and hydrological DEM parameters were calculated by R-package RSAGA [53,54]. The DEM was provided by the DFG RU 816 database [55]. Its cell size was adapted to the landsat image's cell size of 30 m.

The DEM parameters are listed in Table 2. Please be aware of their ID numbers 15 to 236. Predictors 37 to 236 were calculated with different computing window sizes of 3x3 to 41x41 cells corresponding to different degrees of terrain smoothing and search radii of 45 to 615 m.

Table 2. Predictors obtained from DEM

ID	Terrain Parameter	Module Library	Module	Reference	SAGA author/ year
15	Altitude (DEM)	Grid—Spline Interpolation	Multilevel B-Spline Interpolation		Conrad/ 2006
16	Slope		Morphometric Features	[56]	Conrad/ 2013
17	Aspect				
18	Mass Balance Index (MBI)		Mass Balance Index	[57–59]	Conrad/ 2008
19	Slope Height				
20	Valley Depth	Terrain Analysis -	Relative Heights		Böhner &
21	Normalised Height	Morphometry	and Slope Positions		Conrad/ 2008
22	Standardised Height				
23	Wind Effect		Wind Effect		Böhner & Ringeler/2008, Conrad/ 2011
24	Hill Index		Valley and Ridge Detection	[60]	Conrad/ 2013
25	Direct Insolation		Potential Incoming	[61–63]	Conrad/ 2010
26	Diffuse Insolation	Terrain Analysis	Solar Radiation		
27	Visible Sky	- Lighting,	Sky View Factor	[61,62,64]	Conrad/ 2008
28	Sky View Factor	Visibility			
29	Positive Openness		Topographic Openness	[65–67]	Conrad/ 2012
30	Negative Openness				
31	Catchment Area		Catchment Area (Flow Tracing/		
32	Catchment Height	Terrain Analysis -	Kinematic routing algorithm)	[68]	Conrad/ 2001
33	Catchment Slope	Hydrology			
34	SAGA Wetness Index (SWI)	(with pre-processed DEM)	SAGA Wetness Index	[69]	Böhner & Conrad/ 2001
35	Topographic Wetness Index (TWI)		Topographic Wetness Index	[70–72]	Conrad/ 2003
36	LS Factor		LS Factor	[71,73,74]	Conrad/ 2003
37–56	Convergence		Convergence Index (Search Radius)	[75]	Conrad/ 2003
57–76	Terrain Ruggedness Index (TRI)		Terrain Ruggedness Index	[76]	Conrad/ 2010
77–96	Terrain Surface Texture		Terrain Surface Texture	[77]	Conrad/ 2012
97–116	Terrain Surface Convexity		Terrain Surface Convexity	[77]	Conrad/ 2012
117–136	Plan Curvature	Terrain Analysis -			
137–156	Profile Curvature	Morphometry			
157–176	Longitudinal Curvature		Morphometric Features	[78,79]	Conrad/ 2013
177–196	Cross-Sectional Curvature				
197–216	Minimum Curvature				
217–236	Maximum Curvature				

doi:10.1371/journal.pone.0153673.t002

Regression Models

Five machine learning algorithms were compared in their performance to predict the organic layer carbon stocks:

- Random Forest (RF)
- Artificial neural network (ANN)
- Multivariate adaptive regression splines (MARS)
- Boosted regression trees (BRT)
- Support vector machine (SVM)

Random Forest

RF is a recursive partitioning method which grows a number of regression trees [80] and averages the results. In a regression tree the data is subsequently partitioned by the predictor variables into preferably homogeneous subsets regarding the response variable. The mean of each data subset is then used as predicted response value. A partition point or so called node is always defined by that predictor and threshold in its range, which achieves the most homogeneous partition into two subsets (tree branches). At each node all predictors with all possible threshold values are tested.

The RF ensemble model's stability is obtained through varying the trees' input dataset and the subset of predictors used to subsequently split the data in each tree node. According to Hastie et al. [24], RF does not require much tuning. The RF model was adapted by R-package "randomForest" [81], growing 1000 trees (ntree) and the models default parameters for the size of the predictor subset (mtry = p/3) and the minimum amount of data in the terminal nodes (nodesize = 5). The size of the data subset selected to grow each tree was set to the size of the training set, sampling was done with replacement.

Artificial Neural Network

ANNs consist of a number of neurons, the processing units of the algorithm, which are organized in layers. The input–the predictor vectors at each sampling point–has to pass these processing units to relate to the output: the response variable at each sampling point. The neurons of the input layer send data via synapses to the neurons of the first internal layer and these pass it via other synapses to the neurons of the next layer. The synapses store parameters called synaptic weights, which guide the learning process [82]. In general, an ANN is defined by (1) the interconnection pattern between the layers of neurons, (2) the learning rule which updates the weights of these connections, (3) the propagation function that converts a neuron's weighted input to its output, and (4) the activation function, which determines how the state of the neuron at point of time t+1 is calculated based on its state at point of time t and the new input [83]. Finally, the number of layers and neurons is a crucial decision to be made [36]. Generally one hidden layer is enough to approximate continuous functions, whereas two hidden layers might be necessary in the case of discontinuities [84]. A network with too few neurons cannot differentiate between complex data patterns, whereas too many neurons would lead to overfitting [36]. Data normalization before ANN training, is essential to prevent larger values from dominating smaller values, which otherwise might lead to premature saturation of neurons [36].

A two layer feed-forward ANN with Bayesian regularization [85] was trained with function "brnn" of R-package "brnn" [86]. The optimal number of neurons (1 to 20) was selected by five repetitions of an external tenfold cross-validation.

For the all-predictor model the maximum number of possible neurons was reduced to 13 (instead of 20) to combat computation time. With a number of 238 predictors, 13 neurons resulted in the estimation of 3094 parameters (weights and biases) and tuning the number of neurons with a range of 1 to 13 neurons in five times repeated tenfold cross-validation with parallel computing still resulted in a computation time of c. 30 hours. In this particular case the tuning resulted in best performance with only 1 neuron.

Multivariate Adaptive Regression Splines

The model algorithm is constructed by a weighted sum of piecewise linear basis functions of the form $(x\text{-}t)_+$ and $(t\text{-}x)_+$. Each of these so called linear splines is defined by a predictor (x) and a node (t), with the latter separating the first linear part of the function from the second linear part [24]. Starting with the intercept (the mean of the response variable), the algorithm repeatedly adds such a spline or a product of two or more splines. To define a new spline, at each step, the algorithm searches all predictors and all values of each predictor. This process of adding splines continues until a threshold value for the change in residual error or the maximum number of terms is reached. Due to its construction of splines with their internal node, MARS resembles recursive partitioning algorithms. However, in contrast to the latter, MARS produces continuous models [87]. Alltogether, the model algorithm also resembles stepwise forward linear regression, but instead of using the predictors themselves as input, it also permits to make use of already included model terms and products thereof [24]. MARS was implemented by R-package "earth" [88], which is based on Friedman's manuscripts "Multivariate Adaptive Regression Splines" and "Fast MARS" [87,89]. No pruning of the forward pass was applied, but the number of added terms, 1 to 50, was tuned by 5 repetitions of an external tenfold cross-validation. The degree of interaction between predictors was set to 2.

Boosted Regression Trees

Despite their many benefits, such as intuitive model structure, handling of input data of different measurement level and scale, insusceptibility to irrelevant predictors, insensitiveness to outliers etc., tree models have also disadvantages limiting their performance, particularly in modelling smooth functions, and the dependence of their structure from the input data [90].

Boosting sequentially applies a learning algorithm to repeatedly modified data versions and thereby produces a sequence of simple models, whose predictions are finally combined to make the overall prediction [24]. In BRT, at first a regression tree is constructed by assigning all training observations the same weight. Then at each iteration step a new regression tree is trained by giving those observations which impaired model performance in the previous step a higher weight [20]. In this way a sequence of models is build. The resulting BRT model is similar to an additive regression model with the subsequently fitted regression trees as individual terms [90].

BRT was implemented with R-package "gbm" [91], which is based on Freund and Schapire's AdaBoost [92], using a maximum number of 500 iterations. The optimal number of iterations was then selected by five repetitions of a tenfold cross-validation. Shrinkage, a crucial parameter referring to the learning rate was set to 0.001. The data subsampling rate for each of the trees which incorporates a random effect into the BRT algorithm and provides, therefore, the chance to improve BRT model performance [93], was set to 0.90. From the two parameters determining the size of each of the regression trees, interaction depth was set to 2 and the minimum number of samples in each of the final subsets was set to 5.

Support Vector Machine

SVMs were originally developed as an algorithm for classification, which separates the classes like other machine learning algorithms by a hyperplane. However, in contrast to other algorithms it searches for that hyperplane which leaves the largest possible margin between classes free of data, leading to a better generalization probability [94]. Each object to be classified is represented by a vector; the distance of those vectors closest to the hyperplane is maximised. Consequently, only the vectors close to the hyperplane are important for defining it, and are, therefore, called support vectors [94]. Points on the correct side and far away from it are ignored in the process of finding the best hyperplane.

In support vector regression (SVR) there is an analogy in that these "low-error-points" are those with small residuals [24]. The margin of a regression line shall contain all points with errors smaller than ε, i.e. no importance is given to these errors as long as they are smaller than ε [95]. SVR searches for a function which fulfils this ε-criterion and is as flat as possible [96]. Points outside the margin are allowed while a penalty weight C is introduced. It determines the trade-off between allowing points outside ε and the flatness of the regression function [96]. This is particularly important to decrease the impact of outliers. Non-linear regression is typically achieved by applying the so-called Kernel trick. Here, the training data is first transformed into a higher-dimensional feature space by applying a non-linear kernel function. Then a linear model is adapted to this new feature space [97]. Finally, the linear regression in the new feature space is equivalent to a non-linear regression in the original predictor space.

Data normalization before training a SVM is very important to avoid attributes of higher ranges dominating those with smaller ranges and to avoid problems when calculating the kernel function [37]. The support vector regression model parameters C and ε and the kernel parameters must be tuned to obtain sensible results [37,38]. SVR with a gaussian radial basis function kernel $f(x) = e^{(-\gamma|x-v_i|^2)}$ was applied by R-package "e1071" [98]. The width parameter γ, the inverse of the radius of influence of the selected support vectors v_i, determines the amount of generalization of the model. According to Cherkassky and Ma [38] $\varepsilon = 3\sigma\sqrt{\ln(n)/n}$ gives good performance. C and the radial kernel parameter γ were adapted by five repetitions of a tenfold cross-validation. For C a range of 0.1 till 1.0 with a step size of 0.1 was chosen, following Mattera and Haykin [99] and Cherkassky and Ma [38]. For γ a range of 0.2 till 0.5 with a step size of 0.05 was chosen [38].

Model Validation and Comparison

Five- or tenfold cross-validation are usually recommended to test model performance (e.g. [100,101]). The decision on how the subsets are formed remains important as the formed subsets may have an impact on the cross-validated error [100]. Subsets were formed by random selection. However, to account for the impact of subset membership, tenfold cross-validation was repeated five times with different subsets. Parallel computing of the five repetitions was employed with r-packages "doParallel" and "foreach" [102,103] to combat computation time. Predictors as well as response variable where normalized prior to model training for all five model algorithms to allow for comparison regarding their predictive performance; the scaling default setting, which exists in some of the models (ANN, MARS and SVM), was turned off.

Predictor Selection

Wrapper methods developed for multiple regression models include forward selection, backward elimination, stepwise selection and best subsets. The forward selection procedure usually starts with the predictor most highly correlated with the response variable and tests if the resulting model is significant. It then repeatedly adds predictors at each step, testing for model improvement due to the significance of the calculated F-statistic [19]. We applied a slightly different approach to forward selection: At first all individual predictor models of the five machine learning algorithms were compared in performance by their repeated cross-validated RMSE resulting in a predictor ranking. The single predictor included in the best individual predictor model was then used to start the forward selection procedure. At the end of step 1 it was tested whether the various algorithms' parameter ranges to select the optimal parameter via cross-validation were reasonable or needed to be adapted.

The model including the ten best predictors of the predictor ranking obtained after step 1 (Fig 3) was compared to two forward selection procedures: Simple forward selection starting with the best predictor (as selected by the best individual predictor model) and a three step forward selection procedure. Fig 3 includes all three procedures:

- Predictor ranking at the end of step 1 (Fig 3) based on all individual predictor models and their repeatedly cross-validated RMSE. Build model of the 10 best predictors as selected by the ten best individual predictor models: 10bestPR.
- Simple forward selection (sFS) includes steps 1 and 2 (Fig 3).
- Three-step forward selection (3stepFS) included all three steps and repeats steps 2 and 3 until no further improvement in repeated cross-validated RMSE is achieved.

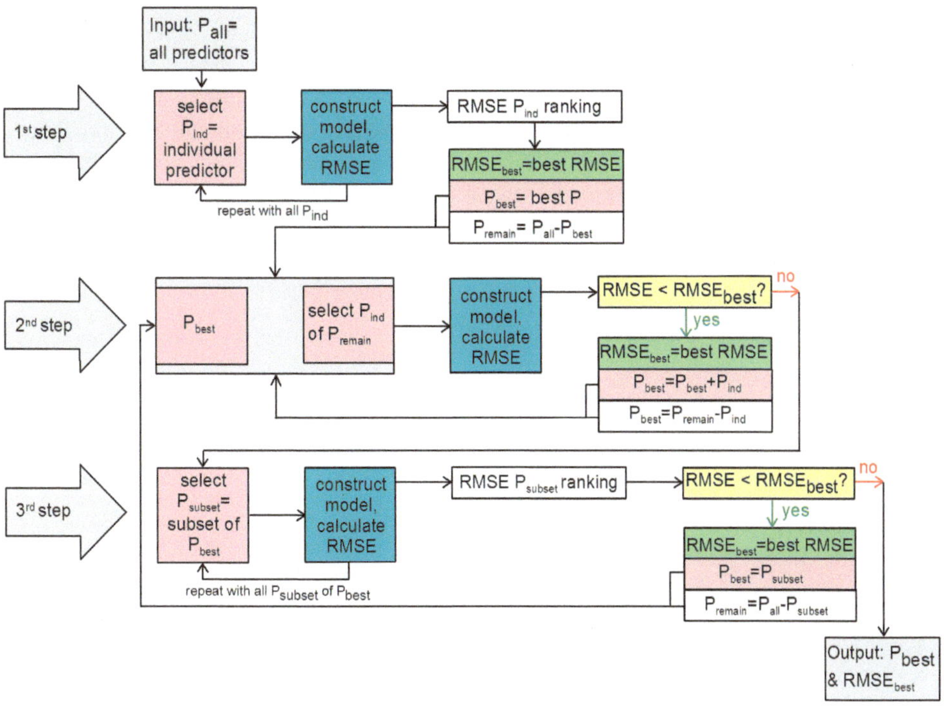

Figure 3. Forward selection procedure for predictor subset selection.

RESULTS AND DISCUSSION

Selected Predictors

Table 3 gives an overview of the selected predictors for each machine learning algorithm (please compare Tables 1 and 2 for predictor IDs). The selected predictors are ordered according to their individual importance (predictor ranking) or order of inclusion during the selection procedures. The best individual predictor is predictor 22 "Standardized Height" in all five model algorithms. As the sFS procedure starts with the best individual predictor, this first predictor is the same as in the model ranking (10bestPR). In two cases this best predictor is removed while applying 3stepFS, the two tree algorithms: RF and BRT. Interestingly, most of the ten single most important predictors (10bestPR) are not included in sFS: only one predictor for RF and MARS, two for ANN and BRT, and three for SVM (indicated by surrounding boxes and bold style in Table 3). For MARS and SVM the 3stepFS procedure did not lead to any improvement compared to sFS, which is why the selected predictors are the same. In the three other model algorithms 3stepFS resulted in an improvement in predictive performance and a reduced number of predictors.

Table 3. Selected predictors in order of importance

Model		**RF**			**ANN**			**MARS**			**BRT**			**SVM**		
Selection Proc.		**1**	**2**	**3**	**1**	**2**	**3**	**1**	**2**	**3**	**1**	**2**	**3**	**1**	**2**	**3**
Predictors	1	**22**	**22**	236	**22**	**22**	22	**22**	**22**	22	**22**	**22**	26	**22**	**22**	22
	2	174	236	15	18	21	217	30	97	97	**26**	**26**	198	37	**26**	26
	3	37	15	198	97	148	26	18	36	36	15	198	148	18	148	148
	4	119	198	169	37	**217**	20	153	145	145	5	148	47	**26**	217	217
	5	97	20	26	38	26	225	20	156	156	37	47	118	**15**	**15**	15
	6	153	169	103	33	20	65	197			175	118	219	12	119	119
	7	5	26		235	225	10	33			119	219	73	174	121	121
	8	175	103		236	65	167	198			174	73	178	198		
	9	178	149		39	10	124	23			30	178	171	38		
	10	155			**217**	40		37			155	171	150	217		
	11					138						150				
	12											24				
	13											14				

Predictor selection: 1 = 10bestPR, 2 = sFS, 3 = 3stepFS, predictor IDs are listed in order of importance from 1 to 13

doi:10.1371/journal.pone.0153673.t003

Table 4 includes only those predictors of Tables 1 and 2 which have been selected at least once. To summarize, similar predictors were grouped. This shows that there are many predictors which describe terrain curvature in one way or another, such as Hill Index, Convergence, TS Convexity and the various types of curvature. Filled table cells indicate whether a certain predictor was selected by the predictor selection procedure in the particular machine learning algorithm. The numbers in some of the cells refer to the search radii of the selected predictors. The table shows that sometimes a particular predictor was included in different radii in the forward selection process compared to the 10best predictor ranking. All together convergence and curvature parameters were always among the selected predictors. Another often chosen predictor was "Diffuse Insolation". The common best predictor for all machine learning algorithms was "Standardized Height", a predictor referring to the slope position.

Table 4. Overview of selected predictors

Algorithm		**RF**			**ANN**			**MARS**			**BRT**			**SVM**		
Predictor selection		**1**	**2**	**3**	**1**	**2**	**3**	**1**	**2**	**3**	**1**	**2**	**3**	**1**	**2**	**3**
group	predictor															
	NIR	X									X					
satellite	TIRS2					X	X									
data	PVI													X		
	NDMI											X				
altitude	altitude		X	X							X			X	X	X
MBI	MBI				X			X						X		
slope	Valley Depth		X			X	X	X								
position	Norm. Height															
	Stand. Height	X	X		X	X	X	X	X	X	X	X		X	X	X
lighting,	Diff. Insolation		X	X		X	X				X	X	X	X	X	X
visability,	Neg. Openness							X			X					
wind	Wind Effect															
hydrology	catchm. Slope				X			X								
	LS factor								X	X						
TRI						285	285					525	525			
	hill index											X				
		45			45	135		45			45	345	345	45		
	Convergence				75									75		
					105											
	TS Convexity	45	225	225	45				45	45						
	Plan Curv.	105					255				105	75	75		105	105
conver-															165	165
gence/	Prof. Curv.	525	405			75		17	285	285	585	375	375		375	375
curvature		585				375			615	615		435	435			
	Long. Curv.	555	405	405			345				555	465	465	555		
		585									585					
	Cross.-Curv.	75										75	75			
	Min Curv.		75	75				1				75	75	75		
								2								
			615	615	45	45	45					105	105	45	45	45
	Max Curv.				585	285	285									
					615											

predictor selection: 1 = 10bestPR, 2 = sFS, 3 = 3stepFS

filled cells indicate selected predictors, numbers in cells indicate the applied GIS search radius in meters

doi:10.1371/journal.pone.0153673.t004

Altitude was only important for RF, BRT and SVM, but not for ANN and MARS. Satellite data was important for all algorithms except MARS. The predictor "diffuse insolation" was among the final predictor set (3stepFS selection) in four of the five algorithms. Interestingly, MBI was among the 10bestPR in ANN, MARS and SVM but never selected in any of the two selection procedures in any of the five algorithms. From the hydrological parameters only catchment slope and LS Factor were of any relevance for ANN and MARS.

Model Performance

Fig $4A_1$–$4E_1$ compare the results of the three selection procedures (1, 2, 3) for the five machine learning algorithms to the all-predictor model (all) results. The figures show the RMSE boxplots of the repeated cross-validation. Please be aware that the RMSEs refer to the normalized C stock data.

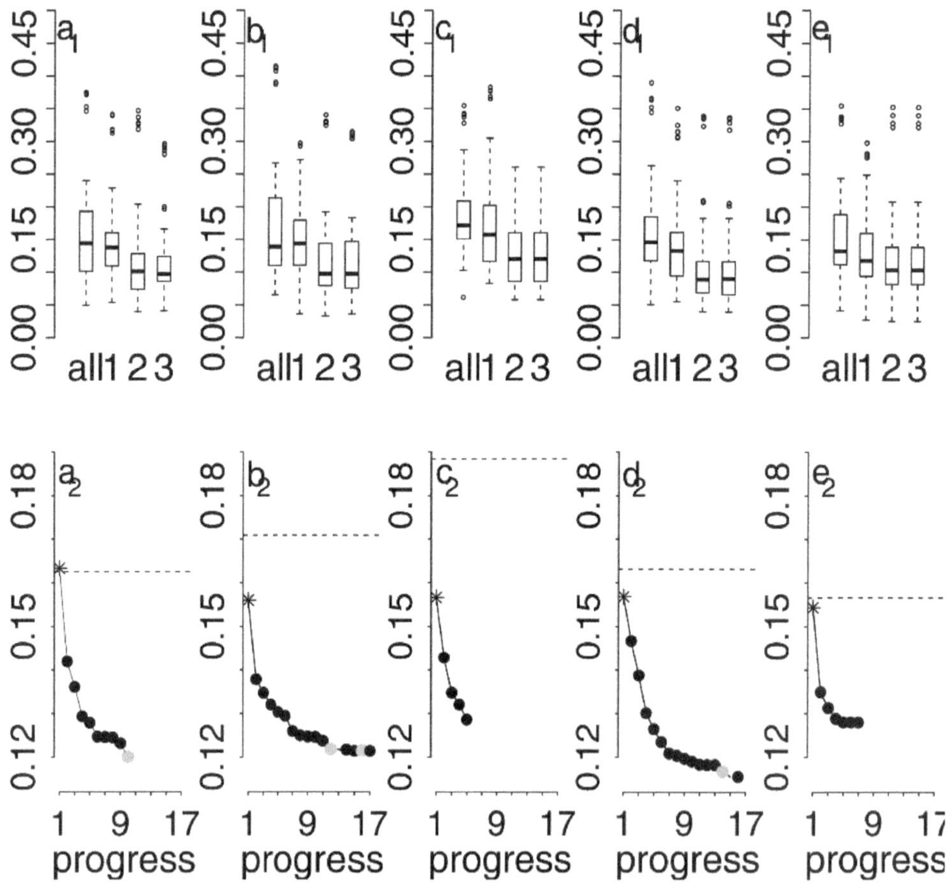

Figure 4. RMSE boxplots of repeated cross-validation (a_1-e_1) and development of the mean RMSE of rpeated cross-validation during the predictor selection process (a_2-e_2). a) RF, b) ANN, c) MARS, d) BRT, e) SVM. In a_1-e_1: "all" refers to the all-predictor model, "1"refers to the 10bestPR model, "2" to the sFS model, and "3" to the 3stepFS model. In a_2-e_2: The star refers to the mean RMSE of the best individual predictor model of step 1, black points refer to added predictors and the resulting mean RMSE during step 2, and grey points refer to the mean RMSE after step 3. The dashed line represents the mean RMSE of the all-predictor model.

The 10bestPR predictor selection improved predictive performance compared to the all-predictor model in all but the ANN algorithm. Applying the sFS procedure improved it even further and in all five algorithms. As mentioned before, in two algorithms, MARS and SVM the subset selection (step 3 in Fig 3) did not improve model results. Therefore, the RMSE boxplots 2 and 3 of Fig 4C_1 and 4E_1 are exactly the same. But even for ANN and BRT there was only a very slight improvement (Fig 4B_1 and 4D_1); the median and the interquartile range of the RMSE boxplots are very similar. This improvement is somewhat more pronounced for RF, particularly while comparing the interquartile ranges of RMSE boxplot 2 and 3 (Fig 4A_1).

Fig 4A_2–4E_2 show the development of the repeatedly cross-validated mean RMSE during the three-step selection procedure. Here the dashed line represents the repeatedly cross-validated mean RMSE of the all-predictor model. The star represents the same error measure regarding the best individual predictor model (step 1, Fig 3). The hereby selected first predictor is then entering the simple forward selection procedure (step 2, Fig 3). The first black point refers to the error measure of the model with the first additional predictor (step 2, Fig 3). The number of this first sequence of black points corresponds to the number of predictors added in step 2 before no further improvement can be achieved in sFS. Then the grey point represents the error measure of the best subset selected in step 3 (Fig 3). Black points after the grey point indicate that another sequence of simple forward selection (step 2) is applied and so on. For RF (Fig 4A_2) it was enough to run the three steps once, BRT (Fig 4D_2) repeated step 2 a second time, ANN needed to repeat step 2 a third time (Fig 4B_2). Finally, applying the 3stepFS to BRT resulted in the best predictive performance with a mean RMSE of 0.116, corresponding to a ten-predictor-model (please compare Table 3). However, RF and ANN performed only slightly worse with a mean RMSE of 0.120 and 0.121, respectively. Particularly, the RF model has the benefit of being much less complex with a number of only six predictors (Table 3). Interestingly, the one-predictor model improved model performance compared to the all-predictor model in all but one machine learning algorithm (RF). For MARS this improvement was comparatively high. In all five considered machine learning algorithms sFS resulted in an improvement in the predictive performance. This improvement was particularly pronounced for the three recursive partitioning algorithms: RF, MARS and BRT. Usually, recursive partitioning methods are said to be rather resistant to non-informative predictors [23]. They choose the best predictor at each split and should, therefore, theoretically be resistant to irrelevant predictors. However, adding irrelevant predictors to a classification tree algorithm also resulted in a deterioration by 5 to 10% on datasets tested by Witten and Frank [18]. Witten and Frank [18] also explain why even relevant attributes can cause harm in recursive partitioning algorithms due to the possibly highly unbalanced subdivision into two subsets. According to Kuhn and Johnson [23], ANN and SVM are even affected to a much larger extent by irrelevant predictors, which could not be confirmed by our results.

With a high number of non-informative predictors, sFS will always result in an improvement even after few iterations steps. And a number of non-informative predictors is likely to be tested in situations were the predictor-response correlation is weak. During model parameter tuning we further realized that the level of improvement is controlled by the algorithms' fine tuning. BRT for example resulted in the overall worst model while only 500 iterations were computed, but in the best with the final 10,000 iterations. At the end of step 1 it was tested whether the various algorithms' parameter ranges to select the optimal parameter via cross-validation were reasonable or needed to be adapted. Accordingly, the upper and lower limit of the SVM gamma and C parameter ranges needed to be adapted with the new range from 10^{-3} to 10^{4}. A higher range of neurons might have resulted into a slightly better RMSE. However, we refrained from doing so since it tremendously increased computation time of the already most time consuming machine learning algorithm. To allow for practicability, the amount of parameter tuning was set to reduce the very time-consuming forward selection (step 2) to a maximum of 12 hours (with parallel computing). This means we do not know if particularly ANN and SVM could be improved even further by testing smaller steps in the C and gamma ranges of SVM and a higher number of neurons in ANN.

Predicted Carbon Stocks

Fig 5A shows the median carbon stocks of the soil organic layer under tropical mountain forest and páramo vegetation as predicted by the overall best model: BRT with 3stepFS. The area north of the interprovincial road (Fig 1) was not predicted due to lack of data for this area of high disturbance by fire and non-natural vegetation (pasture). The majority of studies assessing SOC in tropical forests have been conducted in lowland forests, while the SOC stocks of tropical mountain forests are less well known [104]. Moser et al. [104] investigated SOC in two plots of 20 x 20 m within our research area (at 1890 and 2380 m a.s.l.) and report 3.7 to 4.8 Kg m^{-2} SOC in the organic layer which lie within the lower range of our findings. The corresponding interquartile range of our predictions (Fig 5B) is below 1 kg m^{-2} for 90% of the area indicating a low prediction uncertainty due to the underlying position specific density function in areas of high SOC stocks. Accordingly, the relative prediction uncertainty in areas of low SOC is higher; rainfall induced landslides, frequently occurring within the area, certainly reduce predictability. The apparently lower organic carbon stocks above an altitude of about 2500 m a.s.l. coincide with a change from mountain forest to shrub and páramo vegetation and the corresponding shallower organic layers (1 to 15 cm) observed during sampling: Transect M4 to M7 (Fig 1). In general, median organic layer thickness varies between c. 20 to 70 cm throughout the area, with organic layer thickness being highest on mid-slope positions [10].

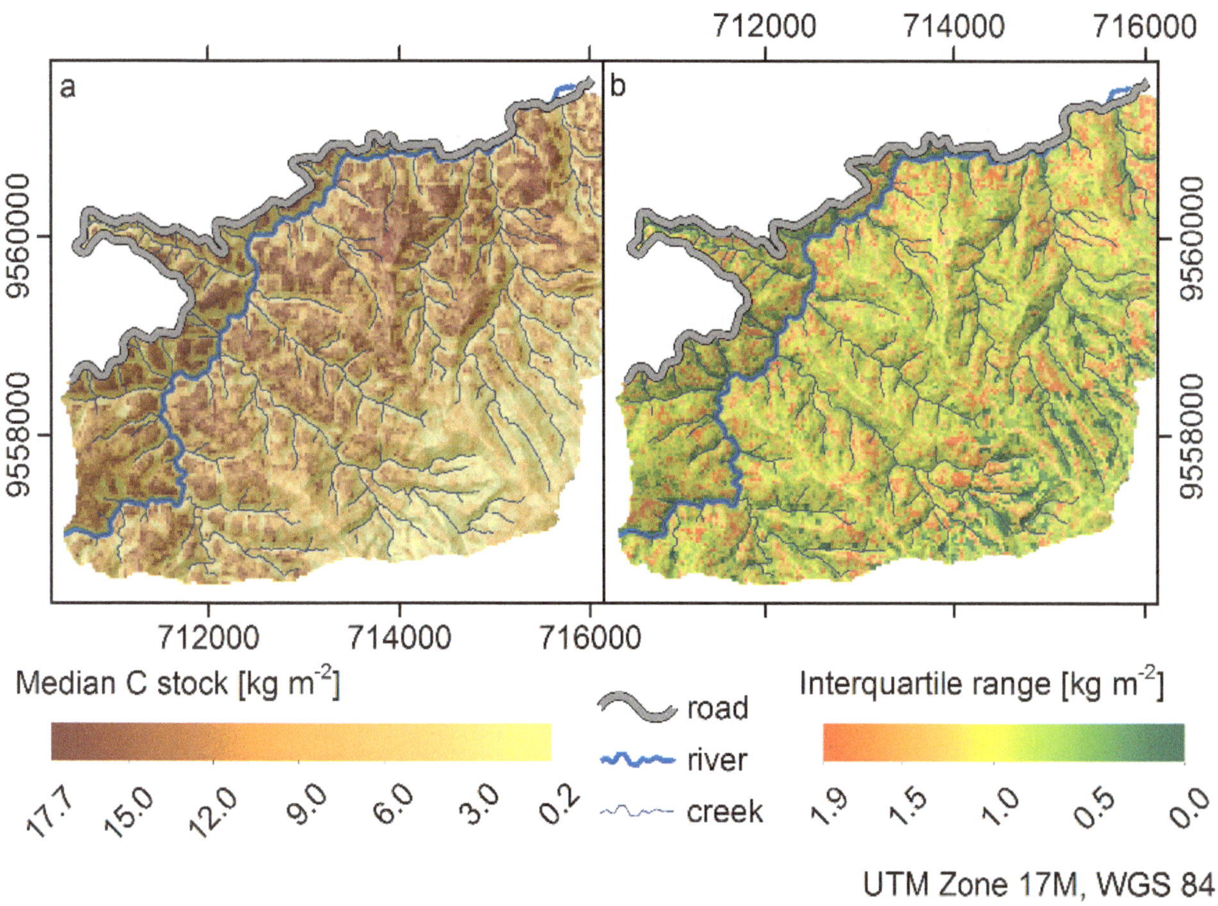

Figure 5. SOC stock prediction with best model BRT 3stepFS. a) Median prediction value, b) Interquartile range (overlaid hillshading with light source from north).

According to Roman et al. [105], the high SOC stocks of tropical montane cloud forests cannot only be explained by low soil organic matter turnover due to slow rates of litter decomposition, soil acidity and reduced rates of nutrient cycling. They conclude that the organic soil layer of these ecosystems is highly variable. Furthermore, soil water logging and the altitudinal gradient alone cannot explain the complex spatial pattern [10]. SOC stocks in tropical forest soils of Papua New Guinea are reported to vary between 4.8 and 19.4 kg m^{-2} (litter and top 100 cm) with an increase of 5.1 kg m^{-2} per 1000 m increase in altitude, while SOC in litter stayed the same [106]. Don et al. [4] report SOC stocks between 7.3 and 10.5 kg m^{-2} in tropical primary forests. They analyzed 385 studies on SOC from tropical countries but had to exclude data on organic layer SOC due to dataset scarcity.

The selected predictors which influence the spatial variation include diffuse insolation, topographic ruggedness index, convergence and all considered curvatures (some in different search radii). Diffuse insolation and topographic ruggedness index are related to site exposure, a factor termed important by Roman et al. [105] when studying the global and local variations in tropical montane cloud forest soils and particularly the accumulation of soil organic matter. The low search radii for plan curvature, cross-sectional curvature, minimum and maximum curvature (75 and 105 m) indicate the importance of local curvatures, while the rather high search radii for the profile and longitudinal curvature (375 to 465 m) indicate the importance of the larger landscape topography. The influence of the diffuse insolation and the influence of the ridge and valley topography on the spatial carbon stock pattern are clearly visible in Fig 5. Carbon stocks are lower in areas of high diffuse insolation (Summit area in the South-East) and in the concave valley structures. Similarly, Wilcke et al. [107], who studied transects in the lower part of the here investigated area, also report higher humus concentrations on the ridge-tops as compared to the valley structures and explain this with stronger acidification and nutrient leaching on the exposed sites. Gessler et al. [27] report flow accumulation being a good predictor for SOC. According to them, ignoring the effect of topography presents a serious weakness in approaches to regional and global carbon modelling, as water distribution between convex and concave landscape positions may account for differences in SOC similar to dramatically different climatic zones. According to Grimm et al. [25], topographical parameters approximate water and solute transport, relate to solar insolation and determine the micro-climate. In that effect, they have the potential to explain large parts of SOC variation.

Predictors obtained from satellite images were not among the selected predictors for the best model: According to Waring and Running [108] multi-spectral reflectance data obtained from satellite images can provide valuable information on forest structure and productivity. However, in forested areas, remote sensing methods are of limited utility for soil studies due to the variable effects of vegetation [12].

CONCLUSIONS

It was shown, that particularly in the complex soil-landscape setting of tropical mountain areas with possibly low predictor-response correlations, the applied methodology for spatial prediction is of high importance to improve the expected poor model results. Considering different spatial settings/scales for the predictors as well as applying predictor selection and model tuning are important to improve predictive performance.

Concerning predictor selection, choosing predictors individually was not the best option. The forward selection resulted in better predictions than the 10bestPR, emphasizing the importance of predictor interaction. Among the forward selection procedures, the tested 3stepFS algorithm did not result into much improvement compared to the sFS. However, the amount of improvement in predictive performance due to predictor selection depended on model tuning. Finally, even a single- or two-predictor model was better than the all-predictor model independent of the applied machine learning algorithm. Predictor selection is, therefore, also important for recursive partitioning algorithms which are often reported to be immune to non-informative predictors. To save computation time in the case of many (>100) predictors we recommend to rather apply forward instead of backward selection. Choosing the initial best starting predictor by running the model and not due to correlation seems favorable to make the approach more consistent.

Predictors obtained from satellite images were not among the selected predictors for the best overall model. We assume that satellite images are most probably of higher importance in predicting SOC in areas with contrasting land use types. The highest C stocks appear at the ridges and mid-slope positions at altitudes below 2500 m a.s.l.. Various curvatures in different search radii and the diffuse insolation are best suited to describe this pattern. The interquartile range show slightly bigger uncertainties in the transition zone from montane forest to paramo vegetation at 2500 m a.s.l.. One might argue that other predictors would lead to even better results. However, in most tropical mountain regions it will be difficult to obtain additional predictors apart from the here presented.

ACKNOWLEDGMENTS

This research was funded by the German Research Foundation (DFG) as part of the Platform for Biodiversity and Ecosystem Monitoring and Research in South Ecuador (PAK 825, LI 2360/1-1). Logistic support by the NGO Nature and Culture International (NCI) and the municipal public agency ETAPA is gratefully acknowledged.

AUTHOR CONTRIBUTIONS

Conceived and designed the experiments: ML. Performed the experiments: ML JS. Analyzed the data: ML JS. Contributed reagents/materials/analysis tools: ML JS BG. Wrote the paper: ML JS BG.

REFERENCES

1. Cerri CEP, Easter M, Paustian K, Killian K, Coleman K, Bernoux M, et al. Predicted soil organic carbon stocks and changes in the Brazilian Amazon between 2000 and 2030. Agric Ecosyst Environ. 2007;122: 58–72. doi: 10.1016/j.agee.2007.01.008.
2. Pan Y, Birdsey R, Fang J, Houghton R, Kauppi PE, Kurz W, et al. A large and persistent carbon sink in the world's forests. Science. 2011;333: 988–993. doi: 10.1126/science.1201609. pmid:21764754

3. Post WM, Kwon KC. Soil carbon sequestration and land-use change: processes and potential. Glob Chang Biol. 2000;6: 317–327. doi: 10.1046/j.1365-2486.2000.00308.x.
4. Don A, Schumacher J, Freibauer A. Impact of tropical land-use change on soil organic carbon stocks—a meta-analysis. Glob Chang Biol. 2011;17: 1658–1670. doi: 10.1111/j.1365-2486.2010.02336.x.
5. Watson RT. Land Use, Land-Use Change, and Forestry [Internet]. 2000. p. 392. Available: http://books.google.com/books?id=EyXQW-gRey0C
6. FAO. State of the World's Forests. 2014.
7. Tapia-Armijos MF, Homeier J, Espinosa CI, Leuschner C, De La Cruz M. Deforestation and forest fragmentation in south Ecuador since the 1970s - Losing a hotspot of biodiversity. PLoS One. 2015;10: 1–18. doi: 10.1371/journal.pone.0133701.
8. Beck E, Hartig K, Roos K. Forest clearing by slash and burn. In: Beck E.; Bendix J.; Kottke I.; Makeschi F.; Mosandl R, editor. Gradients in a tropical mountain ecosystem of Ecuador Ecological Studies 198. Heidelberg: Springer; 2008.
9. Bahr E, Chamba Zaragocin D, Makeschin F. Soil nutrient stock dynamics and land-use management of annuals, perennials and pastures after slash-and-burn in the Southern Ecuadorian Andes. Agric Ecosyst Environ. Elsevier B.V.; 2014;188: 275–288. doi: 10.1016/j.agee.2014.03.005.
10. Ließ M, Hitziger M, Huwe B. The sloping mire soil-landscape of southern ecuador: Influence of predictor resolution and model tuning on random forest predictions. Appl Environ Soil Sci. 2014;2014. doi: 10.1155/2014/603132.
11. Jenny H. Factors of Soil Formation. A System of Quantitative Pedology. New York: Dover Publications, Inc.; 1941.
12. Ryan PJ, McKenzie NJ, O'Connell D, Loughhead AN, Leppert PM, Jacquier D, et al. Integrating forest soils information across scales: Spatial prediction of soil properties under Australian forests. For Ecol Manage. 2000;138: 139–157. doi: 10.1016/S0378-1127(00)00393-5.
13. Ließ M. Sampling for regression-based digital soil mapping: Closing the gap between statistical desires and operational applicability. Spat Stat. Elsevier Ltd; 2015;13: 106–122. doi: 10.1016/j.spasta.2015.06.002.
14. Cavazzi S, Corstanje R, Mayr T, Hannam J, Fealy R. Are fine resolution digital elevation models always the best choice in digital soil mapping? Geoderma. Elsevier B.V.; 2013;195–196: 111–121. doi: 10.1016/j.geoderma.2012.11.020.
15. Behrens T, Zhu AX, Schmidt K, Scholten T. Multi-scale digital terrain analysis and feature selection for digital soil mapping. Geoderma. Elsevier B.V.; 2010;155: 175–185. doi: 10.1016/j.geoderma.2009.07.010.
16. Maynard JJ, Johnson MG. Scale-dependency of LiDAR derived terrain attributes in quantitative soil-landscape modeling: Effects of grid resolution vs. neighborhood extent. Geoderma. Elsevier B.V.; 2014;230–231: 29–40. doi: 10.1016/j.geoderma.2014.03.021.
17. Samuel-Rosa A, Heuvelink GBM, Vasques GM, Anjos LHC. Do more detailed environmental covariates deliver more accurate soil maps? Geoderma. Elsevier B.V.; 2015;243–244: 214–227. doi: 10.1016/j.geoderma.2014.12.017.
18. Witten I, Frank E. Data mining–Practical machine learning tools and techniques. 2nd ed. Amsterdam: Elsevier; 2005.
19. Larose DT. Data mining methods and models. Hoboken: Wiley-Interscience/ John Wiley & Sons, Inc.; 2006.
20. Xu G, Zong Y, Yang Z. Applied data mining. Booca Raton: CRC Press; 2013.
21. John G, Kohavi R, Pfleger K. Irrelevant Features and the Subset Selection Problem. Proceedings of the Eleventh International Conference on Machine Learning. 1994. pp. 121–129.
22. Kohavi R, John H. Artificial Intelligence Wrappers for feature subset selection. 2011;97: 273–324. doi: 10.1016/s0004-3702(97)00043-x
23. Kuhn M, Johnson K. Applied predictive modeling. New York: Springer Science; 2013. doi: 10.1007/978-1-4614-6849-319.
24. Hastie T, Tibshirani R, Friedman J. The elements of statistical learning. 2nd ed. Springer; 2009. doi: 10.1007/978-0-387-98135-2.
25. Grimm R, Behrens T, Märker M, Elsenbeer H. Soil organic carbon concentrations and stocks on Barro Colorado Island—Digital soil mapping using Random Forests analysis. Geoderma. 2008;146: 102–113. doi: 10.1016/j.geoderma.2008.05.008.
26. Guo PT, Li MF, Luo W, Tang QF, Liu ZW, Lin ZM. Digital mapping of soil organic matter for rubber plantation at regional scale: An application of random forest plus residuals kriging approach. Geoderma. Elsevier B.V.; 2015;237–238: 49–59. doi: 10.1016/j.geoderma.2014.08.009.
27. Gessler PE, Chadwick OA, Chamron F, Holmes K, Althouse L. Modeling soil-landscape and ecosystem properties using terrain attributes. Soil Sci Soc Am J. 2000; 2046–2056. doi: 10.2136/sssaj2000.6462046x
28. Gasparini P, Di Cosmo L. Forest carbon in Italian forests: Stocks, inherent variability and predictability using NFI data. For Ecol Manage. Elsevier B.V.; 2015;337: 186–195. doi: 10.1016/j.foreco.2014.11.012.
29. Zhang S, Huang Y, Shen C, Ye H, Du Y. Spatial prediction of soil organic matter using terrain indices and categorical variables as auxiliary information. Geoderma. Elsevier B.V.; 2012;171–172: 35–43. doi: 10.1016/j.geoderma.2011.07.012.
30. Martin MP, Wattenbach M, Smith P, Meersmans J, Jolivet C, Boulonne L, et al. Spatial distribution of soil organic carbon stocks in France. Biogeosciences. 2011;8: 1053–1065. doi: 10.5194/bg-8-1053-2011.
31. Martin MP, Orton TG, Lacarce E, Meersmans J, Saby NPA, Paroissien JB, et al. Evaluation of modelling approaches for predicting the spatial distribution of soil

organic carbon stocks at the national scale. Geoderma. Elsevier B.V.; 2014;223–225: 97–107. doi: 10.1016/j.geoderma.2014.01.005.

32. Pastick NJ, Rigge M, Wylie BK, Jorgenson MT, Rose JR, Johnson KD, et al. Distribution and landscape controls of organic layer thickness and carbon within the Alaskan Yukon River Basin. Geoderma. Elsevier B.V.; 2014;230–231: 79–94. doi: 10.1016/j.geoderma.2014.04.008.

33. Bou Kheir R, Bøcher PK, Greve MB, Greve MH. The application of GIS based decision-tree models for generating the spatial distribution of hydromorphic organic landscapes in relation to digital terrain data. Hydrol Earth Syst Sci. 2010;14: 847–857. doi: 10.5194/hess-14-847-2010.

34. Dai F, Zhou Q, Lv Z, Wang X, Liu G. Spatial prediction of soil organic matter content integrating artificial neural network and ordinary kriging in Tibetan Plateau. Ecol Indic. Elsevier Ltd; 2014;45: 184–194. doi: 10.1016/j.ecolind.2014.04.003.

35. Were K, Bui DT, Dick ØB, Singh BR. A comparative assessment of support vector regression, artificial neural networks, and random forests for predicting and mapping soil organic carbon stocks across an Afromontane landscape. Ecol Indic. Elsevier Ltd; 2015;52: 394–403. doi: 10.1016/j.ecolind.2014.12.028.

36. Basheer IA, Hajmeer M. Artificial neural networks: fundamentals, computing, design, and application. 2000;43: 3–31. pmid:11084225 doi: 10.1016/s0167-7012(00)00201-3

37. Hsu C, Chang C, Lin C. A Practical Guide to Support Vector Classification. BJU Int. 2008;101: 1396–400.

38. Cherkassky V, Ma Y. Practical selection of SVM parameters and noise estimation for SVM regression. Neural Networks. 2004;17: 113–126. doi: 10.1016/S0893-6080(03)00169-2. pmid:14690712

39. Hitziger M, Ließ M. Comparison of three supervised learning methods for digital soil mapping: Application to a complex terrain in the Ecuadorian Andes. Appl Environ Soil Sci. Hindawi Publishing Corporation; 2014;2014. doi: 10.1155/2014/809495

40. Fries A, Rollenbeck R, Göttlicher D, Nauss T, Homeier J, Peters T, et al. Thermal structure of a megadiverse Andean mountain ecosystem in southern Ecuador. Erdkd—Arch Sci Geogr. 2009;63: 321–335. doi: 10.3112/erdkunde.2009.04.03

41. Rollenbeck R. Variability of precipitation in the Reserva Biólogica San Francisco / Southern Ecuador. Lyonia. 2006;9: 43–51.

42. Litherland M, Aspen JA Jamielita R. The metamorphic belts of Ecuador. Overseas Mem Br Geol Surv. 1994;11: 1–147.

43. Ließ M, Glaser B, Huwe B. Uncertainty in the spatial prediction of soil texture. Comparison of regression tree and Random Forest models. Geoderma. 2012;170: 70–79. doi: 10.1016/j.geoderma.2011.10.010

44. Bauer FC. Water flow paths in soils of an undisturbed and landslide affected mature montane rainforest in South Ecuador. 2010; 150.

45. Ließ M, Glaser B, Huwe B. Digital soil mapping in Southern Ecuador. Erdkunde. 2009;63: 309–319. doi: 10.3112/erdkunde.2009.04.02.

46. Schmidt J. Statistische Modellierung der Kohlenstoff- und Stickstoffvorräte der organischen Auflage unter Bergregenwald in Südecuador. Universität Leipzig. 2012.

47. USGS. Using the USGS Landsat 8 product [Internet]. 2013. Available: http://landsat.usgs.gov/Landsat8_Using_Product.php.

48. Jackson TJ, Chen D, Cosh M, Li F, Anderson M, Walthall C, et al. Vegetation water content mapping using Landsat data derived normalized difference water index for corn and soybeans. Remote Sens Environ. 2004;92: 475–482. doi: 10.1016/j.rse.2003.10.021.

49. Richardson A, Wiegand C. Distinguishing vegetation from soil background information. Photogramm Eng Remote Sens. 1977;43: 1541–1552.

50. Fox G, Sabbagh G, Searcy S, Yang C. An Automated Soil Line Identification Routine for Remotely Sensed Images. Soil Sci Soc Am J. 2004;68: 1326. doi: 10.2136/sssaj2004.1326.

51. Baret F, Guyot G. Potentials and limits of vegetation indices for LAI and APAR assessment. Remote Sens Environ. 1991;35: 161–173. doi: 10.1016/0034-4257(91)90009-U.

52. Baret F, Guyot G, Major DJ. TSAVI: A Vegetation Index Which Minimizes Soil Brightness Effects On LAI And APAR Estimation. 12th Can Symp Remote Sens Geosci Remote Sens Symp. 1989;3: 1355–1358. doi: 10.1109/IGARSS.1989.576128.

53. Brenning A. RSAGA: SAGA geoprocessing and terrain analysis in R [Internet]. 2015. Available: http://cran.r-project.org/web/packages/RSAGA/index.html.

54. Association SUG. SAGA–System for Automated Geoscientific Analysis [Internet]. 2014. Available: http://www.saga-gis.org.

55. Nauss T, Göttlicher D, Dobbermann M, Bendix J. Central data services in multidisciplinary environmental research projects. e-Zeitschrift für Agrarinformatik. e-Zeitschrift für Agrar. 2007;2. Available: http://www.preagro.de/ezai/index.php/eZAI/article/view/28/28

56. Lyle WZ, Colin RT. Quantitative analysis of land surface topography. Earth Surf Process Landforms. 1987;12: 47–56. doi: 10.1002/esp.3290120107.

57. Friedrich K. Digitale Reliefgliederungsverfahren zur Ableitung bodenkundlich relevanter Flächeneinheiten. Frankfurt/M: Frankfurter Geowissenschaftliche Arbeiten D21; 1996.

58. Friedrich K. Multivariate distance methods of geomorphographic relief classification. In: Heinicke H, Eckelmannn W, Thomasson A, Jones J, Montarella L BB, editor. Land information systems–Developments for planning the sustainable use of land resources European Soil Bureau–Research Report 4. EUR 17729. Office for official publications of the European Communities, Ispra; 1998. pp. 259–266.

59. Möller M, Volk M, Friedrich K, Lymburner L. Placing soil-genesis and transport processes into a landscape context: A multiscale terrain-analysis approach. J Plant Nutr Soil Sci. 2008;171: 419–430. doi: 10.1002/jpln.200625039.
60. Rodriguez F. The Black Top Hat function applied to a DEM: A tool to estimate recent incision in a mountainous watershed (Estibère Watershed, Central Pyrenees). Geophys Res Lett. 2002;29: 2–5. doi: 10.1029/2001GL014412.
61. Böhner J, Antonic O. Land surface parameters specific to topo-climatology. In: Hengl T.; Reuter HI, editor. Geomorphometry–Concepts, Software, Applications Developments in Soil Science, Vol 33. Amsterdam: Elsevier; 2009. pp. 195–226.
62. Oke TR. Boundary layer climates. London: Taylor & Francis; 1987.
63. Wilson JP, Gallant JC, editors. Terrain analysis–Principles and Applications. New York: John Wiley & Sons, Inc.; 2000.
64. Häntzschel J, Goldberg V Bernhofer C. GIS-based regionalisation of radiation, temperature and coupling measures in complex terrain for low mountain ranges. Meteorol Appl. 2005;12: 33–42. doi: 10.1017/s1350482705001489
65. Anders N, Seijmonsbergen A, Bouten W. Multi-Scale and Object-Oriented Image Analysis of High-Res LiDAR Data for Geomorphological Mapping in Alpine Mountains. Geomophometry 2009. 2009; 61–65.
66. Prima ODA, Echigo A, Yokoyama R, Yoshida T. Supervised landform classification of Northeast Honshu from DEM-derived thematic maps. Geomorphology. 2006;78: 373–386. doi: 10.1016/j.geomorph.2006.02.005.
67. Yokoyama R, Shlrasawa M, Pike RJ. Visualizing Topography by Openness: A New Application of Image Processing to Digital Elevation Models. Photogramm Eng Remote Sens. 2002;68: 257–265.
68. Lea NL. An aspect driven kinematic routing algorithm. In: Parsons A.J.; Abrahams AD, editor. Overland flow: hydraulics and erosion mechanics. London: CRC Press; 1992.
69. Böhner J, Köthe R, Conrad O, Gross J, Ringeler A, Selige T. Soil regionalisation by means of terrain analysis and process parametrisation. In: Micheli E, Nachtergaele F ML, editor. Soil Classification 2001 European Soil Bureau, Research Report No 7, EUR 20398 EN. 2002.
70. Beven KJ, Kirkby MJ. Physically based, variable contibution area model of basin hydrology [Internet]. Hydrological Sciences Bulletin. 1979. pp. 43–69. doi: 10.1080/02626667909491834.
71. Böhner J, Selige T. Spatial prediction of soil attributes using terrain analysis and climate regionalisation. SAGA—Anal Model Appl Göttinger Aeographische Abhandlungen. 2006;115: 13–28.
72. Moore ID, Grayson RB, Ladson AR. Digital Terrain Modelling: a Review of Hydrological, Geomorphological, and Biological Applications. Rev Lit Arts Am. 1991;5: 3–30. doi: 10.1002/hyp.3360050103
73. Desmet P, Govers G. A GIS procedure for automatically calculating the USLE LS factor on topographically complex landscape units. J Soil Water Conserv. 1996;51: 427–433.
74. Kinnell PIA. Alternative Approaches for Determining the Usle-M Slope Length Factor for Grid Cells. Soil Sci Soc Am J. 2005;69: 674. doi: 10.2136/sssaj2004.0047.
75. Köthe R, Lehmeier F. SARA–System zur automatischen Relief-Analyse. User Manual. unpublished; 1996.
76. Riley S, De Gloria S, Elliot R. A terrain ruggedness that quantifies topographic heterogeneity. Intermt J Sci. 1999;5: 23–27.
77. Iwahashi J, Pike RJ. Automated classifications of topography from DEMs by an unsupervised nested-means algorithm and a three-part geometric signature. Geomorphology. 2007;86: 409–440. doi: 10.1016/j.geomorph.2006.09.012.
78. Wood J. The geomorphological chracteristics of digital elevation models. University of Leicester. 1996.
79. Wood J. Geomorphometry in LandSerf. In: Heng T, Reuter H editors. Geomorphometry: Concepts, Software, Applications Developments in Soils Science, 33. Amsterdam: Elsevier; 2009. pp. 333–349.
80. Breiman L, Friedman J, Olshen R, Stone C. Classification and Regression Trees. Belmont: Wadsworth International Group; 1984.
81. Breiman L, Cutler A. Breiman and Cutler's random forests for classification and regression. Packag "randomForest." 2012; 29. Available: https://cran.r-project.org/web/packages/randomForest/randomForest.pdf
82. Lawrence J. Neuronale Netze–Computersimulation biologischer Intelligenz. München: Systhema Verlag; 1992.
83. Scherer A. Neuronale Netze–Grundlagen und Anwendungen. Vieweg, editor. Braunschweig; 1997.
84. Masters T. Practical neural network recipies in C++. California: Academic Press; 1993.
85. Foresee FD, Hagan MT. GAUSS-NEWTON APPROXIMATION TO BAYESIAN LEARNING ** School of Electrical and Computer Engineering. Network. 1930; 1930–1935. doi: 10.1109/icnn.1997.614194
86. Perez Rodriguez P, Gianola D. Package "brnn." 2015;
87. Friedman J. Multivariate Adaptive Regression Splines. Ann Stat. 1991;19: 1–67. doi: 10.1214/aos/1176347963
88. Milborrow S. Package "earth". Multivariate Adaptive Regression Splines [Internet]. 2015. Available: http://cran.r-project.org/web/packages/earth/index.html.
89. Friedman JH. Fast MARS. Technical Repoert No. 110 [Internet]. 1993. Available: http://www.milbo.users.sonic.net/earth/Friedman-FastMars.pdf.
90. Elith J, Leathwick JR, Hastie T, Elith R. Leathwick J., Leathwick & Hastie A working guide to boosted regression trees—Online Appendices Page 1. J Anim Ecol. 2008;77: 802–13. doi: 10.1111/j.1365-2656.2008.01390.x. pmid:18397250

91. Ridgeway G. Package “gbm”. Generalized Boosted Regression Models. 2015; Available: http://cran.r-project.org/web/packages/gbm/index.html.

92. Freund Y, Schapire RE. A Decision-Theoretic Generalization of On-Line Learning and an Application to Boosting. J Comput Syst Sci. 1997;55: 119–139. doi: 10.1006/jcss.1997.1504.

93. Friedman JH. Stochastic Gradient Boosting. Comput Stat Data Anal. 2002;38: 367–378. doi: 10.1016/s0167-9473(01)00065-2

94. Bennett K, Campbell C. Support vector machines: Hype or hallelujah? SIGKDD Explor. 2000;2. Available: http://www.acm.org/sigs/sigkdd/explorations/issue2-2/bennett.pdf.

95. Vapnik VN. The nature of statistical learning theory. New York: Springer; 1995.

96. Smola A, Schölkopf B. A tutorial on support vector regression. Stat Comput. 2004;14: 199–222. doi: 10.1023/B:STCO.0000035301.49549.88.

97. Olsen D, Delen D. Advanced data mining techniques. Berlin: Springer; 2008.

98. Meyer D, Dimitriadou E, Hornik K, Weingessel A, Leisch F, Chang C, et al. Package “e1071”. Misc Functions of the Department of Statistics (e1071) [Internet]. E1071. 2015. Available: http://cran.r-project.org/package=e1071

99. Mattera D, Haykin S. Support vector machines for dynamic reconstruction of a chaotic System. In: Schölkopf B.; Burges C.J.C.; Smola A.., editor. Advances in kernel methods–Support vector learning. Massachusets Institute of Technology; 1999.

100. Breiman L, Spector P. Submodel selection and evaluation in regression. The X-random case. Int Stat Rev. 1992;60: 291–319. doi: 10.2307/1403680.

101. Kohavi R. A Study of Cross-Validation and Bootstrap for Accuracy Estimation and Model Selection. Int Jt Conf Artif Intell. 1995;14: 1137–1143.

102. Weston S. Package ‘ doParallel ‘. Foreach Parallel Adaptor for the “parallel” Package [Internet]. 2015. Available: https://cran.r-project.org/web/packages/doParallel/doParallel.pdf.

103. Weston S. Package “foreach”. Provides Foreach Looping Construct for R [Internet]. 2015. Available: https://cran.r-project.org/web/packages/foreach/foreach.pdf.

104. Moser G, Leuschner C, Hertel D, Graefe S, Soethe N, Iost S. Elevation effects on the carbon budget of tropical mountain forests (S Ecuador): The role of the belowground compartment. Glob Chang Biol. 2011;17: 2211–2226. doi: 10.1111/j.1365-2486.2010.02367.x.

105. Roman L, Scatena F, Bruijnzeel L. Global and local variations in tropical montane cloud forest soils. In: Brujinzeel LA, Scatena FN HL, editor. Tropical Montane Cloud Forests International Hydrology Series. New York: Cambridge University Press; 2010.

106. Dielemann W, Venter M, Ramachandra A, Krockenberger A, Bird M. Soil carbon stocks vary predictably with altitude in tropical forests: Implications for soil carbon storage. Geoderma. Elsevier B.V.; 2013;204–205: 59–67. doi: 10.1016/j.geoderma.2013.04.005.

107. Wilcke W, Boy J, Goller R, Fleischbein K, Valarezo C, Zech W. Effect of topography on soil fertility and water folw in an Ecuadorian lower montane forest. In: Brujinzeel LA, Scatena FN HL, editor. Tropical Montane Cloud Forests International Hydrology Series. New York: Cambridge University Press; 2010.

108. Waring R, Running S. Forest Ecosystems: Analysis at Multiple Scales. San Diego: Academic Press; 1998.

INDEX

O

P

R

S

T

V

W